# Springer Series in Plasma Science and Technology

Plasma Science and Technology covers all fundamental and applied aspects of what is referred to as the “fourth state of matter.” Bringing together contributions from physics, the space sciences, engineering and the applied sciences, the topics covered range from the fundamental properties of plasma to its broad spectrum of applications in industry, energy technologies and healthcare.

Contributions to the book series on all aspects of plasma research and technology development are welcome. Particular emphasis in applications will be on high-temperature plasma phenomena, which are relevant to energy generation, and on low-temperature plasmas, which are used as a tool for industrial applications. This cross-disciplinary approach offers graduate-level readers as well as researchers and professionals in academia and industry vital new ideas and techniques for plasma applications.

Frederick B. Marcus

# Systems Approaches to Nuclear Fusion Reactors

Springer

Frederick B. Marcus
Rixensart, Belgium

ISSN 2511-2007 ISSN 2511-2015 (electronic)
Springer Series in Plasma Science and Technology
ISBN 978-3-031-17710-1 ISBN 978-3-031-17711-8 (eBook)
https://doi.org/10.1007/978-3-031-17711-8



This Springer imprint is published by the registered company Springer Nature Switzerland AG
The registered company address is: Gewerbestrasse 11, 6330 Cham, Switzerland

# Preface

## Motivation

Major physics and engineering advances provide a strong basis for constructing nuclear fusion reactors as an inexhaustible energy source. Such machines are complex and internally interconnected, requiring systems approaches to design, construction and operation.

Now is the right time to publish a monograph about applying systems approaches. Several more countries have entered the field of fusion research. Successes on toroidal magnetic confinement experiments have led to international construction projects and demonstration reactor design. Other fusion concepts such as stellarators have obtained improved plasma confinement. Laser inertial confinement has demonstrated a fusion burn and near energy breakeven. Private industries are making large-scale investments in innovative reactor designs.

## Overview

Systems approaches for engineering, architecture and analysis of emergent properties such as robustness are currently being applied to proto-fusion machines and demonstration reactor designs. Multi-level methodologies can deal with the complex interactions of physics and engineering in fusion power plants.

The first chapter provides a basic plasma physics background, definitions of systems concepts and a logical structure for their applications. In the following chapters and in order of increasing complexity, most currently operating tokamak experiments are referenced and extensively analysed. In-depth case studies are developed of important existing tokamaks, including fusion experiments on JET, superconducting technology machines and the ITER proto-reactor under construction. The final chapter on tokamaks demonstrates in detail how systems approaches are currently being applied to fusion reactor design. Further chapters apply these principles to other magnetic and inertial confinement machines, including stellarators, linear machines and lasers. In the concluding chapter, a synthesis of the application of systems concepts is provided. A glossary and index are provided.

## Readership

This monograph is structured for teaching systems approaches to students and young researchers interested in plasma physics and nuclear fusion engineering. Practising fusion physicists and engineers can learn how their findings are integrated into systems design of fusion reactors and how they can contribute most effectively. General readers with a technical background will find an accessible survey of the design and operation of almost all existing fusion experiments and concepts, including their results and future prospects as reactor concepts.

Rixensart, Belgium Frederick B. Marcus

# Acknowledgements

I acknowledge and appreciate the help and guidance received from Springer Nature Responsible Editor Hisako Niko, Assistant Editors Nitesh Shrivastava and Malini Arumugam, and Production Assistant Sindhuja Aroumougame. I am grateful for the extremely helpful and challenging comments from reviewers Basil Duval, Gianfranco Federici and Fritz Wagner for Chaps. 4, 8 and 9, respectively.

# Contents

# About the Author

**Fredrick B. Marcus** graduated from MIT in 1969 with high honours for a BS in physics. He won a USA NSF graduate fellowship to attend the University of Oxford in the UK and obtained a DPhil in plasma physics, working at the United Kingdom Atomic Energy Authority (UKAEA) Culham Laboratory. An Oxford postdoc was carried out on the Culham superconducting levitron. At Oak Ridge National Laboratory (ORNL), he worked on the design of the EPR tokamak fusion reactor. At General Atomic (GA), he was a physics group leader and principal scientist on Doublet III. At the Swiss École Polytechnique Fédérale de Lausanne (EPFL), he was co-responsible for the physics and engineering design and initial construction of the TCV tokamak. At the European Commission's fusion project Joint European Torus (JET) at Culham, he was a member of the neutron diagnostics group and was scientific secretary for designing the high-power deuterium-tritium fusion (DTE1) programme. The EURATOM-employed team on JET was then dissolved, and he finished his career as a project officer and expert scientist at the European Commission in Brussels, directing systems biology and bioinformatics projects and publishing two related monographs with Springer. His hybrid interests in systems approaches and the recent progress in fusion research led to this book.

# Part I
# Introductory Systems and Plasma Fundamentals

# Chapter 1
# Introduction to Systems Approaches to Nuclear Fusion

**Abstract** Current experiments and supporting theory have developed much of the knowledge needed to design a nuclear fusion reactor. Since reactors are highly complex and interactive, their optimized and successful design relies on the application of systems approaches and numerical models. This chapter presents basic physics and engineering concepts needed for understanding reactor components. It explains the detailed logical structures of systems approaches, including engineering to examine relationships between elements, architecture to develop the layouts and links between form and function, and emergent properties to evaluate operation. Systems approaches are developed with fusion-specific examples and are summarized by checklists. Successive chapters apply this analysis, starting with existing experiments and finally reaching the complexity needed for fusion reactors.

## 1.1 Fusion Physics and Systems Approaches

### *1.1.1 Fusion Reactors and Systems*

#### 1.1.1.1 Introduction

One of the world's greatest challenges and opportunities is to build reactors that harness the nuclear fusion process which powers the sun, to provide a clean and inexhaustible energy source. The sun uses its own immense size and gravity to compress, confine and heat its hot plasma of ions and electrons to temperatures of millions of degrees. It can fuse protons together by overcoming electrostatic repulsion and burn for billions of years. Here on earth, there are two main methods of achieving controlled fusion by burning the more reactive hydrogen isotopes of deuterium (d) and tritium (t). One method uses strong magnetic fields to confine the hot ionized plasma away from surrounding vacuum vessel walls to produce nearly steady-state confinement. The other method is to use an ultra-high-power rapidly-pulsed energy source, usually lasers, to compress and heat a hydrogen pellet or plasma to very high density and temperature. Inertial confinement must last long enough for the fuel to burn before the pellet is blown apart.



F. B. Marcus, *Systems Approaches to Nuclear Fusion Reactors*, Springer Series in Plasma Science and Technology, https://doi.org/10.1007/978-3-031-17711-8_1

Controlled nuclear fusion research has been advancing steadily over decades. As physics and technical capabilities have increased, so have the complexities of the machines used to produce the plasmas. The results of these experiments have become so impressive that near energy-breakeven fusion conditions (where fusion output equals heating input) have been obtained in both magnetic and inertial confinement experiments. However, the next generations of machines beyond these, leading to a fusion reactor, will become yet more complex and more internally coupled, so that systems approaches will be required to design, construct and operate them.

The word system stands for connectivity between elements, often dynamic in nature. Logical structures and sets of procedures have been developed to deal with this complexity. It is the goal of this monograph to describe systems approaches and apply them to machines of increasing complexity, including potential fusion reactors.

#### 1.1.1.2 Experiencing Fusion Research

An introduction to fusion research should not neglect to mention the excitement of actually participating in experiments. I was fortunate enough to participate in the high power fusion experiments on the world's largest operating magnetic confinement machine, the Joint European Torus (JET). On 9 November 1991, high energy tritium was injected into a deuterium plasma, resulting in the world's first controlled release of fusion energy and a power production of 1.7 megawatts [1]. The experiments were led by Paul Rebut, the director and inspiration for these experiments.

Almost exactly 6 years later, we carried out maximum fusion power experiments. I had been scientific secretary to Alan Gibson for planning the experimental campaign. The need to limit vacuum vessel radioactivity meant we were allowed only two attempts at the highest fusion power. The control room was packed with operators, diagnostic physicists, directors, TV reporters and cameramen.

The plasma discharge in JET was ready to be started, the klaxon sounded, the machine countdown began. The computers controlled the pre-programmed sequences. Suddenly, the plasma viewing camera inside JET, which was sensitive to fusion-produced neutrons, showed a white haze characteristic of a giant neutron flux. The measurement of the total neutron power was 16 megawatts [2], ten times greater than in our first experiments. The fusion power was nearly equal to the plasma loss power.

In February 2022, JET operation was extended to 5 s of fusion power production of over 11 megawatts [3] in nearly steady-state conditions. The highest amount of controlled fusion energy of 59 megajoules was produced, paving the way for successful operation of ITER, a proto-reactor designed to follow on from JET.

#### 1.1.1.3 Types of Systems Approaches

The approaches used are *systems engineering* [4], *systems architecture* [5] and *systems emergent properties* [6, 7]. These procedures are best described using relevant examples from fusion research. Therefore, a brief introduction is provided of essential plasma physics and engineering concepts for fusion machines.

A system comprises parts or elements that have relationships with each other. In this context, *systems thinking* [4] is a way of understanding these relationships and connections. *Systems engineering* [4] involves developing the relationships between elements and the problem solving process. *Forward engineering* is applied to designing a new machine. *Reverse engineering* is the process of studying an existing machine. *Systems architecture* [5] describes the physical layout and relationships between form and function of elements to optimize machine design and its implementation. *Systems modelling* [6] uses mathematics or computations of the interactions of the components of the system. *Systems complexity* [7] leads to *systems emergent properties*, particularly *performance and metrics, robustness and fragility*, which can be optimized or used for mitigation.

The analysis of systems involves choices of the correct levels of detail and simplification. In an ideal world with infinite resources, the dictum of Max Planck might be achieved: "The goal is nothing other than the coherence and completeness of the system not only in respect of all details, but also in respect of all physicists of all places, all times, all peoples, and all cultures [8]." This monograph demonstrates that by breaking down a system into component parts which are analysed at an appropriate level, successful machine designs can be efficiently produced without needing infinite resources.

### *1.1.2 Detailed Monograph Organization*

The overall strategy of this monograph is to start by developing the formalisms and then analysing many of the relevant tokamak (toroidal magnetic chamber) experiments using both systems reverse and forward engineering with the aim of demonstrating ever increasing levels of complexity. Next, systems approaches are shown to be essential for the proto-reactor ITER [9, 10] and the European Union (EU) funded consortium (EUROfusion)-sponsored demonstration fusion reactor (DEMO) [11–14] using systems codes [15, 16]. An EU DEMO case study examines how the designers used a methodology which is consistent with that developed in this book and the best principles of systems engineering. The monograph then goes on to explore stellarator, linear, reversed field and inertial confinement devices and their possibilities for fusion reactors. A synthesis and comparison of various concepts is developed.

#### 1.1.2.1 Part I – Introductory Systems and Plasma Fundamentals

This Chap. 1 presents a basic overview of plasma and tokamak physics [17–25] followed by the development of systems formalisms applied to experiments, with applications summarized in checklists.

Chapter 2 creates the design space for tokamaks and different views of interacting system elements. Medium plasma current experiments are examined to classify machine elements and plasma operational scenarios.

#### 1.1.2.2 Part II – High-Current Tokamaks

Chapter 3 examines Doublet III [26] and its upgrade DIII-D [27] with its flexible shaping system to serve as a case study of systems reverse engineering. Other machines reviewed are Asdex-U, Alcator C-Mod, MAST-U and NSTX-U [28–31].

Chapter 4 discusses TCV [32] as a case study for design and operation, chosen for its maximum flexibility plasma shaping system [33].

Chapter 5 studies JET [34] in depth, including forward engineering of the first high power d-t (deuterium–tritium) fusion experiments [2]. Further d-t experiments were conducted under EUROfusion management [35] and obtained a record fusion energy production. Experiments using d-t on TFTR [36] are summarized.

Chapter 6 makes a case study of the superconducting and long-pulse Tore Supra and WEST and reviews EAST, KSTAR, SST-1, JT-60U and JT-60SA [37–43].

#### 1.1.2.3 Part III – Prototype Tokamak Fusion Reactors

Chapter 7 analyses the role of research, innovation and prototyping in systems integration of the design and construction of ITER [20] with the aim of achieving sustained fusion power.

Chapter 8 considers concepts for tokamak fusion reactors, highlighting the EU DEMO [44] as a case study in the advantages of systems engineering for designing fusion reactors. This is followed by analysis of other fusion reactor concepts and designs, some based on ITER, others exploring compact machines with high magnetic field and/or low aspect ratio.

#### 1.1.2.4 Part IV – Helical, Linear and Inertial Fusion Reactor Concepts

Using the systems formalism already developed for tokamaks, Chap. 9 explores in detail the stellarator, helias, heliotron/torsatron and other concepts, which are considered in terms of robustness and fragilities. A case study is made of the W7-X [45] helias. Reactor concepts are considered based on W7-X and on the LHD torsatron/heliotron [46].

Chapter 10 analyses linear magnetic traps and their toroidal versions and makes a forward engineering case study of MFTF-B [47]. Geometries include simple and tandem mirrors, field reversed configurations and reversed field pinches [48]. Potential reactor developments are examined.

Chapter 11 investigates inertial fusion concepts such as direct-drive and indirect-drive laser implosion, laser shock and fast ignition and linear z-pinch compression. The laser powered National Ignition Facility (NIF) [49] which has had the very impressive recent result of energy breakeven, is presented as a case study in forward engineering. Reactor concepts are examined including HiPER [50].

#### 1.1.2.5 Part V – Synthesis and Conclusions

Chapter 12 provides a synthesis of the conclusions presented in this monograph relating to the advantages of using systems approaches. Comparisons are made of various concepts.

### *1.1.3 A Simplified Description of Plasmas*

The following basic description of plasmas [21, 22, 24] provides background for reading this monograph and clarifies the examples used when discussing systems definitions. Readers who are familiar with plasmas and tokamaks can jump to Sect. 1.2.

- **Plasma:** A plasma is an electrically conducting collection of ions and electrons (ionized atoms) that provide charge (quasi-) neutrality. The ions in fusion reactors are mostly deuterium and tritium.
- **Density and temperature:** A plasma is locally characterized by density per cubic metre ($m^3$) for electrons and hydrogen isotope ions as $n_e$ and $n_i$, impurity ions with charge Z as $n_Z$. A collection of ions and electrons interact and collide with each other via the Coulomb force, establishing a Maxwellian distribution of energies characterized by a temperature T, where 1 keV (kilo electron volt) is equivalent to 11.6 million degrees Kelvin, since they are related by the Boltzmann constant. A Maxwellian velocity distribution is proportional to E times exp(−E/T), where the particle energy is $E = mv^2/2$. Electron and ion temperatures are defined as $T_e$ and $T_i$. A singly charged particle has an energy of 1 keV when accelerated through a potential of 1 kV (kilovolt). The most probable particle energy in kV equals the temperature in keV in a Maxwellian distribution. Fast electrons or ions created or injected externally or by resonant electromagnetic fields, will deposit energy and heat via Coulomb collisions.

- **Current and ohmic heating:** The electrons flowing in a plasma and creating a current encounter a resistivity $\eta = 2.8 \times 10^{-8} / T^{1.5}{}_{e,keV}$ ohm-meters. In a plasma with radially varying electron temperature, plasma current will redistribute until there is constant applied voltage everywhere. At constant applied voltage, the resistivity drops as the plasma heats above 1 keV, so it is difficult to reach fusion conditions with ohmic heating alone.
- **Plasma creation and start-up:** The plasma is usually created by accelerating electrons with electric fields or microwaves in a mostly hydrogen isotope gas. These electrons create a cascade which results in full ionization. During ionization, other processes are important, such as plasma resistivity, atomic and molecular physics, charge exchange, ionization, and radiation from electron bremsstrahlung (slowing-down radiation).
- **Plasma frequency and waves:** Approximate charge neutrality of the plasma is maintained by electric fields that appear when separation begins to occur. These fields cause oscillations at the electron plasma frequency (in radians/sec) $\omega_{pe}^2 = n_e\, q_e^2 / (\, m_e\, \varepsilon_0)$, where the quantities are electron density ($n_e$), charge ($q_e$) and mass ($m_e$) plus the permittivity of space ($\varepsilon_0$) in SI units. At $10^{20}$ $m^{-3}$, the plasma frequency ($f_{pe} = 2\pi\omega_{pe}$) is 90 GHz (gigahertz). Plasmas can oscillate and carry waves. Microwaves propagating through a plasma are cut-off below this frequency. When ions are better confined than electrons, a confining electrostatic potential in kV is created of the order of a few $T_e$/e.
- **Magnetic field:** When a plasma is immersed in a magnetic field (B), each ion and electron has velocity components parallel and perpendicular to the magnetic field. Ions and electrons rotate perpendicularly to the magnetic field in Tesla (T) at a cyclotron circular frequency of $\Omega_{ce,\ ci} = q_{i,\ e}\ B\ /\ m_{e,\ i}$ . At a field of 1 T, the electron cyclotron frequency $f_{ce}$ is 28 GHz. The perpendicular radius ($\rho$) of the orbit of the particle with velocity (v) around the field line is given by $\rho_{e,i} = v_{\mathbf{ei\ perp}}$ **/** $\boldsymbol{\Omega}_{\mathbf{ce,ci}}$ **.** For a 15 keV (kiloelectrron volt) ion in a field of 1 T, $\rho_i$ is 0.025 m.
- **Instabilities:** An instability can result when a plasma can move from a higher to a lower energy state. Instabilities can be macroscopic at the plasma fluid level description, or microscopic at the particle velocity distribution level, and can propagate and grow via waves in the plasma. Sources of instability can be gradients in density and temperature, the plasma, its current and high energy particles. Instabilities can result in plasma motion that destroys plasma magnetic confinement or leads to enhanced energy losses either inside or at the plasma edge. Much of fusion research is devoted to creating machine configurations that suppress instabilities.
- **Nuclear fusion:** Fusion reactions occur when fast ions collide with enough energy to tunnel through the repelling Coulomb barrier. The most energetically favourable fusion reaction is d + t → n (14.1 MeV) + $He^4$ (3.5 MeV). The reaction rate per unit volume is given by the product of the cross section ($\sigma$) and ion relative velocity (v) averaged over the velocity distribution as $\langle \sigma v \rangle$. The peak value for deuterium and tritium is $0.9 \text{x} 10^{-21}$ $m^3 s^{-1}$ at about 60 keV temperature,

but is nearly half of the maximum at 20 keV. The reaction rate per unit volume is $n_d n_t \langle \sigma v \rangle$. The $He^4$ (α-particle) ion deposits heat in the plasma.

- **Radioactivity:** Fusion product alpha particles and neutrons deposit their energy in structures surrounding the plasma and generate energy as heat, which could be processed into electricity. The neutrons also induce radioactivity into the structure, and can be used to breed tritium. Tritium itself is radioactive.
- **Fusion gain Q:** An important metric is Q, the energy gain of a plasma, defined by the total fusion power (or energy) produced by the plasma, divided by the heating power (or energy) applied to it. Therefore, $Q = 1$ is referred to as the breakeven condition. The ignition condition $Q =$ infinity($\infty$) occurs when no external heating is required, i.e. when contained alpha particles heat the plasma and maintain a fusion burn.

### *1.1.4 Tokamak Magnetic Field Coils and Geometry*

In this chapter, examples of systems concepts are applied to the nuclear fusion concept called the tokamak, an acronym for (toroidal chamber magnetized).

In early 1969, Lev Artsimovich was in charge of the T-3 tokamak, which had reported record results [51]. He presented four lectures at the Massachusetts Institute of Technology (MIT) in April 1969 (which I attended). MIT soon proposed its own high magnetic field version of a tokamak based on Bitter magnet technology [52]. Many other tokamaks were constructed around this period [53]. Tokamaks are currently the most investigated fusion concept, shown schematically in Fig. 1.1.

The main elements of a tokamak are:

- **Vacuum chamber:** The torus-shaped vacuum chamber is surrounded by poloidally wrapped coils which produce a toroidal magnetic field pointed along the axis of the inside of the vacuum vessel, so that plasma particles flow along the directions of the field inside the torus and are contained.
- **Toroidal magnetic field and its gradient:** The toroidal field is stronger than the poloidal field and provides particle confinement. However, because of the toroidal geometry, the toroidal magnetic field is stronger nearer the inner part of the torus than the outer part, decreasing with the major radius (R) as 1/R. This magnetic gradient causes electrons and ions to drift upwards and downwards producing an electric field,. Without confining fields, the particles would escape and hit the walls of the vessel.
- **Plasma current:** In order to stop particles drifting out, an electrical current is induced in the plasma, which creates a magnetic field circling around the plasma current, the so-called poloidal field. The individual plasma particles follow the combined toroidal and poloidal field, and spiral around the plasma. Since particles are then drifting upwards half the time and downwards half the time, the drift is cancelled and the particles are confined.

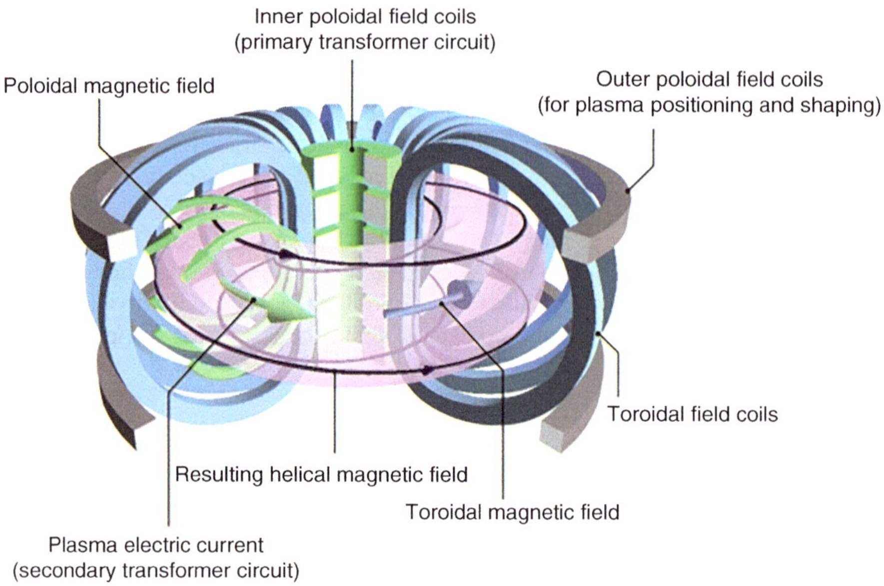


**Fig. 1.1** Schematic of a tokamak chamber and magnetic profile. (From [54], Fig. 1)

- **Poloidal field coils:** These are fields that are symmetric around the torus and generated by coils that are wound in the toroidal direction. They can be wired to produce special multipole fields (see below) or they can produce a superposition of these fields by individual coil current control to provide plasma positioning and shaping and confinement.
- **Transformer:** The plasma current itself is generated by transformer action either by a coil with changing current or a coil plus an iron core going through the hole in the vacuum vessel torus. A magnetic flux change in this primary winding creates a one-turn voltage that acts on the plasma to ionize the gas and create a flowing current which ohmically heats the plasma by collision with the accelerated electrons.
- **Vertical field:** As the plasma current and pressure from ohmic and external heating increase, the toroidal plasma expands radially outwards. A vertical magnetic field perpendicular to the plasma current, generated by external coils, produces a radially inward force which holds the plasma in place.
- **Shaping field:** Although the simplest plasma shape is a circular torus, it is often desirable to further shape the plasma with additional coils in the toroidal direction to generate shaping poloidal magnetic fields.
- **Horizontal field:** Vertically elongated plasmas can be up-down unstable. A horizontal field can raise or lower the plasma and be used with feedback for positional stability.
- **Plasma exhaust:** At the same time as the plasma is being heated, particles and energy are also escaping in the form of electromagnetic radiation, energetic ions,

electrons and fusion products. Some power is going either to a plasma limiter structure and the walls, or to a magnetic configuration that removes plasma energy (divertor). The plasma escape needs to be controlled to avoid damaging surrounding structures.

### 1.1.5 Plasma Current and MHD Fluid Equilibrium

There are essential physics elements [21] that are central to tokamak design. Current in the ionized plasma flows along the toroidal magnetic field $B_{tor}$, which creates a poloidal magnetic field $B_{pol}$ (see Fig. 1.1) that is perpendicular to the plasma current. Closed surfaces of constant poloidal magnetic flux are formed on which there is a force equilibrium between magnetic fields and the plasma pressure. Magnetic field lines lie on a flux surface, so particles following field lines will stay on that surface. The shape of these nested flux surfaces can be calculated by using Magneto-Hydro-Dynamic (MHD) fluid equations developed by Grad and Shafranov [21], which describe the force balance in equilibrium conditions.

The so called safety factor (q) is defined as the number of times a magnetic field line goes around toroidally for each time it goes around poloidally on a flux surface, and it is a function of minor radius. q may be non-integral and is calculated in a circular plasma with major radius R and minor radius r by the formula:

$$q = \frac{rBtor}{RBpol} \tag{1.1}$$

It is valid in the large aspect ratio A = R/r regime. In a tokamak, the lowest value of q at the plasma edge r = a is kept above 2 and often 3, otherwise resonant instabilities may occur.

The current in MA in a plasma with toroidal field $B_{tor}$ and vertical elongation $\kappa$ is approximately given by:

$$I_{MA} = \frac{5a\kappa Btor}{\left(\frac{R}{a}\right)q} \tag{1.2}$$

The number of volt-seconds of magnetic flux required from the primary transformer and ohmic heating coil is given by $LI_{plasma}$, where L is the plasma self inductance. Additional voltage used to sustain plasma resistance is calculated as resistivity times plasma circumference divided by cross section. As an example, the self inductance of an unshielded current loop of major radius R and cross section a, for uniform current, is given by.

$$L_{circular} = \mu_0 \, \mathrm{R} \left( \mathrm{In} \left( \frac{8R}{r} \right) - 2 \right) \tag{1.3}$$

For an aspect ratio of 3, L becomes 1.5 microhenries times R. Hence to produce a plasma current of 1 MA at R = 0.8 m requires about 1.2 volt-seconds for the magnetic energy of an unshielded plasma, and less if nearby shaping coils reduce the effective inductance. The resistive losses in forming the plasma need to be added into the volt-second flux requirement from the ohmic heating transformer coil.

### *1.1.6 Plasma Pressure and Confinement Time*

Plasma stability requirements limit the ratio beta (β) of the plasma pressure (nT) integrated over particle species and space to the magnetic field pressure,

$$\beta = nT / \left( \frac{B_{tor}{}^2}{2\mu_0} \right) \tag{1.4}$$

The Troyon Limit [21] in a tokamak on β was determined from computation and experiment to be:

$$\beta(\%) limit < 2.8 \; x \; I_{MA} / aB_{tor} \tag{1.5}$$

The Lawson criteria is the value of plasma density (n) times energy confinement (τ) at which a fusion reactor reaches energy breakeven, and is approximately equal to an nτ product of $10^{20} m^{-3} s^{-1}$ for deuterium and tritium at 10 or more keV temperature. Combining the above formulas (with MKS units), the Lawson criteria for plasma self-ignition becomes

$$\frac{\tau \; Btor^2 \; \kappa}{\left( \frac{R}{a} \right) q} > 17 \tag{1.6}$$

Using suggested reactor parameters of $B_{tor} = 5$ Tesla, elongation $\kappa = 2$, Aspect ratio R/a = 3, and q = 2, an energy confinement time of about 2 seconds is required, which is the confinement time estimated for the so-called L-mode confinement in ITER. With divertor plasmas, enhanced confinement modes become possible with even higher confinement times.

Wesson [21] notes that for a parabolic radial distribution of deuterium and tritium total ion density with peak density values $n_i(0)$, ion temperature $T_i(0)$ and global energy confinement time $\tau_E$, the triple product for plasma ignition by self-heating from fusion product alpha particles is:

$$n_i(0)T_i(0)\tau_E > 5 \times 10^{21} m^{-3} keVs \tag{1.7}$$

A group of scientists developed a scaling law of energy confinement time for the ITER project (see Chap. 7), the ITER ELMy H-mode confinement time [55, 56]. It is valid operating in a high confinement mode (H-mode) when the plasma has Edge Localized Mode (ELM) instabilities and is given by:

$$\tau_{H98(y,2)} = 0.144 I^{0.93}\ B^{0.15} \bar{n}_{20}^{0.41}\ P^{-0.69}\ R^{1.97}\ \kappa_a^{0.78}\ \varepsilon^{0.58}\ M^{0.19} \tag{1.8}$$

Various tokamak designs have taken different approaches to maximizing the figures of merit, either by emphasizing an increase in the toroidal field or an increase in elongation, or a reduction in the aspect ratio as in spherical tokamak design. JET (see Chap. 5) tried to optimize all three of these, with R/a = 2.4, $B_{tor}$ = 3.5 tesla and elongation κ = 1.7. Ignition in JET would have required a confinement time of about 4 seconds, whereas only about 1 second was reached. Even this lower confinement time allowed the approach to Q = 1, where the plasma loss power from auxiliary heating equals the plasma loss power.

### *1.1.7 Individual Particle Effects in Toroidal Systems*

In fusion reactors with spatially dependent magnetic fields, the mirror effect occurs when a particle follows a magnetic field line. The gyroradius-averaged motion of a particle is constrained by adiabatic invariants, meaning constants of motion where external fields change slowly. During a time much shorter than the time to significantly collide with other particles, the total particle energy E is constant, and the magnetic moment involving the energy perpendicular to the field M = $E_{perp}$/B is an adiabatic constant of the motion. As the particle moves into a higher magnetic field, the perpendicular energy increases and conservation means the parallel energy and velocity decreases. If the field variation is large enough, the parallel energy becomes zero, and the particle is reflected (mirrored) back in the opposite direction. The mirror ratio is defined as the ratio between the maximum and minimum magnetic field along this trajectory.

In linear mirror machines, fast particles bounce back and forth between two intense magnetic fields. In toroidal tokamaks and stellarators, particles are trapped between the inside and outside of the spatially varying toroidal field. Since particles drift vertically in opposite directions on one bounce and on the return bounce, the shape of their orbits is that of a banana, hence the term banana orbits, which can have a significant effect on plasma transport. These orbits and the rate of collision with other particles determine the rate of transport labelled as neoclassical in a torus, which is enhanced compared to classical losses of non-trapped particles.

Banana orbits can contribute to a current referred to as bootstrap, which is generated by plasma pressure radial gradients.

### *1.1.8 Auxiliary Heating and Current Drive*

Auxiliary heating methods can be chosen to preferentially heat ions or electrons, and also to impart momentum and drive spin and plasma current. When one species is heated, it can transfer energy to the other species by Coulomb interactions, which are stronger at high density and weaker at high electron temperature.

- **Neutral beam injection (NBI):** An external source is used to inject ions into a high voltage static electric field which accelerates them to high energy. They are then neutralized by charge exchange with a cold gas and travel into the vacuum vessel, where they are then ionized by the plasma, then trapped by the magnetic field. These injected ions slow down and heat the plasma. They can contribute to fusion via beam-plasma and beam-beam collisions.
- **Electron cyclotron resonance heating (ECRH):** A microwave source such as a high power gyrotron injects microwaves into the plasma where they are resonant with the electron cyclotron frequency at its first or higher harmonics. These waves can heat the electrons and also induce current drive.
- **Lower hybrid current drive (LHCD):** Lower hybrid current drive is produced by injecting microwaves via waveguides at a frequency intermediate between the electron and ion cyclotron frequencies. It has proved to be relatively efficient for current drive by interaction with electrons.
- **Ion cyclotron resonance heating (ICRH):** Antennas inside the vacuum vessel are used to launch radio frequency waves resonant with the cyclotron frequency of ions. ICRH is most efficient when heating minority helium or impurity species of a few percent which in turn heat the main plasma.
- **Alfven Wave Heating (AWH):** Alfven waves are created at low radio frequency by antennas inside the vacuum vessel. Heating seems to be accompanied by continuous plasma density rise.
- **Alpha particle heating (APH):** When d-t fusion occurs, the alpha particles are trapped by the magnetic field, if strong enough. The plasma is heated as the alphas slow down, with heat transfer typically to electrons.

### *1.1.9 Elements of a Tokamak Fusion Reactor*

Tokamak reactor designs were initiated during the 1970’s, including the Experimental Power Reactor (EPR) [57], based upon previous general reactor considerations [58]. This design was superseded, but it highlighted several features illustrating the complexity of a fusion reactor compared to existing tokamaks without neutron production. In a reactor, plasma self-heating and ignition require confinement of charged alpha particles from d-t fusion, while neutrons escape the plasma into the surrounding structure. A plasma current of at least 1 MA is required for some alpha confinement and 2.5 MA is a minimum for confining most alpha particles. Shielding was essential to protect components against sustained 14 MeV neutrons. A breeding

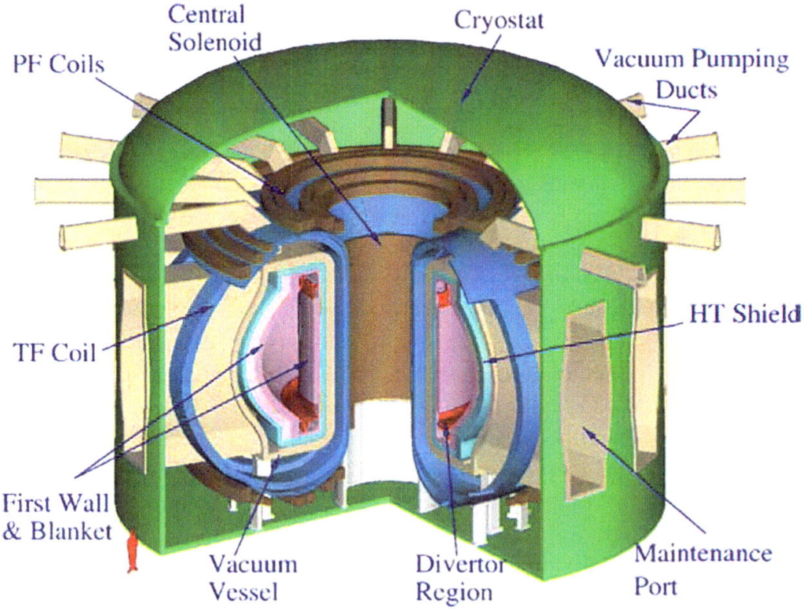


**Fig. 1.2** The ARIES-AT fusion power core. (From [62], Fig. 1)

blanket with lithium-6 and lithium-7 was needed to breed tritium fuel from their reactions with neutrons.

The EPR parameters were derived from systems scoping studies [59]. The nuclear blanket and shield were sufficiently thick so as to protect the superconducting coils. The result was a tokamak with minor radius of 2.25 m, major radius of 6.75 m, magnetic field on axis of 4.8 T and plasma current of 7.2 MA. The poloidal magnetics system [60] was designed both to reduce the total volt-second energy requirements and to reduce the field time-variation at the toroidal field coils.

Reactor studies in the USA continued as a community effort that produced the ARIES series, starting with ARIES-I, ARIES-II and ARIES-III [61]. A design shown in Fig. 1.2 is for the ARIES-AT Advanced Tokamak fusion reactor [62]. This particular 1000-MWe ARIES-AT design had a major radius of 5.2 m, minor radius of 1.3 m, toroidal $\beta$ of 9.2% ($\beta_N = 5.4$) and on-axis toroidal field 5.6 T. The plasma current was 13 MA and the current-drive power was 35 MW.

These concepts evolved and demonstration reactors are being designed [63].

## 1.2 Systems Engineering and Architecture Principles

### *1.2.1 Elements, Relationships and Systems Thinking*

Let us begin by looking at the essentials of systems engineering [4] and how they apply to fusion experiments and reactors. The word *system* stands for connectivity and in this context refers to a fusion experiment or reactor installation.

Systems consist of *elements*, which have properties and functions, and can themselves be seen as systems, with their own subsystems. Such an element could be an ionized plasma with properties such as density and temperature, and functions such as producing fusion products such as neutrons and alpha particles. Another element might be the poloidal field shaping coils in a tokamak.

Elements are linked to each other through *relations*. For example, poloidal field shaping coils influence the plasma shape, but a moving or disrupting plasma can induce currents in the shaping coils. Relations can be in one direction or in both directions and can be complex, involving several types of interactions.

Elements and relations form a construct which demonstrates an order, forming the *structure* of a system. Examples of structural types include *hierarchical*, *network* and *structures with feedback.*

*Subsystems* are found by taking a system and determining its constitutive elements at a lower level and then connecting them by means of relations. Two plasma subsystems might be the electron temperature radial profile (determining plasma resistivity), and the plasma current radial profile. Their steady-state relationship is determined by Ohm's law, where the voltage (V) is the product of current (I) and resistance (R) as $V=IR$ when there is a constant voltage across the plasma.

Systems have emergent properties [7] of robustness and fragility, for example in the context of dynamic systems biology [64]. These properties and metrics allow the designer to evaluate the stability and probable success of systems.

These elements and their relations are discussed in detail in Chap. 2 in the context of the design space available for nuclear fusion experiments and reactors.

### *1.2.2 Hierarchy*

A system divided into multiple levels results in a *systems hierarchy*. Under the system, there is a level of subsystems of order 1 and below are subsystems of order 2, etc., and the lowest level is the element level, with relations acting as connectors.

A typical hierarchy might involve the following descending levels: tokamak; power supplies; poloidal coil power supplies and transformer; coil #5 power supply; solid state thyristor rectifiers. The lowest element would require a controlling systems relationship with the power supplies.

The elements can be seen as *black boxes*, so only the function (the purpose) or desired inputs and outputs (results) are of significance. For a thyristor, we do not need to know the detailed semiconductor physics, only that a control signal will open a gate and produce a voltage and current in a power supply and coil.

A *white box*, for example a plasma in a tokamak, is when the exact connection between output and input is of great significance, such as whether heating a fusion plasma produces enough fusion self-heating for near-ignition and enough neutrons to breed tritium and to heat coolant to power an electrical generator.

A *grey box* is roughly or partially structured. As system design is developed, there is a transition from black to grey to white boxes.

### 1.2.3 Aspects

Every system consisting of elements and relations can be described from different viewpoints, called *aspects*. For example, in its interaction with other parts of a fusion reactor, a plasma can be regarded from various aspects; a source of neutrons for tritium breeding and electricity generation; energy flow; a component in a plasma-coil electromagnetic system; a danger to the vacuum vessel in case of disruption.

Different views of a system are possible, depending on the level of complexity required to study a problem. These different views include a *simple system, massively interconnected complicated system, dynamic complicated system, complex system*. One of the goals of systems engineering is to look for the simplest view and structure that allows analysis. The choice of level of complexity is central to whether useful insights can be found.

Another possibility is the *input/output view*, where the relation between input and output is governed by a transfer function, for example a function that related the neutron production from a plasma to the heating power applied.

### 1.2.4 Models

An essential principle of systems thinking involves systems models and model-based illustrations. Models are abstractions of reality, hence they reveal only partial aspects, and thus the question of relevance depends on choosing the correct level of complexity. Simple graphic and tabular models can demonstrate relationships and promote problem awareness, and provide the basis for development of quantitative models. Aids for illustrating relations or structures include graphs and matrices. In the graphs, elements are shown as rectangles which are connected by directional or multi-directional arrows representing relationships.

In a matrix, the elements are listed in rows and columns, and the intersections can be text or quantitative numbers representing the strength of the relation. The two row-column and column-row intersections on opposite sides of the diagonal can represent relations in two directions. In the simplified Table 1.1 concerning a tokamak, a sample function matrix in the electromagnetic view is shown.

**Table 1.1** Sample function matrix for tokamak poloidal field system- electromagnetic view

| Functional relations | Plasma | Vacuum vessel | Poloidal coils |
|---|---|---|---|
| Plasma | Plasma is internally self connected by inductance, diffusion, instability. | Vacuum vessel stabilizes instabilities in plasma. | Poloidal coils induce plasma current and shapes and stabilize plasma. |
| Vacuum vessel | Plasma current changes and disruptions induce currents in vessel. | Vacuum vessel currents redistribute within the shape and thickness. | Poloidal coils create currents in vacuum vessel which resistively decay. |
| Poloidal coils | Plasma disruption induces currents in coils. | Vacuum vessel currents induced by disruption will decay and induce currents in poloidal coils. | Poloidal coils can induce currents in each other. |

**Table 1.2** Sample form matrix for tokamak poloidal field system – Shape View

| Form relations | Plasma | Vacuum Vessel | Poloidal Coils |
|---|---|---|---|
| Plasma | X | Vacuum vessel toroidally surrounds plasma. | Poloidal coils are outside vacuum vessel, but near to plasma. |
| Vacuum vessel | | X | Poloidal coils are outside vacuum vessel and inside toroidal field coils. |
| Poloidal coils | | | X |

Matrix numerical algorithms could optimize the relationships, for example by using mutual inductances between coils and plasma. All elements have both form (what it is) and function (what it does). The *architecture* also has the goal of determining the functional and spatial relationship of each element, see Table 1.2.

An example of the use of function and form studies might be to compare two fusion systems, such as tokamak and laser-driven inertial confinement reactors. In the relationship connecting external heating systems with the fusion plasma, both would have a large number of interactions concerning the effect of heating system on the plasma. Inversely, there is an influence of the plasma on a radio frequency heating system due to neutrons hitting the antenna. A laser system would have fewer inverse interactions, since most of the laser can be relatively well isolated from the neutron source. Thus a matrix reflecting the neutron damage aspect of a fusion system would favour a laser system. The structure-oriented view of a system under different aspects allows observation of the system through different filters.

Using system hierarchical thinking, it is possible to analyse a problem, such as the design of a new machine, by a simple model involving a scoping study incorporating the basic elements of a system. For a tokamak, the elements include: toroidal field, plasma current, aspect ratio, elongation, divertor configuration, and auxiliary heating type and power. Then other levels with subsystems can be added until the appropriate level of detail and analysis is achieved.

Many models for analysing nuclear fusion plasmas and the surrounding machinery have been developed and will be discussed in Chap. 2. There are also dedicated systems engineering and architecture languages such as SysML described in OMG Systems Modelling Language [65] and Object Process Methodology (OPM) [66]. Examples of plasma models include the tokamak simulation code (TCS) [67] and the analysis code TRANSP [68].

### *1.2.5 Agility*

An agile system should be capable of being variable and modified, which can be an important concern in a rapidly changing environment. Agile fusion experiments have been able to incorporate new features which greatly improve their performance. Examples include the addition of auxiliary heating of the plasma, or the installation of divertor coils that improve plasma performance and impurity and gas control.

Some machines are designed to be highly flexible from the beginning; others have enough flexibility to allow important upgrades. Plasma physics concepts and new technologies are being developed that could result in machine upgrades allowing significant improvements to the performance of ITER.

## 1.3 Systems Engineering Process Model

Systems engineering [4] involves systems thinking about the relationship between elements and the problem solving process shown in Fig. 1.3.

There are four basic systems principles [4] for solving a problem:

- Proceed from the general to the detailed.
- Think in variants: consider alternative solutions to each problem.
- Use a phased approach in developing ideas.
- Develop a problem solving cycle as a general work logic.

The first stage of a project is the planning of the general objectives of the project and of the general means to carry out the objectives. For a nuclear fusion experiment, it is necessary to look at all of the previous knowledge and experience of related experiments, and then assess what needs to be done to achieve the project objectives. Important inputs are the capabilities of the group planning the experiment and the amount of financing available to carry it out.

Proceeding from the general to the detailed involves using the hierarchical method already discussed. It starts with a simple model using black boxes, and evolves to a detailed treatment of more complex white boxes. General objectives and a general solution framework need to be established. Concepts at higher levels serve as guidance for the detailed design. In the context of nuclear fusion experiments, the basic goals need to be decided, the approximate budget applied as a constraint and

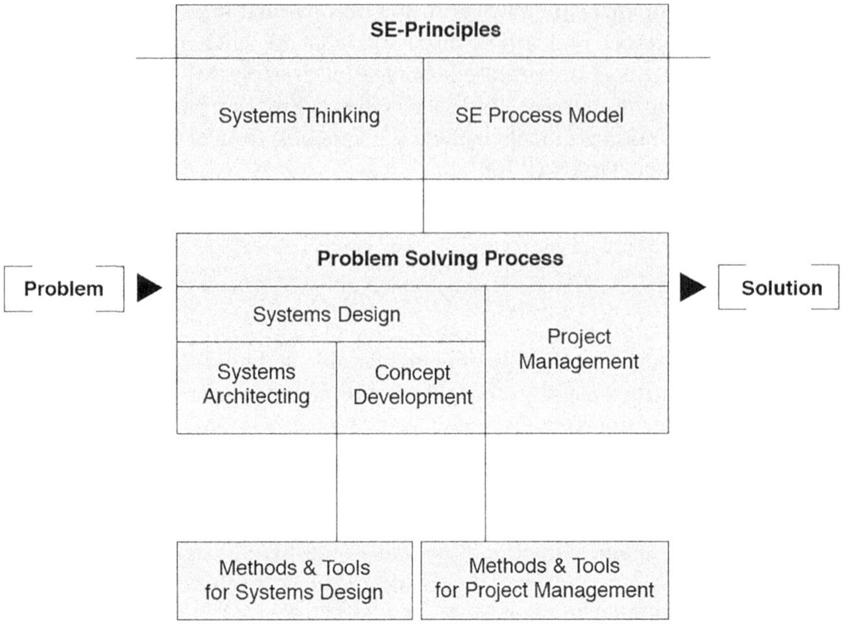


**Fig. 1.3** Systems Engineering (SE) concepts. (From [4], Fig. 1)

the overall machine parameters initially decided. Then basic choices need to be made such as superconducting or resistive coils, with or without divertor, auxiliary heating, etc.

Variant creation involves studying several solutions to the same problem. For example, if the goal is creating a high current tokamak, variants might involve high or low toroidal field, plasma aspect ratio, degree of elongation, type of divertor. Extreme variants could be scoped for feasibility. During the detailed design of the optimum choice, small variations can be explored for optimization. Variants can involve different paradigms, even an entirely different fusion concept.

Dividing the development of a project into phases that proceed from one to the next in a logical and timely fashion coincides with moving from the general to the specific. The number of phases and approaches depend on the complexity of the project. Phase I might involve a general specification of machine design suitable for preliminary external evaluation for funding, and Phase II might involve a more detailed design suitable for actual application for funding approval. Phase III might be the production of highly specific details needed for manufacturing.

The problem-solving cycle requires a chain of logic for each problem, which involves setting objectives and goals, searching for a solution, and selecting the most appropriate approach. This can include prototyping, where a low cost test of a component or concept may be made to determine its feasibility.

### 1.3.1 Problem Solving Process, Systems Architecture and Design

The systems engineering process model is implemented using the tools of systems architecture and concept development. *Systems architecture* [5] goes beyond structure and involves the *allocation of function* to the *elements of a structure*. Systems design involves *systems architecture* and *concept development*.

Systems architecture is defined as:

- the allocation of function to elements;
- the arrangement of these elements in a structure;
- the definition of interfaces between these elements and the system environment; and
- creation of a defined value.

An example for a tokamak involves the architecture of the plasma, vacuum vessel, toroidal field coils, and poloidal field shaping coils. Two different architectures could involve having the poloidal shaping coils either inside or outside the toroidal field coils. In the two cases, the functions assigned to each element would differ. The plasma function would be to demonstrate the value of a complex shape while giving the shaping coils the function of producing that configuration, made easier if inside the toroidal field coils. The vacuum vessel would have the obvious function of producing a vacuum for the plasma, but its conducting walls would also have the function of stabilizing the plasma by image currents induced by plasma motion. Each subsystem has its own architecture, but is part of the overall tokamak system. The definition of interfaces between the systems involves different aspects, including mechanical structure, cooling, electromagnetics, etc.

The definition of good architecture [5] is adapted here for fusion reactors:

- The experiment's architecture should be able to make multiple and significant contributions towards a fusion reactor.
- It should fulfil the project objectives and goals.
- It ensures compliance with regulations, especially health and safety.
- It can be operated and serviced efficiently and is sustainable.
- It is scalable and adaptable with minimal effort.
- Further machines and experiments can be developed from it.
- It is elegant.

### 1.3.2 Concept Development

*Concept design and development* involves producing a selected architectural design in a more concrete and detailed fashion. The process includes searching for objectives and possible solutions, then making a selection of the best approach.

A *design space* of existing experiments provides a basis for fixing goals and for deciding on the best options for a systems architecture to take forward. A design space is developed for tokamaks in Chap. 2, where lessons learned from existing tokamaks are summarized both for construction choices and for demonstrated operational scenarios obtained in particular systems architectures. The tokamak design space is relevant to other fusion concepts, which have their own design spaces developed in the relevant chapters.

A key element of concept design is working with models. These models advance from general concepts to specific designs, with numerical constructs to verify and quantify the concepts. Plasma physics involves developing a mathematical framework and numerical solutions for plasma behaviour and its interaction with external systems, including the best theoretical descriptions. Fusion reactor design additionally requires very extensive neutronics and other calculations relevant to a fusion product environment.

Three additional principles of good design are:

- minimizing constraints by preferring solutions that allow the most room for further developments;
- minimizing interfaces so that each subsystem can have the maximum robustness;
- using modular structures so that components could be used in other situations. In this context, a tokamak design would include the option of reusing the toroidal field coils and power supplies and motor generator in a machine upgrade or replacement.

### *1.3.3 Constraints*

Any design is subject to constraints, and choices between variants can be made by applying metrics. Primary among the constraints are *logical constraints*, which have the character that they involve choices that cannot or should not be done, or that obviously will not work as conceived. There are also constraints of *reasonableness*, where you probably would not want to exercise a certain option. Externally imposed constraints include a maximum budget for spending on the project, time and manpower limitations, and the level of complexity that can be handled.

### *1.3.4 Metrics and Risk*

Metrics allow the quantification of aspects of the design and are an excellent way of choosing between design variants and allowing the evaluation of the success of a system once built and operating. Metrics can include projected performance of the various systems, cost, time to completion, technological difficulty, amount of new innovation required, risk and performance. The metrics of actual performance are an emergent property of the entire system (see below).

Risks can be evaluated at the individual element level or at the project level. A risk on a component might depend on whether or not the component will fail during the lifetime of a project, and is related to the difficulty of repair or maintenance. Risk can also involve failure of the entire project. The risk associated with each element of a system can be quantified, and the overall risk to the system calculated as the product of the individual risks. A typical quantification of a multi-element system in an important final product would be risk level probabilities: high (0.90); medium (0.95), low (0.98), very low (0.99). The project management can then decide what level of risk is acceptable. There will often be trade-offs between total risk and total cost, for example with machining tolerances.

## 1.4 Systems Emergent Properties, Robustness and Dynamics

### *1.4.1 Emergent Properties of Systems*

Emergent properties of systems such as robustness are revealed by analysis, computational modelling and experience. Once the interactions are known, they can be integrated into a dynamic simulation. During the design of a fusion machine, the operation phase can be simulated and optimized. During the operations, analysis and simulation can be incorporated for real time control.

Emergent properties of systems need to be considered during these optimizations. In a dynamic system, robustness refers to the ability to tolerate perturbations. Robustness includes the ability of a system to resist change without damaging the main system functions. Perturbations can be small at an anticipated level, or large enough to represent a danger to the system. Disruptions in tokamaks are a prime example. Robustness therefore has two dimensions, which are resistance and avoidance.

As in the analysis of biological organisms [64], the properties for developing a **robust system** [7, Chap. 17] in fusion reactors must be optimized, including:

- systems control
- fault-tolerance
- modularity
- decoupling

### 1.4.2 Mechanisms for Robustness

Systems control involves negative or positive feedback on the desired control variable, which can make a system stable around the desired equilibrium. For example, if a motion is detected in the plasma by a magnetic field sensor, a vertical magnetic field can be strengthened to reduce the displacement to zero via negative feedback. There can also be switches to drive systems from one stable state to another, for example to transform a molecular gas into a plasma.

Fault tolerant mechanisms or control systems increase resistance to component failure and external changes, often by having alternate components or methods that maintain system function. If control of a plasma is lost, there need to be mechanisms to minimize the danger to surrounding systems. Fault tolerance can also involve redundancy or diversity. Redundancy occurs when a similar system can take over from another. Diversity involves having several different ways of accomplishing the same goals. When a plasma is in contact with a limiter, there can be other limiters at different locations that protect the vacuum vessel.

Modularity provides isolation of a perturbation from the rest of the system. It is perhaps best achieved when each individual subsystem has internal protections and controls designed to maintain it in a state, with interconnections between the subsystems that are monitored and controlled. If the plasma makes a rapid motion, it should not damage the external magnetic field coils.

Decoupling isolates low-level noise and fluctuations from functional structures. Surrounding a plasma with a conducting metal surface will produce image currents in the metal that counteract plasma motion and help to maintain its stability.

### 1.4.3 Robustness Trade-Offs

Robust systems resist fluctuations that attempt to move the system away from its desired state. Robustness involves tradeoffs:

- robustness itself
- fragility
- resource demands
- performance

Avoidance of fragilities may be achieved by designing a system such that certain fluctuations simply cannot occur. A plasma is subject to instabilities. It is possible to create magnetic geometries where plasma fluctuations are suppressed.

### 1.4.4 Fragilities

A fragility is any event that can damage or destroy a system, ranging from small perturbations during operation to a systematic failure mode. Fragilities can occur even in a system that is designed to be robust against many perturbations [69]. Not all perturbations can be anticipated.

For example, a plasma experiment would be highly robust against many plasma instabilities, but could be destroyed by a crack in a vacuum window. Several superconducting magnets in the Large Hadron Collider at CERN were destroyed not by a quench, but by a helium valve not designed for high pressure.

A certain level of performance is expected in systems. The goal of a new plasma experiment may be to push performance to new records. Components would be tested to their limits. Obviously, if operations were conducted short of the design limits, then system fragility would be reduced. Hence the size of safety margins affects performance and robustness.

### 1.4.5 Resource Demands

Resource demands can range from the quantity and complexity of components to funding for the project. Quantification of resource demands includes metrics such as expense per element, the cost of systems integration and of the total system itself. Consumption of resources is also important during an experiment such as volt seconds from a transformer or power from a motor-generator.

### 1.4.6 Performance

Performance is an emergent property of a system and is perhaps the most important metric for evaluating original project goals. It can include quantification of the progress made towards a higher level goal. Metrics of success in a tokamak can include maximum toroidal magnetic field and plasma current. Other metrics are the triple product of plasma density, temperature and energy confinement time achieved, as a measure of progress towards building a fusion reactor.

## 1.5 Examples of Systems Engineering, Architecture, and Emergent Properties

### *1.5.1 Building a House*

House building is given in [4] as an example of systems engineering. A study was presented to set goals and to develop variants and choose among them. Systems and subsystems were then identified so as to provide assessments and choice of variants so as to proceed to building. The goals altered strongly during the study and illustrated some important lessons:

- Objectives direct the solution search and enable decision making.
- Thinking in variants is essential and decisions channel the next stages.
- Incremental implementation clarifies goals, since a project focus can change.
- Working in phases helps and exit possibilities are necessary.
- Use of a problem solving cycle (PSC) is essential, where systems logic is applied to each project phase of increasing content.

### *1.5.2 Landing on the Moon and Returning*

The design of the lunar mission spacecraft of 1969 was analysed [5] in terms of developing systems architecture. Even very complicated systems can be deconstructed into simpler systems and subsystems, until a level of detail is reached that allows development of variants and optimization and decisions. The goal was to get astronauts to the moon and back, but the choice of means was wide open. The mission plan was divided into nine major decisions, and variants were determined for each, including earth orbit rendezvous, launch type, lunar orbit, etc. To decide among variants, the designers used constraints and metrics including performance, costs, developmental and operational risk. The analysis led to a series of successful missions. Even when one mission experienced major faults, the astronauts returned alive.

### *1.5.3 Guiding an Airplane in Flight*

An airplane provides an example of emergent properties [7] such as robustness, and how systems work together to achieve optimum performance. The airplane course is maintained by a pilot and an automatic flight controller which uses control surfaces and negative feedback. Fault tolerance is provided by the pilot, several redundant control surfaces and backup automatic flight controllers. Modularity comes from the steady thrust of multiple and redundant jet engines. Decoupling is achieved by the large size of the aircraft wings and the minimum size of an analogue fluctuation necessary to activate the digital controls. Overall robustness could be destroyed by the fragility of electronic control systems to a lightning strike, unless properly mitigated.

## 1.6 Systems Reverse Engineering

### *1.6.1 Identifying Elements and Relationships in Existing Devices*

There are several steps in the reverse engineering of a system. Key concepts are the form an element takes and the function defining the relationship between elements. The architecture is not only the way elements are put together spatially, but also how other interconnections are made.

The stages of system reverse engineering are:

- Identify form, functions and relationships of elements within the architecture.
- Identify boundaries and elements of highest level system.
- Identify formal view of spatial relationships.
- Determine the functional interactions at the highest level.
- Repeat the exercise for subsystems of appropriate elements, down to the level necessary for analysis.

### *1.6.2 Analysing Metrics and Emergent Properties*

Once the key elements of the experimental system are determined, an analysis can be carried out to evaluate the machine and especially to determine the knowledge acquired and the lessons learned for future machines and fusion reactors. These methodologies include:

- Using constructional, operational and experimental evidence to determine the outcomes of these functional relationships;
- Developing computational models of function, where appropriate, to aid in interpreting the results;
- Examining the emergent properties of the system in light of the goals of the system;
- Evaluating the emergent properties in terms of robustness and fragilities, and implications for future operations and machines and performance.

### *1.6.3 Contributing Lessons Learned to the Design Space*

A full analysis of existing machines and those in construction that illustrates design innovations and operational results provides crucial contributions to a design space for that class of fusion reactors. A very wide range of construction philosophies for tokamaks and other concepts have been explored. An equally broad knowledge base has been developed for physics and engineering understanding and of operational scenarios developed that can be applied.

The additional analysis of emergent properties also contributes to the design space. Sometimes the failures and lack of optimum operation can be as useful as successes in guiding future designs.

## 1.7 Systems Forward Engineering – From Concept to Architecture

This section describes how logically to apply the principles of systems engineering and systems architecture in the forward engineering of a new system. It involves setting goals and implementing them through a systems architecture procedure [5]. Specific steps are described for determining basic goals, evolving several concepts and producing a systems level design and then a full design.

### *1.7.1 Developing a Solution-Neutral Function*

A *solution-neutral function* is defined as the function of a system without reference to how it is achieved. It involves an *operand* and a *value-related process*, with the aim of meeting *value-related goals*.

The value-related goal of nuclear fusion research is to continuously generate electricity suitable for a power grid by burning light elements in a reactor. The operand is light isotopes for fuel, the value-related state is whether the light elements are burnt. The value related process is burning light isotopes in a fusion reactor and the function is changing the fuel (d, t and Li) into fusion products and energy (He + n + 17.6 MeV).

### *1.7.2 Developing and Implementing Concepts*

Concepts are the highest level of mapping of function to form, and represent a bridge towards detailed architecture. These concepts involve determining the form of objects that could be used to deliver the solution-neutral function.

For nuclear fusion, two major concepts are involved:

- One concept is to create a hot plasma with deuterium and tritium ions and electrons, burning continuously and confined by magnetic fields.
- The other is to create a stream of small d-t pellets, each of which is ignited by lasers or other means, thereby causing a rapid burn of the fuel.

The next step is to develop a hierarchy of concepts, where the next level down in turn becomes the solution-neutral function. For example, having chosen the magnetic confinement solution as the operand, the next level value-related goal is whether the light elements can be burnt or not. The next step down is to choose the most suitable magnetic configuration among toroidal and linear systems, and then for toroidal systems, among the tokamak, stellarator and reversed field pinch, and so on.

Graphic and tabular models can demonstrate relations, promote problem awareness and are the basis for possible development of quantitative models. Mathematical models can involve the full range of theoretical derivations and/or numerical analysis.

### *1.7.3 Organization of Project Management*

Project management is a key area for implementing systems thinking, and the complexity of the management clearly depends on the complexity of the project. The problem solving process is divided into systems design and project management [4] which includes measures for planning, leading, steering and monitoring a project. Usually, there is a project manager and deputy responsible for both design and management, and personnel often work on both aspects. Management involves decision making and implementation and is composed of four dimensions: functional, institutional, instrumental and personnel.

Large management structures with extensive external consultation have been established for projects such as JET [19] and ITER [20]. In large scale collaborative research projects [6, Chap. 10], the process of science management is described for projects supported by the European Commission. Overall management goals are provided from collaboration in the EUROfusion Roadmap [35].

In planning experiments on existing machines, a key aspect is to look at project goals and to make maximum use of the expertise of the scientific and engineering team working on the project.

## 1.8 Summary Checklists of Systems Strategies for Fusion

The system strategies discussed in this chapter are organized into checklists developed below for several general types of analysis. They are indicative and need to be adapted to each individual project. The checklists are first described and then listed together in Sect. 1.8.9.

### *1.8.1 Systems Concepts Summary – Checklist 1.1*

The main considerations for systems thinking are summarized in Checklist 1.1:

overall structure, definitions, systems, elements, relations, systems aspects, systems agility and project management

### *1.8.2 Analyse Existing Plasma Physics and Nuclear Fusion Machines – Checklist 1.2*

Analysing existing machines and experiments makes full use of the methodology developed for reverse engineering and evaluation, summarized in Checklist 1.2. The checklist will be used in a case study in Chap. 3 on Doublet III /DIII-D and in Chap. 5 on JET.

Chapters 9, 10 and 11 deal with fusion concepts using both magnetic and inertial confinement, thus distinct from the tokamak design space, although there are some common features. Therefore, a new design space will be developed in each chapter for these concepts. Existing experiments will be reverse engineered and analysed on the basis of Checklist 1.2.

### *1.8.3 Design, Fund, Build and Commission New Fusion Machines – Checklists 1.3A and 1.3B*

#### 1.8.3.1 Challenges of Designing a Reactor-Relevant Fusion Experiment

No fusion reactor has yet been built, so fusion experiments need to contribute prototyping information towards a reactor. In planning a new experiment in nuclear fusion, a key decision is needed on the level of ambition for each experiment, and how close the experiment would be to modelling or investigating projected reactor conditions. The initial years of difficulty in fusion research provided a sentiment that first it was necessary to produce a plasma that had a good energy confinement time, high temperature and high density sufficient for a reactor. The burning plasma aspects of neutrons and tritium and the need for a positive fusion power balance could be considered later for the whole reactor plant, plus tritium breeding for fuel and neutron shielding. There was also the hope that new and future technologies such as high temperature superconductors would help to resolve present design difficulties.

The consequence of this approach was that earlier tokamak experiments were built with water cooled or nitrogen cooled pulsed toroidal field coils for medium plasma pulse lengths, rather than superconducting magnets that had effectively zero power consumption so that a power plant could produce net electricity output. Earlier experiments were sometimes conducted with hydrogen rather than with deuterium and/or radioactive tritium, which would have produced high energy neutrons necessitating personnel and component shielding.

In experiments that injected deuterium neutral beams into plasmas, the relatively low level of neutron flux was still an important design consideration. In the first tritium experiments in plasma, the tritium was purchased from fission reactors rather than produced in breeding blankets converting neutrons into fuel, thus saving space and cost. As experiments approached closer to being reactors, machines such as JET and ITER incorporated tritium and shielding to become more reactor relevant. An important design consideration was whether experimental features were relevant to reactors.

#### 1.8.3.2 Detailed Checklists

The two checklists give general systems principles for forward engineering in 1.3A and detailed steps using these principles in 1.3B. These checklists will be used in a case study in Chap. 4 on the TCV tokamak, using the toroidal confinement systems design space in Chap. 2. They will also be used in Chap. 9 on W7-X, Chap. 10 on MFTF-B, and Chap. 11 on NIF.

### *1.8.4 Design and Carry Out Fusion Experiments on Existing Machines – Checklist 1.4*

To plan experiments on already existing facilities, detailed implementation strategies incorporate features of systems analysis to maximize success.

After machine construction, commissioning up to first plasma is completed with start-up diagnostics and operating systems. Next, more diagnostics are added. Finally, the machine reaches full performance in toroidal field and plasma current, and then auxiliary heating and a full set of plasma diagnostics are added, and detailed plasma physics studies can begin.

A more challenging experimental environment occurs when deuterium and tritium fusion is involved, because of radioactive tritium and induced radiation and safety hazards from 14 MeV neutrons. Isotope mass effects may also occur in particle, energy and gas transport and neutral beam charge-exchange efficiencies.

Using the forward engineering principles of Checklist 1.3A, Checklist 1.4 is developed as a guide for devising experimental programmes in general, and specifically on deuterium-tritium experiments. The checklist will be used in a case study in Chap. 5 on the deuterium-tritium PTE, DTE1 and DTE2 experiments on the JET tokamak.

### *1.8.5 Analyse Prototype Reactor Machines and Experiments – Checklist 1.5*

Taking the ITER project as a model of a proto-fusion reactor, several experiments were constructed around the world with one of their goals being to create a prototype for ITER by using superconducting coil systems with appropriate structural components and heat shields. Part of this prototyping is to explore very long plasma pulse operation. These experiments contribute towards making ITER and future demonstration fusion reactors a success.

The Checklist 1.5 will be used in a case study in Chap. 6 on superconducting machines Tore-Supra and WEST that are prototyping ITER and fusion reactors.

### *1.8.6 Planning and Building Next-Step Fusion Reactor Machines – Checklist 1.6*

The next-step proto-fusion reactor ITER is the subject of Chap. 7. Several systems approaches and models are being used in its design construction. It is a highly documented machine, with a large number of options already considered. Many new innovations have produced solutions to difficult problems [70]. Therefore, ITER is a machine whose construction parameters are fixed, but there are no experimental results, and there are more years of prototyping results on other tokamaks to consider. Hence we use a modified reverse engineering checklist to illustrate the design choices made for ITER.

### *1.8.7 Designing and Building a Fusion Reactor – Checklist 1.7*

Several demonstration tokamak reactors and fusion engineering devices are being designed whose main features include very long plasma duration or even steady state operation, and high and sustained fluxes of 14 MeV neutrons. The checklist 1.7 will be used in Chap. 8 for a case study on EUROfusion's EU DEMO and on tokamak demonstration and engineering test fusion reactors to look at the choices that are being made in the reactor design space and compare them for their robustness and fragility properties.

## *1.8.8 Comparing Approaches to Diverse Concept Fusion reactors – Checklist 1.8*

Comparisons are made in a structured manner (Checklist 1.8) in Chap. 12 of goals, achievements and prospects of toroidal, linear, and inertial fusion concepts.

## *1.8.9 Detailed Checklists 1.1 to 1.8*

### 1.8.9.1 Checklist 1.1 Summary of the Main Considerations for Systems Thinking

Overall Structure – Systems Engineering Principles

- Use systems thinking and a systems engineering process model.
- Develop a problem solving process which involves systems design, systems architecture and concept development, plus project management, and modeling tools.

Definitions – Systems, Elements and Relations

- A system such as a fusion device involves connectivity between elements.
- Elements, which have properties and functions, can themselves be seen as systems, with their own subsystems.
- Relations can be in one direction or in both directions.
- Structures include hierarchical, network and feedback.

Systems Hierarchy

- Systems hierarchy is a system divided into multiple levels.
- Under the system, there is a level of subsystems of order 1, 2, etc. down to elements, seen as:
  - black boxes, where only the function or desired inputs and outputs (results) are of significance.
  - white box, when the exact connection between output and input is of great significance
  - grey box, which is roughly or partially structured.

- Systems design involves a transition from black to white boxes.

Systems Aspects

- Aspects can be used to describe every system from different viewpoints.
- System views can range greatly in complexity.
- The goal of systems engineering is to look for the simplest view and structure that allows analysis.

Systems Agility

- Agile systems are variable and modifiable in a changing environment.

Project Management

- Project management should be used in the context of systems thinking.

#### 1.8.9.2 Checklist 1.2: Systems Reverse Engineering and Evaluation

Identify Form and Function and System Architecture

- Review the machine's design and scientific and engineering publications and laboratory reports on experimental design.
- Identify elements of highest level system.
- Identify form of elements and formal view of spatial relationships.
- Determine the functional interactions at the highest systems level.
- Repeat the exercise for subsystems of appropriate elements, down to the level necessary for analysis.
- Use constructional, operational and experimental evidence to determine the outcomes of these functional relationships.
- Where appropriate, develop functional models to aid in interpreting the results.
- Examine the emergent properties of the system in light of the goals of the system and of new goals developed during operation.

Identify and Evaluate Metrics and Performance

- Evaluate the robustness and fragilities of the entire system and of operational modes.
- Evaluate performance of overall system and of components.
- Make use of a historical design space for comparing construction choices and operational scenarios to those of other machines.
- Determine new knowledge acquired and lessons learned for other experiments and for future machines.
- Identify new contributions to a common machine design space in terms of construction choices and developed operational scenarios.

#### 1.8.9.3 Checklist 1.3A: General Systems Considerations of Forward Engineering for Constructing a Fusion Device or Designing Experiments

Process-Model Design Activities Should Include:

- Going from general to detailed is the correct progression.
- Variants and alternative solutions to each problem should be developed.
- Phased approaches should be used in developing ideas and projects.
- Problem solving cycles should be used with objectives, various solutions and choice of approach.
- Project management is a key area for initiating these ideas.

Systems Architecture

- Systems architecture is defined as:
  - Allocation of function to elements;
  - Arrangement of these elements in a structure involving relations;
  - Definition of interfaces between these elements system environment;

Good Architecture Involves:

- Multiple and significant contributions towards a fusion reactor.
- Project objectives and goals should be achieved.
- Compliance with laws and regulations should be assured.
- Operation and servicing should be efficient and sustainable.

Concept Development

- Concept design involves developing an architecture in a more concrete and detailed fashion.
- Use the design spaces of existing experiments such as the one in Chap. 2, or those of other concepts, to identify goals, construction choices and operating scenarios as a basis for creating initial conceptual designs.
- Identify innovations in construction and operating scenario options that widens the choice of goals and variants to be studied.

Systems Models

- Systems thinking involves system models and model-based illustrations.
- Models are a key element of concept design.
- Models can demonstrate relations and promote problem awareness.

Systems Dynamics Applied to Fusion Reactors

- Dynamics involve the mathematical and computational modeling of time-dependent interactions of a system's components.

- Design phase modeling simulates the operation phase dynamically so as to optimize the design of each system.

Project Management

- Project management should be developed in the context of systems thinking and be appropriate to the scale and ambition of a project.

#### 1.8.9.4 Checklist 1.3B: Systems Forward Engineering for Constructing a Fusion Device

Getting Started

- Establish preliminary goals.
- Review other machines and lessons learned.
- Examine options from available design space and identify innovations needed.
- Choose a small team to rapidly develop primary ideas.
- Choose main machine options: minor upgrade, major upgrade, new machine, re-use existing hardware, group expertise, maximum budget, level of ambition.

Review and Prioritize Goals and Evaluate Variants

- Review project goals and reactor relevance and most optimistic outcomes.
- Review goals for a partially equipped machine at start-up, and prospects for later equipment or capability added as project progresses.
- Scope the highest system level of major variants, especially upgrade or new machine.
- Develop or find models for primary level systems, estimate performance.
- Produce first report and discuss and present with wider group and laboratory or project management.

Produce Definitive Proposals and Engineering Design

- Choose one or more of the most promising variants, expand project team, and begin detailed design of level 1 and level 2 subsystems of key areas.

Phase I Proposals (Euratom Levels)

- Produce Phase I proposal with specific design, appropriate physics and engineering modeling, cost and time estimates.
- Review at team, laboratory or project level, submit to funding source.

Phase II Proposal

- If Phase I application is successful, incorporate outside suggestions, proceed to full scale engineering and physics design, being ready to revise design in case of problems encountered.

- Carefully decide what if any innovation and development is needed for construction,
- Allow for future upgrades.
- Assess robustness and fragilities in advance, and take precautions, especially the ability to repair faults.
- Produce and internally review Phase II proposal.
- Submit for definitive funding approval.

Final Engineering Design and Construction After Final Funding Approval

- If successful, incorporate any suggested changes, proceed to detailed engineering design suitable for calls for proposals.
- Prepare design and specifications for calls for proposals, and apply the same level of detailed design to simple and complex systems, because it is often the simplest systems that fail from lack of attention.
- Develop detailed management and construction planning, and most especially a quality control and interface system.
- Submit calls for proposals, and if possible, have as wide a range of bidders as appropriate.
- Continuously plan for additional possible funding and machine, plasma heating and diagnostic upgrades.

#### 1.8.9.5 Checklist 1.4: Systems Forward engineering for Designing Experiments and Operational Procedures on Fusion Machines

Getting Started

- Establish preliminary goals.
- Review other machines and lessons learned (see scenarios in design spaces).
- Examine options from available machine operation space and identify new experiments needed using Checklist 1.3A.

Prepare for Preliminary Tritium Experiments

- Determine goals for preliminary experiment in terms of hardware, tritium systems, gas puffing, pellets and neutral beam injection, and target plasmas.
- Develop or find models for primary level systems, estimate performance.
- Obtain target plasmas using deuterium-deuterium plasmas to produce test 14 MeV neutrons from d-d produced MeV tritons to calibrate diagnostics
- Perform technology upgrades necessary for working with tritium.
- Revise goals, examine options, and carefully plan tritium and 14 MeV budget.

Perform Preliminary Tritium Experiments

- Design and carry out experiments.
- Coordinate publications.

Prepare for Full Fusion Power Experiments

- Analyse results, revise goals.
- Perform necessary machine upgrades, develop new scenarios over years of operation in preparation.

Planning Full Power Experiments with Significant (Tens of %) Tritium.

- Assemble DTE task force with project manager, scientific secretary, experimental project leaders, and scientific group leaders.
- Review overall machine goals, determine technical and time constraints: total neutron and tritium use budget with recycling, machine limitations.
- Ask project and task leaders to propose experiments and neutron and tritium budgets in terms of primary goals, secondary goals and supplementary experiments that can be carried out at the same time.
- Use group discussions and preferences to rank experiments and to revise overall experimental goals and neutron, tritium and time budgets.

Carry Out Full Power Tritium Experiments

- Perform prototype experiments of the program in deuterium.
- Start the main experiment with trace tritium tests.
- Carry out the main program, adjusting according to results.
- Carefully document the tritium cleanup.
- Coordinate publications.

Upgrade Machine to Reactor Prototype and Repeat

- Observe progress on all machines and prepare prototype experiments for proto-reactor experiments such as ITER.
- Update procedures and plan for DTE2.

#### 1.8.9.6 Checklist 1.5: Systems Reverse Engineering for Analyzing Superconducting Machines with Long-Pulse plasmas for Prototyping Reactors

Develop Systems Analysis Strategy for Prototyping Reactor Subsystems

- Determine goals and systems architecture choices for superconducting and steady-state machines.
- Review parameters and technological choices for existing and under-construction machines.
- Identify key elements and constraints of steady-state operation.

Develop Design Space for Superconducting and Steady-State Machines

- Establish design space for superconducting and long-pulse machines
- Review goals for the design space to be established.
- List technical aspects of the design space for superconducting magnets.
- Carry out the systems architecture of choices, trade-offs and operational space options in the superconducting and steady-state design space.
- Determine the methods of systems architecture choices of form and function.

Examine in Detail the Design Choices Implemented in Several Machines

- Review design.
- Study main operating scenarios developed.
- Determine robustness, fragilities, and relevance to steady state operation.

#### 1.8.9.7 Checklist 1.6: Systems Reverse Engineering and Evaluation of a Proto-Fusion Reactor Under Construction

Identify Form and Function and System Architecture

- Review general goals and basis parameters of ITER and put them in the context of past and future machines.
- Review detailed systems architecture, design space innovations
- Determine the emergent properties and expected performance in light of the system goals.

Identify and Evaluate Metrics and Performance and Lessons Learned

- Evaluate the expected robustness and fragilities of the entire system and subsystems and of operational modes.
- Determine new knowledge to be acquired and lessons to be learned for future machines.
- Speculate on future directions and upgrades for ITER in the context of future needs of fusion reactor systems.

#### 1.8.9.8 Checklist 1.7: Systems Strategy for Planning and Building a Demonstration Fusion Reactor

Identify Goals and Essential Elements of a Fusion Reactor

- Review general goals and basis parameters of proto-reactors and demo reactors to put them in the context of past and future machines.
- Identify key plasma physics, magnets, blanket physics and technology.

- Review detailed systems architecture, design space innovations and expected performance and compare to proto-reactors such as ITER.

Use Systems Codes to Develop Design Analysis Methods

- Identify form and function and system architecture of various subsystems.
- Identify and evaluate metrics and performance.

Plan Major Steps for Designing a Fusion Reactor

- Carry out initial design planning.
- Use systems modeling tools to fully explore and optimize the relationship between elements in the design space.
- Prepare detailed design at optimum scoping point.
- Develop comparison strategies for choosing technologies and optimizing performance by analysing design variants.

Examine Fusion Reactor Designs and Their Variations

- Identify key design features of each proposed machine.
- Reverse engineer specific designs to highlight their basis for a fusion reactor.
- Evaluate expected performance of overall system and of components.
- Evaluate the expected robustness and fragilities of the entire system and of operational modes.
- Determine new knowledge to be acquired and lessons to be learned for future machines.
- Compare the various approaches to a fusion reactor.

#### 1.8.9.9 Checklist 1.8: Systems Strategy for Comparing Different Fusion Reactor Concepts

Develop Overall Systems Analysis Strategy for Treating Concepts by Systems Architecture

- Use a systems strategy for comparing different fusion reactor concepts.
- Evaluate systems level metrics in existing and planned machines

Examine Progress Towards and Operating Fusion Reactor

- Determine the stages of progress towards a working fusion reactor and characterize degree of progress of each concept.
- Analyse the robustness and fragilities of different concepts.
- Extend the analysis of robustness and fragility to reactor concepts.
- Integrate methods for improving robustness and countering fragilities.

**Summarize optimization of fusion reactors by systems approaches**

# References


1. F.B. Marcus et al., Neutron emission profile measurements during the first tritium experiments at JET. Nucl. Fusion **33**(9), 1325 (1993)
2. M. Keilhacker et al., High fusion performance from deuterium-tritium plasmas in JET. Nucl. Fusion **39**(2), 209 (1999)
3. EUROfusion NEWS. https://www.euro-fusion.org/news/2022/european-researchers-achieve-fusion-energy-record/. 9 February 2022
4. R. Haberfellner et al., *Systems Engineering: Practice and Applications* (Imprint Birkhäuser, Springer, Cham, 2019)
5. E. Crawley et al., *Systems Architecture: Strategy and Product Development for Complex Systems* (Pearson, Boston, 2016)
6. F.B. Marcus, *Bioinformatics and Systems Biology* (Springer, Heidelberg, 2008)
7. H. Kitano, Chapter 17: Cancer robustness and therapy strategies, in *Cancer Systems Biology, Bioinformatics and Medicine: Research and Clinical Applications*, ed. by A. Cesario, F. B. Marcus, (Springer, Dordrecht, 2011)
8. J.H. Heilbron, *The Dilemmas of an Upright Man: Max Planck and the Fortunes of German Science, with a New Afterword* (Harvard University Press, 1996)
9. P. Barabaschi et al., System engineering and configuration management in ITER, presented by S. Chiocchio, Proceedings of 24th SOFT Conference, Warsaw (2006). https://slideplayer.com/slide/1514227/. Last accessed 18 June 2022
10. S. Chiocchio et al., Integrated systems engineering approach to manage the installation and start up of a fusion reactor, Semantics Scholar (2008). https://www.semanticscholar.org/paper/TO-MANAGE-THE-INSTALLATION-AND-START-UP-OF-A-FUSION-Chiocchio-Kuehn/11662dc0a6a04efce1eac4891a801a2f7924c3a5?p2df. Last accessed 18 June 2022
11. G. Federici et al., Overview of the DEMO staged design approach in Europe. Nucl. Fusion **59**, 066013 (2019)
12. I. Jenkins et al., A SysML model of the tokamak subsystems involved in a DEMO pulse, report EUROFUSION WPPMI-CP(16) 15445, Proceedings of 29th Symposium on Fusion Technology (SOFT) (2016)
13. G. Grossettia et al., Systems engineering perspective to the integration of the heating and current drive system in the EU DEMO: analysis of requirements and functions. Fusion Eng. Des. **136A**, 53 (2018)
14. G. Federici et al., The plan forward for EU DEMO. Fusion Eng. Des. **173**, 112960 (2021)
15. M. Kovari et al., 'PROCESS': a systems code for fusion power plants – part 2: engineering. Fusion Eng. Des. **104**, 9 (2016)
16. C. Reux et al., DEMO design using the SYCOMORE systems code: influence of technological constraints on the reactor performance. Fusion Eng. Des. **136**, 1572 (2018)
17. G. McCracken, P. Stott, *Fusion: The Energy of the Universe*, 2nd edn. (Academic/ Elsevier, Oxford, 2013)
18. S.O. Dean, Search for the ultimate energy source: a history of the U.S, in *Fusion Energy Program*, (Springer, New York, 2013)
19. E.N. Shaw, *Europe's Experiment in Fusion: the JET Joint Undertaking* (North-Holland, Amsterdam, 1990)
20. M. Claessens, *ITER: the Giant Fusion Reactor: Bringing a Sun to Earth* (Springer, Cham, 2020)
21. J. Wesson, *Tokamaks*, 4th edn. (Oxford University Press, Oxford, 2011)
22. F.F. Chen, *Introduction to plasma physics and controlled fusion*, 3rd edn. (Springer, Cham/ Heidelberg, 2018)
23. Tokamak, Tokamak Wikipedia. https://en.wikipedia.org/wiki/Tokamak. Last accessed 17 June 2022
24. E. Morse, *Nuclear Fusion* (Springer, Cham, 2018)

25. V. Glukhikh, O. Filatov, B. Kolbasov (eds.), *Fundamentals of Magnetic Thermonuclear Reactor Design* (Elsevier/Woodhead Publishing, 2018)
26. K.H. Burrell et al., Attainment of reactor level volume-averaged beta in doublet III. Nucl. Fusion **23**(4), 536 (1983)
27. C.C. Petty and the DIII_D Team, DIII-D research towards establishing the scientific basis for future fusion reactors. Nucl. Fusion **59**, 112002 (2019)
28. H. Meyer et al., Overview of physics studies on ASDEX upgrade. Nucl. Fusion **59**, 112014 (2019)
29. M. Greenwald et al., 20 years of research on the Alcator C-Mod tokamak. Phys. Plasmas. **21**, 110601 (2014)
30. J.R. Harrison et al., Overview of new MAST physics in anticipation of first results from MAST upgrade. Nucl. Fusion **59**(11), 112021 (2019)
31. J.E. Menard et al., Overview of NSTX upgrade initial results and modelling highlights. Nucl. Fusion **57**(10) (2017)
32. S. Coda and the TCV Team, Physics research on the TCV tokamak facility: from conventional to alternative scenarios and beyond. Nucl Fusion **59**, 112023 (2019)
33. F.B. Marcus et al., Controlled evolution of highly elongated tokamak plasmas. Phys. Rev. Lett. **55**, 2289 (1985)
34. X. Litaudon et al., Overview of the JET results in support to ITER. Nucl. Fusion **57**, 102001 (2017)
35. A.J.H. Donné, Eurofusion roadmap, Symposium on Fusion Technology, Sicily, Italy, 17 September 2018
36. R.J. Hawryluk et al., Results from D-T experiments on TFTR and implications for achieving an ignited plasma. Philos. Trans. A: Math. Phys. Eng. Sci. **357**(1752), 443–469 (1999) The Approach to Ignited Plasma
37. A. Isayama et al., Achievement of high fusion triple product, steady-state sustainment and realtime NTM stabilization in high-βp ELMy H-mode discharges in JT-60U. Nucl. Fusion **43**(10), 1272 (2003)
38. R. Aymar et al., *Tore Supra: Basic Design Tokamak System, Report EUR-CEA-FC-1068* (Fontenoy-aux-Roses, France, 1980)
39. J. Bucalossi et al., The WEST project: testing ITER divertor high heat flux component technology in a steady state tokamak environment. Fusion Eng. Des. **89**(7–8), 907 (2014)
40. X. Gao et al., Key issues for long-pulse high-βN operation with the experimental advanced superconducting tokamak (EAST). Nucl. Fusion **57**, 056021 (2017)
41. G. Lee, The KSTAR project: an advanced steady state superconducting tokamak experiment. Nucl. Fusion **40**(3Y), 575 (2002)
42. Y.C. Saxena, SST-1 team present status of the SST-1 project. Nucl. Fusion **40**(6), 1069 (2000)
43. JT60-SA. http://www.jt60sa.org/b/n1/introduction.htm
44. G. Federici et al., DEMO design activities in Europe: progress and updates. Fusion Eng. Des. **136**, 729 (2018)
45. G. Beidler et al., Physics and engineering design for Wendelstein VII-X, IAEA 7th International Workshop on Stellarators, Oak Ridge, Tennessee, pp. 479, 10–14 April 1989
46. G. Motojima, Recent design considerations for the large helical device, IAEA 7th International Workshop on Stellarators, 10–14 April 1989, Oak Ridge, Tennessee, Report IAEA-TECDOC-558, 471 (1989)
47. V.N. Karpenko, The Mirror Fusion Test Facility (MFTF-B) – Status, Lawrence Livermore Laboratory Report UCRL-92247, 6th Topical Meeting on the Technology of Fusion Energy, Knoxville, Tennessee, Paper CONF-850310 – 107 (1985)
48. M. Gobbin et al., Ion Heating and Energy Balance During Magnetic Reconnection Events in the RFX-Mod Experiment, 28th IAEA Fusion Energy Conference, Nice, France (virtual) 10-15 May 2021, Presentation, Paper IAEA–CN–286-1052, Submitted to Nuclear Fusion (2021)

49. P. Patel, Progress in the U.S. Inertial Confinement Fusion Program, 28th IAEA Fusion Energy Conference, Nice, France (Virtual) 10–15 May 2021, Presentation OV/3–1, Paper IAEA–CN–286-1168, Submitted to Nuclear Fusion (2021)
50. B. Le Garrec, HiPER laser: from capsule design to the laser reference design, Proceedings of SPIE – The International Society for Optical Engineering 7916, February 2011
51. L.A. Artsimovich et al., *Plasma Physics and Controlled Nuclear Fusion Research*, vol I (IAEA, Vienna, 1969)
52. Alcator design group, in Proceedings of the IVth European Conference on Controlled Fusion and Plasma Physics, Rome, p. 20, 1970, Design of Alcator – The MIT high field torus (1970)
53. List of fusion experiments (Wikipedia). https://en.wikipedia.org/wiki/List_of_fusion_experiments. Last accessed 17 June 2022
54. S. Li et al., Optimal tracking for a divergent-type parabolic PDE system in current profile control, Abstract and Applied Analysis, Volume 2014, Article ID 940965, Hindawi Publishing Corporation (2014)
55. ITER Physics Expert Group, Chapter 2: Plasma confinement and transport. Nucl. Fusion **39**(12), 2175 (1999)
56. V. Mukhovatov et al., Overview of physics basis for ITER. Plasma Phys. Control. Fusion **45**, A235 (2003)
57. M. Roberts et al., *Oak Ridge Tokamak Experimental Power Reactor Study Reference Design, Report ORNL-TM-5042* (Oak Ridge National Laboratory, Oak Ridge, 1975)
58. D.J. Rose, M. Clark Jr., *Plasmas and Controlled Fusion* (MIT Press, Cambridge, 1961)
59. D. McAlees, F. B. Marcus (eds.), *Oak Ridge Tokamak Experimental Power Reactor Study-1976-Part 2 – Plasma Engineering in a Deuterium-Tritium Fueled Tokamak, Report ORNL-TM-5573* (Oak Ridge National Laboratory, Oak Ridge, 1976)
60. F.B. Marcus, Air Core Poloidal Magnetic Field System for a Toroidal Plasma Producing Device. U S Patent Number 4,087,322 (1978)
61. F. Najmadabi et al., The ARIES tokamak fusion reactor study, IEEE Thirteenth Symposium on Fusion Engineering, Knoxville, TN, USA, 2, 1021 (1989)
62. F. Najmabadi et al., The ARIES-AT advanced tokamak, advanced technology fusion power plant. Fusion Eng. Design **80**, 3 (2006)
63. V. Chan, M.Q. Tran, G. Wurden, Preface to the special issue: creativity and innovations on the path to fusion energy. J. Fusion Energy **38**, 1–2 (2019)
64. A. Cesario, F.B. Marcus, *Cancer Systems Biology, Bioinformatics and Medicine: Research and Clinical Applications* (Springer, Dordrecht, 2011)
65. OMG Systems Modelling Language, Version 1.6, An OMG Systems Modeling Language Publication. https://www.omg.org/spec/SysML/1.6/ (2019). Last accessed 17 June 2022
66. D. Dori, *Model-Based Systems Engineering with OPM and SysML* (Springer, New York, 2016). https://doi.org/10.1007/978-1-4939-3295-5. ISBN 9781493932955
67. S.C. Jardin et al., Dynamic modeling of transport and positional control of Tokamaks, J. Comput. Phys. **66**, 481 (1986)
68. R. Budny et al., *Simulation of DT Experiments in TFTR*, Report PPPL-2808 (Princeton Plasma Physics Laboratory, Princeton, 1991)
69. J.M. Carlson, J. Doyle, Highly optimized tolerance: a mechanism for power laws in designed systems. Phys. Rev. E: Stat. Phys **60**, 1412–1427 (1999) Plasma Fluids related Interdisciplinary topics
70. D.J. Campbell et al., Innovations in technology and science R&D for ITER. J. Fusion Energy **38**, 11–71 (2019)

# Chapter 2
# Systems Design Space for Tokamak Physics and Engineering

**Abstract** A design space is developed which contains the architectural elements and experimental scenarios needed for systems approaches. Detailed views are developed including time views of plasma operation and spatial views of form. Component and interface options are catalogued, including the interactions between elements. The designs and key experiments of several tokamaks are examined from machines with less than 700 kA plasma current. Their operational scenarios are then summarized along with those of larger tokamaks discussed in Chaps. 3, 4, 5 and 6. Theoretical and simulation models are an essential part of systems design and are discussed and classified. The design space is integrated into systems approaches to machine design and their analysis.

## 2.1 Developing a Design Space

The systems architecture approach to developing a tokamak design space requires looking at the integrated system and subsystems from several views including:

- the time dependent view of tokamak operation;
- the form view of the placement and interfaces of various machine and plasma elements;
- the subsystem view of various critical elements, especially plasma and molecular physics;
- the electromagnetic interaction viewpoint.

Chapter 1 provided a brief summary of tokamak and reactor physics [1, 2]. In order to develop a tokamak design space for either reverse engineering of existing machines or forward engineering for building new ones, it is necessary to review the full range of tokamak machines and their experiments. The tokamak design space is constructed from lower current tokamaks examined in this chapter plus scenarios from Chaps. 3, 4, 5 and 6.

In a design space, each component is viewed as an option for the machine or as an element to be controlled to achieve a desired goal. A plasma instability can either be a nuisance, a major danger, or an important tool for plasma confinement and for

 
F. B. Marcus, *Systems Approaches to Nuclear Fusion Reactors*, Springer Series in Plasma Science and Technology, https://doi.org/10.1007/978-3-031-17711-8_2

eliminating heat from the plasma. A confinement vessel can simply produce a vacuum or it can also be a major element in stabilizing the plasma against positional and other instabilities. A poloidal coil optimally positioned for plasma shaping can be a major obstacle to machine repair in case of a fault. Hence it is vital to look at all the interactions of the various systems and consider their implementation within the overall architecture.

## 2.2 Detailed Views of Tokamak Design Space

### *2.2.1 The Time View*

One approach to the time view [3] uses the SysML systems language to extrapolate from JET operation to a demonstration fusion reactor (DEMO) plasma pulse, as shown in Fig. 2.1. They model a plasma discharge as a state machine, going from one state to the next. These states include:

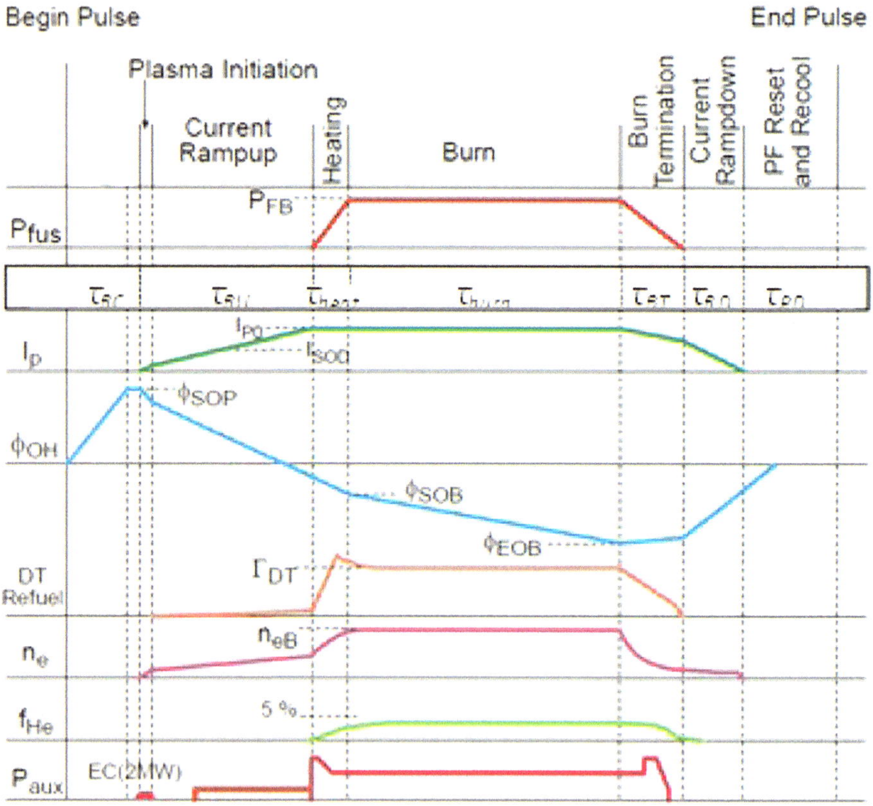


**Fig. 2.1** Schematic showing sequence of events in a plasma pulse. (From [3], Fig. 1)

- vacuum vessel conditioning
- pre-pulse
- initiating plasma
- ramping up plasma current on limiter
- ramping up plasma current with divertor x-point
- plasma heating
- starting fusion burn
- maintaining burn
- terminating burn
- cooling plasma
- ramping down plasma current
- decaying plasma
- dwelling
- end pulse, prepare for next discharge

The plasma at gas breakdown is very different from a burning plasma. It is created in high vacuum conditions obtainable by baking and conditioning of the vessel. Molecular gas or pellets are injected into the vessel, and an initial plasma is formed by heated metal filaments, electron cyclotron breakdown or applied toroidal electric field. The gas breaks down and undergoes an ionization cascade until a highly ionized plasma at medium electron temperature is ready for the plasma current to be increased or maintained. During the rise of the plasma current, external magnetic fields are continuously varied to encourage plasma breakdown and then to maintain plasma position and shape.

The different phases require different combinations of equipment to perform tasks at various times and to be used not at all during some or most phases. For example, special vacuum vessel internal wall preparation might be used only during the vacuum vessel conditioning whereas vacuum pumps would operate continuously. Each tokamak experiment could pose very different requirements for different elements. A toroidal field coil required to run continuously for years has a different design from a machine that runs for a few seconds only.

A tokamak composite view of the elements is used to develop the time view along with more detailed subsystems. Each model needs to be appropriate to the level of detail required and the problems to be solved. Far more detailed models with spatially dependent plasma physics and magnetohydrodynamics (MHD) such as the Tokamak Simulation Code can compute plasma equilibrium, stability and the full current increase to a highly elongated plasma [4].

## 2.2.2 *Time Scales and Design Options*

In choosing design goals for a tokamak, it is necessary to examine the time scale on which these aims are to be achieved. A truly steady-state tokamak fusion reactor with current drive would be able to provide continuous electrical power to the grid for months or years between scheduled maintenance. All equipment must be capable of operating in steady state, even if certain aspects are cyclical or dynamic, including not least the plasma itself.

For experiments serving as a prototype of some aspect of a fusion reactor, all relevant time scales need to be examined to see if the machine goals can be met, and if the machine performance is consistent with these goals. For example, a machine might aim to transiently demonstrate high values of plasma parameters. However, these achievements may be dependent on the particular state of the plasma, which might affect the generality of the result. There are relevant time scales both for the plasma and for the tokamak machine itself:

#### 2.2.2.1 Tokamak Plasma Time Scales

Plasma time scales range over several orders of magnitude and include:

- Plasma breakdown time
- Plasma current rise time
- Alfven time
- Vertical instability time
- Plasma energy confinement time
- Fuel (d-d or d-t) particle confinement time
- Fusion product confinement time
- Impurity particles confinement time
- Plasma profiles internal redistribution times
- Plasma profiles induced redistribution times
- Plasma internal current density redistribution time
- Plasma instability or disruption time scales

#### 2.2.2.2 Tokamak Machine Time Scales

There are also machine time scales:

- Vacuum vessel 'skin' time for field penetration
- Power supply response time
- Coil inertial heating time (for normal conductors)
- Plasma facing components inertial heating time
- Time limit for power from stored energy devices (capacitor, motor generator)
- Cryopumping surfaces saturation and refreshing time
- Time limits on auxiliary heating equipment
- Tritium breeding and reprocessing time
- Time to reach true steady-state equilibrium for all components

### 2.2.3 Plasma View

The plasma is the central element of the tokamak, and its design space needs to be developed first. The various sub-elements within the plasma are highly time-dependent during formation. At high temperature, the plasma will have the following sub-elements, all of which need to be controlled. Derivations and discussions of plasma quantities are available in Chap. 1 and [1, 2].

- Plasma composition: The plasma is composed of fully ionized deuterium and tritium isotopes, and full or partly ionized impurities, electrons, electric and magnetic fields, current rotation, and fast ions and electrons with energies induced by auxiliary heating.
- MHD equilibrium involves:
  - external fields
  - equilibrium and flux functions
  - q, beta, aspect ratio
  - electric fields
  - particle orbits
  - rotation and current drive
- Particle and energy confinement depends upon:
  - confinement modes and empirical scaling
  - H-modes and other high confinement regimes
  - internal transport barriers
  - neoclassical ion thermal conductivity
  - Pfirsch-Schluter currents and bootstrap currents
  - fluctuations and turbulence
  - radiation losses
- MHD stability and related instabilities types are:
  - kink
  - tearing modes
  - neoclassical tearing modes
  - internal kink
  - resistive mode
  - localized modes and ELMs
  - ballooning
  - axisymmetric
  - beta-limit
- Microinstabilities include:
  - electron drift wave
  - passing and trapped particle instabilities
  - micro-tearing mode

- Plasma-surface interactions involve:
    - plasma sheath and scrape-off layer
    - recycling
    - atomic and molecular processes and impurities
    - wall conditioning
    - sputtering
    - arcing
    - limiters
    - divertors
    - heat flux, evaporation and heat transfer
    - deuterium and tritium

### *2.2.4 Spatial View of Form*

In a tokamak experiment or fusion reactor, as the design moves out from the main machine axis, there needs to be room for the toroidal field coil, the ohmic coils, the shaping coils, the vacuum vessel, plasma wall gaps and the size of the plasma itself. In a reactor, almost all the coils need to be shielded from the neutron flux and space needs to be provided for breeding tritium.

With water-cooled or nitrogen-cooled copper coils, the coils heat up rapidly during a plasma pulse and cooling is only effective between pulses. Hence the current density in copper is limited by the maximum allowable temperature rise during operation. Designs are constrained by magnetic torque forces between coil currents and transverse magnetic fields generated by other coils and the plasma.

In superconducting coils, the conductors have a maximum current density above which a highly undesirable current quench can occur. Quenches can be suppressed by surrounding the superconductor with non-superconducting stabilizers such as copper, but their heating must not cause the superconducting helium temperature to rise above a certain level. These tradeoffs are constraints on performance optimization.

In a low aspect ratio (R/a) tokamak reactor, there is much hardware to fit into a small major radius, making it difficult to properly shield inboard coils from neutrons. The plasma major radius grows as the design proceeds with calculations about realistic constraints. It is a considerable challenge to extend present spherical tokamak experiments to a working fusion reactor.

## 2.3 Engineering and Interface Analysis

### *2.3.1 Engineering Aspects*

Fusion machines are complex systems, and their design and assembly involves a wide range of mechanical and nuclear engineering aspects. The process of systems architecture to determine form and unify it with function includes the following aspects:

- spatial – The shape, size and composition of each element and its relationship to the others is chosen.
- tolerances – Elements can only fit together and operate correctly if they are built within tolerances. Even after construction, each element needs to stay within further tolerances during extreme heating (as with plasma impacted surfaces) or extreme cooling (down to superconducting temperatures) or under extreme forces (full and rapidly changing magnetic fields).
- stresses and strains and structure – All components are subject to stresses and strains and various deformations, including structural elements. These stresses are gravitational, electromagnetic, mechanical, thermal, nuclear, chemical, etc.
- thermodynamic – Plasma experiments involve very large heat flows, which need to be managed by either inertial heating or active cooling.
- particle flow – tokamak experiments involve significant gas flows with gas and neutral injection into the plasma and gas removal. If tritium is involved, especially when tritium is bred from other elements in surrounding fluids or structure, it must be extracted and recycled and impurities removed.
- nuclear – Fusion machines generate neutrons and fast particles, which can heat, damage, transmute surrounding materials and breed tritium.
- diagnostics, measurements and control – The machine involves a wide range of diagnostics for both the plasma and the machine itself, all requiring their own detailed engineering. The control systems work for both plasma and machine.
- quality control – The unappreciated hero of machine construction is quality control during design, procurement, construction, assembly and operation. All machine construction is subject to error, which can only be prevented, or detected and fixed, with a highly detailed quality control system in operation.

These aspects need to be developed for each machine element, taking full account of all interfaces. Some aspects of configuration management have already been applied to ITER [5, 6].

### 2.3.2 Interface Analysis

The interfaces between elements can act in both directions. A simplified example is shown in the interaction matrix in Table 2.1 for the electromagnetic view of some major systems elements of smaller tokamaks.

Opposite sides of the matrix diagonal show bi-directional relationships. The tokamak described corresponds to TCV which is discussed in Chap. 4. It has a continuous vacuum vessel with a relatively long time constant to stabilize the plasma. Several shaping coils are positioned outside the vacuum vessel and inside the toroidal field coils. These coils provide plasma current increase, ohmic heating and plasma shaping. The bottom poloidal coils create an x-point divertor

**Table 2.1** Function matrix for tokamak poloidal field system- electromagnetic view

| Relationships | Plasma | Vacuum vessel | Central ohmic solenoid | Inner shaping coils | Outer shaping coils | Divertor coils | Fast stabilization coils | Error and ELM coils |
|---|---|---|---|---|---|---|---|---|
| Plasma | Plasma is internally self connected | Stabilizes instabilities | Induces current in plasma and heats | Shape inner plasma | Shape outer plasma | Create x-point and divertor | Stabilize kHz vertical instability | Stabilize ELMs at plasma edge |
| Vacuum vessel | Current and flux change induce currents in vessel, especially disruption | X | Induces currents in vacuum vessel from applied voltage | Induce image currents which decay in vacuum vessel | Induced image currents in vacuum vessel slow response | Induce eddy currents during plasma ramp up | Effectiveness reduced by inducing image currents in vessel | Response is slowed by image currents in vacuum vessel |
| Central ohmic solenoid | Induces voltages, especially disruptions | Shields the coils | X | Can add to volt seconds from central solenoid | Can add volt seconds and shield stray field | Current ramp up requires more volt-sec | Has no effect since wired in series opposition | Has no effect since creates radial fields |
| Inner shaping coils | Induces voltages + disruptions | Shields the coils | Induces voltages in coils | x | Induce voltages | Induce voltages | Have little effect | Have little effect |
| Outer shaping coils | Induces voltages, especially disruptions | Shields the coils | Induces voltages in the coils | Induces voltages in the coils | x | Induce voltages and shaping flux | Create fast radial field | Effects are averaged out |
| Divertor coils | Induces voltages, especially disruptions | Shields the coils | Induces voltages in the coils | Induces voltages in the coils | Induces voltages in the coils | x | Have little effect | Effects are averaged out |
| Fast stabilizing coils | Induces voltages, especially disruptions | Reduces effectiveness of coils | Has no effect on coils | Have little effect on coils | Compensate to zero coil current | Have little effect | x | Effects are averaged out |
| Error correction and ELM coils | Has little effect | Slows coil response | Has no effect on coils | Have no effect | Has no effect | Have no effect | Have no effect | x |

configuration. Fast stabilizing coils are located inside the vacuum vessel, and ELM stabilizing and error correction coils are located outside the toroidal field coils.

Figure 2.2 [7] illustrates how interfacing of the heating and current drive systems with the plasma can be analysed by systems methods in the EU DEMO design.

Heating and current drive are major tools for influencing all aspects of the plasma. All these interfaces need to be carefully considered during the design phase, and controlled in real time during machine operation as shown in Fig. 2.1.

## 2.4 Operational Scenarios from Tokamaks with Less Than 700 kA Plasma Current

### *2.4.1 Strategy for Developing a Catalogue of Operational Scenarios*

In order to present a full design space for tokamaks, it is necessary to review the scenarios developed over the full range of tokamak operation in different machines. The strategy chosen for this monograph for compiling a reasonably full catalogue is to first review below some key scenarios from two tokamaks (TOSCA and PLT) from the 1970’s and several recently operated tokamaks in the less than 0.5 MA range. The catalogue also includes the scenarios of the tokamaks discussed in more detail Chaps. 3, 4, 5 and 6 of this monograph.

Some tokamaks in the past were designed with a limited goals such as:

- study a particular plasma shape
- create beam-beam fusion
- investigate wall materials
- learn about divertor physics

Tokamaks have proven to be very versatile and a wide range of operating scenarios have been developed that can be applied in new machines or upgrades. These scenarios often integrate a wide range of machine systems and require operation in very particular ways to obtain results or physics knowledge.

In the following sections, the procedure is to briefly describe the tokamaks in the approximate order going from lower plasma current to higher, often correlated with construction dates. Some of their key scenarios are listed to add to the scenario design space. Systems analysis is not attempted, since the goals are to present an introduction to tokamaks and to contribute to the design space.

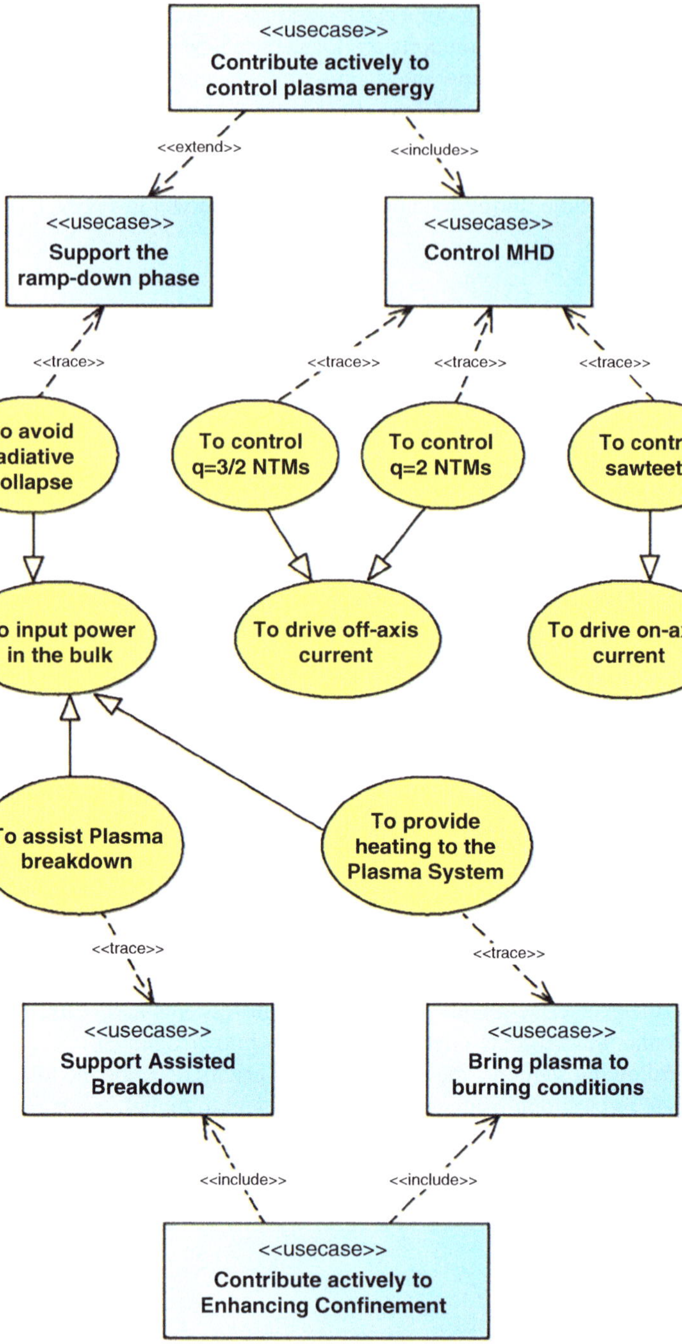


**Fig. 2.2** Relationships between the Use Cases considered in the analysis and main functions for the HCD system. (From [7], Fig. 4)

## 2.4.2 *TOSCA*

Small tokamaks have made important contributions, providing excellent value for money. Built by Derek Robinson, TOSCA's parameters were $R = 0.30$ m, $a = 0.10$ m, $B_T = 1.2$ T, and $I_p$ up to 20 kA. Its coils were driven by capacitor banks, and the 16 poloidal field windings were inside the toroidal field coils, acting as a copper shell to provide plasma stabilization and as a flexible shaping system. Four vertical field coils were located outside the toroidal field coils.

#### 2.4.2.1 Toroidal Field Compression

One method of plasma heating is by compression. TOSCA was able to rapidly increase the toroidal field from 0.6 T up to 1.2 T [8]. Results indicated that the increase in plasma energy from compression was consistent with adiabatic heating.

#### 2.4.2.2 Toroidal Beta Limits from Ohmic Heating

The plasma pressure limit in a tokamak is a key design feature. It was calculated [9] that maximum toroidal beta from ohmic heating is obtained on small aspect ratio devices with small minor radius. The heated plasma axis in TOSCA was highly shifted to larger major radius, indicating high pressure. The ratio of plasma to toroidal field pressure (beta) on axis reached 5%, thereby slightly exceeding the ballooning instability limit. The plasma was also able to operate with inward or outward pointing triangular-shaped plasmas. The lowest q obtained on the outer surface of the plasma was 1.4.

#### 2.4.2.3 Electron Cyclotron Heating at the Second Harmonic

External plasma heating by microwaves is an efficient method of auxiliary heating. Electron cyclotron heating at the second harmonic was successfully applied to TOSCA [10] to increase beta, control the current distribution and investigate current drive. It was separately reported [11] that off-axis current drive was relatively low in the plasma regimes studied. Other experiments stimulated an $m = 2$ instability in low q plasmas due to modification to the current profile [12].

#### 2.4.2.4 Instability Control with Active M = 2 Saddle Coils

MHD instability control is a necessary part of plasma operation. The ($m = 2$, $n = 1$) instability could be controlled [13] by externally applied helical fields. Large perturbations at the plasma surface could be tolerated.

## 2.4.3 *TCABR, Formerly TCA*

Auxiliary plasma heating with Alfven waves was investigated as a low frequency alterative to ion cyclotron heating. TCABR (Tokamak Chauffage Alfven BResilean), formerly TCA, studied auxiliary heating with Alfven waves. TCA [14] started operation in the early 1980's, with parameter R = 0.605 m, a = 0.18 m, $B_T = 1.5$ T, $I_p = 135$ kA. Interpretation of Alfven heating experiments were complicated by a rapid density rise. Electron cyclotron heating was installed for plasma breakdown and initiation studies. The TCA tokamak was later transferred to Brazil and became TCABR.

TCABR participated in large collaborative IAEA programmes [15] on "Joint Research Using Small Tokamaks" (2004–2008) and on "Utilisation of a Network of Small Magnetic Confinement Fusion Devices for Mainstream Fusion Research"(2011–2015) on the tokamaks CASTOR, COMPASS and GOLEM, T-10, ISTTOK, TCABR and STOR-M.

### 2.4.3.1 Plasma Initiation with Electron Cyclotron Assisted Breakdown

It is complicated and expensive to initiate toroidal plasma breakdown with high voltage electric fields induced by external coils, and alternatives have been investigated. Electron cyclotron heating to assist plasma breakdown and start-up was demonstrated in TCA [16]. For both X-mode and O-mode polarized electron cyclotron waves, it was possible to achieve plasma breakdown at zero applied volts, and current rise at 1–2 volts per turn. With ohmic transformer breakdown, only 8–10 volts per turn were required for start-up.

### 2.4.3.2 Plasma Heating with Alfven Waves

Plasma heating antennas operating at high voltage can be bombarded by plasma and require shielding to avoid energy loss and damage. TCABR extended the TCA work on Alfven heating by using antennas with Faraday shields [17]. It was found that electron density rise occurs in all cases, but local AW electron heating was demonstrated by radiometry measurements of electron cyclotron emission. The heating was related to local Alfven wave resonances.

### 2.4.3.3 Ohmic H-Modes with Electrostatic Biasing of Plasma Edge

Many methods have been studied to create a transition in the plasma from a low (L-mode) to a high (H-mode) confinement regime. The transition from the ohmic heating L-mode to the improved confinement H-mode in TCABR was induced by external electrode biasing [18]. A gradual improvement in the plasma confinement with the biasing voltage was observed as a result of the enhancement of the E × B shear flow at the plasma edge. The electron density profile became steeper as the voltage on the electrode increased.

### 2.4.4 *Neutral Beam Heating with PLT and COMPASS*

Auxiliary heating of the plasma from externally generated neutral beams offers a method of injecting both particles and energy. High power neutral beam heating experiments were conducted on the Princeton Large Torus (PLT) [19], with $R = 1.32$ m, $a = 0.4$ m, $B_T = 4$ T and plasma currents of less than 700 kA in a circular plasma defined by a physical limiter. With neutral beam powers of 2.4 MW injected into the plasma, ion temperatures of 7 keV were achieved, with $T_e/T_i = 0.5$ at densities of $5 \times 10^{19} m^{-3}$.

COMPASS, formerly COMPASS-D, was transferred from the UK to the Czech Republic [20]. The COMPASS tokamak was a compact experimental device ($R = 0.56$ m, $a = 0.2$ m) operated in a divertor plasma configuration with an ITER-like plasma cross-section designed for neutral beam heating. COMPASS has operated [21] with plasma current up to 400 kA, toroidal magnetic field 0.8–2.1 T and elongation up to 1.8, with plasma configurations that are circular, elongated or single null X-point.

#### 2.4.4.1 H-mode Experiments with Neutral Beam Heating in Dee Shaped Plasmas with an Open Divertor Configuration with Carbon Tiles

Plasmas with an x-point separatrix and a divertor are able to transition to the H-mode, usually with auxiliary heating. The COMPASS vacuum vessel has an open divertor covered by carbon tiles. Neutral beam injection was used to create and study high-confinement H-modes in D-shaped plasmas [21]. H-mode experiments were carried out at $B_T = 1.2$ T, $I_p = 200$ to 340 kA, $\kappa = 1.8$, and $q_{95} = 2.5$ and 300 kW of neutral beam heating into a lower single-null divertor configuration.

Carborane vapour deposition on the walls produced conditioning that minimized radiation losses from the plasma. The H-mode was accompanied by small high-frequency Type-III ELMs close to the L–H transition or large Type-I ELMs.

H-mode experiments were extended before the final shutdown in 2020 [22] to investigate the dependence of the L-H power threshold on the X-point height above the divertor. The results showed a clear linear trend where the threshold increases by ~18% per 1 cm of the height above the divertor assembly.

#### 2.4.4.2 High Field Side (HFS) Error Correction Coils (ECC)

Error fields in large tokamaks need to be avoided to obtain good energy and particle confinement. COMPASS used its localized 3D coils to investigate the correction of errors in the toroidal field [23]. An artificial error field that simulated a tilt of the central solenoid was applied to low-beta plasma in scenarios relevant to the ITER baseline ($q_{95} = 3$, aspect ratio=3.23). The plasma density $n_e$ was varied to study the influence on locked modes (non-rotating magnetic island) and the L-H transition. It was possible to compensate for central solenoid displacements and recover plasma performance.

### 2.4.5 *Tokamak T-10, Successor to T-3 and T-4*

The tokamak T-10 was the successor to the highly successful tokamak T-3 [24] (and also T-4) which had encouraged much of the world to shift its major nuclear fusion research to this type of confinement system. The parameters of T-10 [25] are $R = 1.5$ m, $a = 0.3$ m, $B_T = 2.5$ T and circular plasma current up to 400 kA, using lithium gettering. T-10 had 2 MW of ECRH heating at 130 and 140 GHz.

#### 2.4.5.1 Electron Internal Transport Barrier (e-ITB)

Plasma confinement can be improved by the production of internal transport barriers. An electron internal transport barrier (e-ITB) is a toroidal surface in a plasma which improves energy confinement inside. Its formation was observed near the $q = 1.5$ surface [25]. It was produced with current ramp-up and off-axis ECRH, which was used to suppress sawteeth oscillations. A flatter q profile was produced in the vicinity of rational surface $q = 1.5$, leading to the formation of the e-ITB.

The influence of ECR-heating on the anomalous transport of tungsten ions was later investigated [26]. Tungsten was removed from the plasma centre mainly due to an increase in its anomalous diffusion.

#### 2.4.5.2 ITER-Grade Heat Flux to a Tungsten Limiter

The flow of heat out of the plasma should not damage external surfaces. A tungsten poloidal limiter was used in T-10 [27] to investigate extremely high heat loads as expected in ITER. The "ecton" mechanism was observed, involving non-ambipolar plasma flow toward the limiter surface which led to surface overheating and melting. The interior tungsten plates of the limiter were heated up to temperature exceeded 2000 ° C, at an estimated local thermal load were of more than 40 MW/m$^2$ on the plate edges. This led to surface melting and therefore concern for ITER.

### 2.4.6 *HL-2A, Formerly ASDEX, and HL-2 M*

The high confinement H-mode was originally discovered [28, 29] in Asdex. High confinement regimes are a central feature of toroidal confinement in machines with a divertor [30, 31]. Asdex had a poloidal divertor for removing plasma exhaust and separatrix contained plasma. Auxiliary heating was by neutral beam injection. It was subsequently shipped to China and became HL-2A [32]. ASDEX / HL-2A has R = 1.65 m and a = 0.4, with plasma current up to 430 kA, and is shown schematically in Fig. 2.3 [33].

The numbers on the figure refer to:

- 1 Ohmic coils to create driving voltage in plasma;
- 2 Vertical field coils to counteract outward plasma motion;
- 3 Compensation field coils for field correction;
- 4 Horizontal field coils for positioning and to give stability against vertical plasma movement;
- 5 Multi-pole field coils to produce the divertor separatrix and pathway for plasma exhaust;
- 6 Vacuum vessel to maintain high vacuum;
- 7 Plasma shown by constant surfaces of poloidal magnetic flux;
- 8 Fixed limiter to define plasma during startup and to protect vessel;
- 9 Movable limiter to adjust outside plasma boundary.

Recent experiments [34, 35] explored physics relevant to ITER and HL-2 M.

#### 2.4.6.1 ELM-Control RMP, LHCD, LBO-Seeded Impurities (Al, Fe, W) and Impurity SMBI (Ar, Ne)

Edge localized modes (ELMs) are associated with H-modes and can be beneficial by expelling particles and heat through a transport barrier, or damaging if the heat loss is too great or too concentrated. Experiments for controlling ELMs [34] have used Rotating Magnetic Perturbations (RMP), Lower Hybrid Current Drive (LHCD), Laser Blow Off (LBO)-seeded impurities (Al, Fe, W) and impurity Supersonic Molecular Beam Injection SMBI (Ar, Ne). The observed ELM mitigation can be simulated by a turbulent heat transport model.

#### 2.4.6.2 Strategies to Control Toroidal Alfven Eigenmodes (TAE) Driven by Energetic Ions

Toroidal Alfvén eigenmodes (TAE) are relevant to alpha-particle heated reactors, since they can lead to enhanced losses of fast particles. These modes have been observed on HL-2A by using high energy neutral beams. Progress has been made in

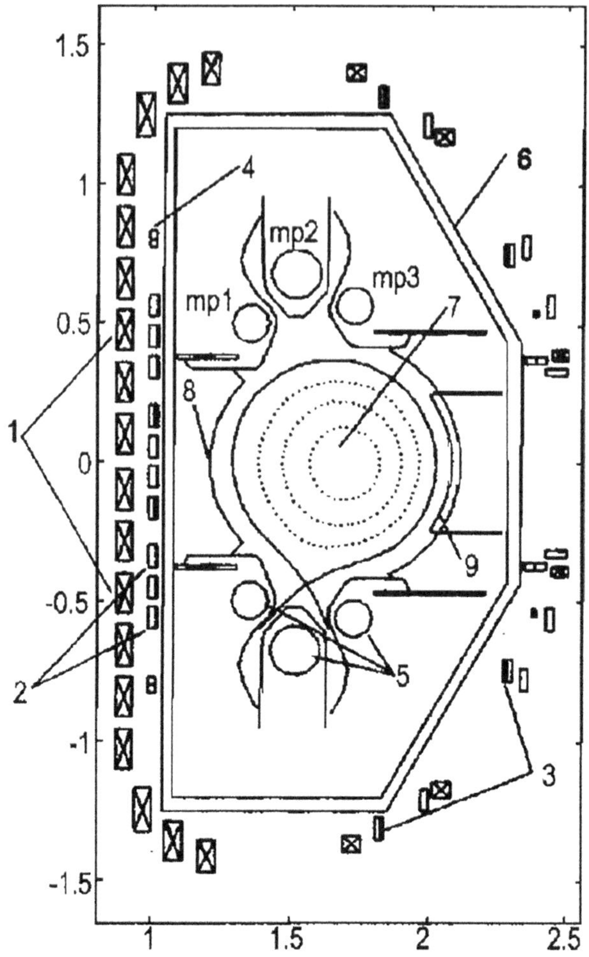


**Fig. 2.3** A schematic cross section of HL-2A, formerly ASDEX, with position of coils, vacuum vessel, and MHD plasma equilibrium surfaces of constant flux and divertor x-point surface. (From [33], Fig. 1)

developing strategies to control instabilities including neo-classical tearing modes and core-localized Alfvén eigenmodes [34].

#### 2.4.6.3 Stabilization of Fishbone Instabilities with ECRH

The m/n = 1/1 ion fishbone instability was stabilized by ECRH on HL-2A [34]. The stabilization depends not only on the neutral beam injected power but also on the radial deposition of ECRH.

#### 2.4.6.4 Formation of an Internal Transport Barrier with NBI in Advanced Tokamak Operation

In HL-2A [34], an ion internal temperature barrier was created with auxiliary heating (mainly NBI) during the current rise phase. There was a weak magnetic shear near the plasma centre, where kinetic electromagnetic instabilities were observed.

Two neutral beams were applied for high performance operation. Hybrid scenarios with double transport barriers were achieved [35]. High $\beta_N$ of greater than 2.0 was maintained for a duration of about 15 energy confinement times.

#### 2.4.6.5 First Plasma in HL-2 M

The first plasma in the much larger HL-2M [35] was obtained in 2020. The design parameters of HL-2M tokamak were $R = 1.78$ m, $a = 0.65$ m, $B_T = 2.2 \sim 3$ T, $I_p = 2.5 \sim 3$MA. The goals were to produce and study:

- high performance plasmas for next-step devices;
- flexible divertor configurations (snowflake, tripod, etc.);
- plasma facing components under high heat and particle flux;
- key issues such as mitigation of ELM, disruption, VDE, etc.

### 2.4.7 *START*

Extremely low aspect ratio tokamaks were predicted to have many advantages [36] for reaching high plasma current and pressure. MHD equilibrium calculations showed that spherical torus plasmas with an edge safety factor $q_a > 2$ are characterized by:

- high toroidal beta ($\beta_t > 0.20$)
- low poloidal beta ($\beta_p < 0.30$)
- naturally large vertical elongation ($\kappa \approx 2$)
- large normalized plasma current with $I_p/(aB_{t0})$ up to about 7
- strong paramagnetism ($B_t/B_{t0} > 1.5$)
- strong magnetic helical pitch ($\Theta$ comparable to F)

To test these ideas, the START spherical tokamak [37] was constructed. It is shown in Fig. 2.4 with the flux surfaces showing a high $\beta = 32\%$ discharge.

START operated at aspect ratio $R/a = 1.35$, with a plasma minor radius of 32.5 cm, toroidal field of 0.3 T, and a plasma current above 200 kA. The plasmas had natural vertical stability at higher elongation, up to 2/1, and operated with x-points top and bottom. Auxiliary heating was provided by neutral beam injection of 0.5 MW of hydrogen at 30 keV into deuterium plasmas.

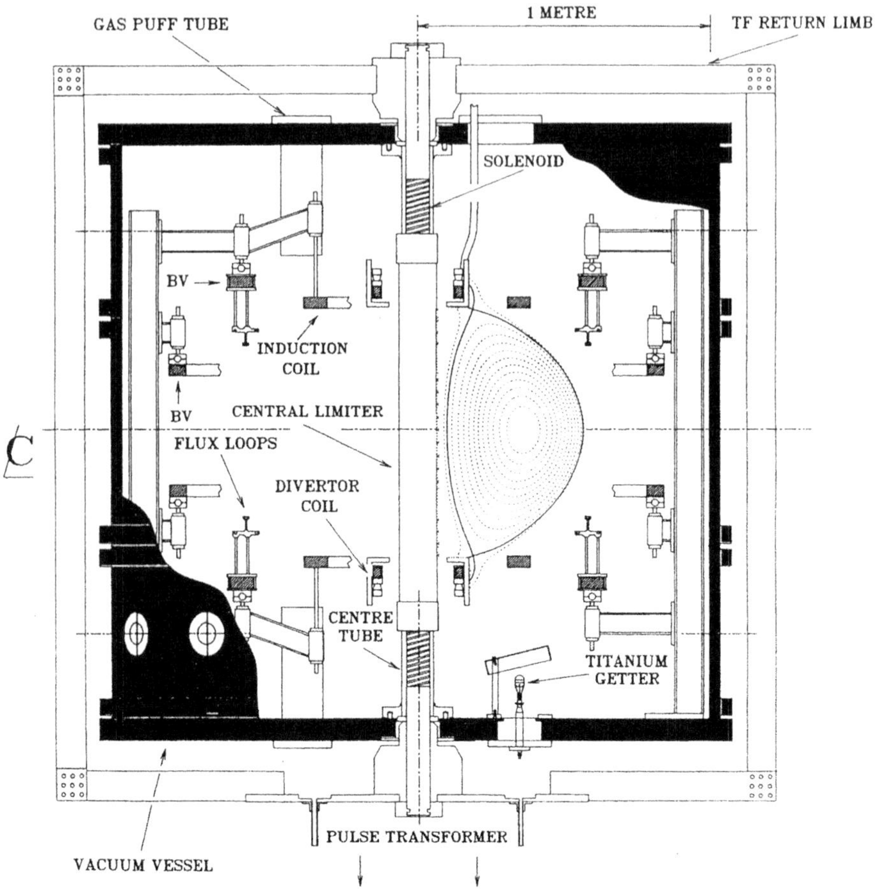


**Fig. 2.4** Schematic of START with β = 32% equilibrium. (From [37], Fig. 1)

#### 2.4.7.1 High Toroidal Beta at Low and Extremely Low Aspect Ratio

The low aspect ratio tokamak DIII-D with R/a = 2.8 had achieved a toroidal beta of 12.5% [38]. START extended these experiments to a much lower aspect ratio and obtained high-β of 32%. These plasmas reached normalized beta values of 4 at values of auxiliary heating power comparable to the ohmic power.

### *2.4.8 GLOBUS-M and GLOBUS –M2*

Globus-M was a spherical tokamak [39] with parameters a = 0.21 m, R/a = 1.67, plasma current ≈ 200 kA, elongation 1.9 and lower null x-point.

#### 2.4.8.1 H-Mode in Spherical Tokamaks

GLOBUS-M was used to study energy confinement scaling in spherical tokamaks. They were able to establish internal transport barriers characteristic of standard H-mode tokamaks [39].

#### 2.4.8.2 High Current and LHCD Experiments in Globus-M2

Lower Hybrid waves are an efficient method of driving plasma current in tokamaks. Experiments were performed in the upgraded Globus-M2 [40], with the toroidal magnetic field up to 0.8 T and plasma current up to 0.4 MA. Temperatures of electrons and ions above 1 keV and densities of $10^{20}$ $m^{-3}$ were achieved with neutral beam injection of 850 kW at 28.5 keV.

The unique use in a spherical tokamak of LHCD at 2.45 GHz for non-inductively driven current exceeded 70% of total current of 0.2 MA. The achieved values of efficiency $\eta = (0.15–0.4) \times 10^{19}$ A $m^{-2}$ $W^{-1}$ were comparable with standard tokamak results.

### *2.4.9 QUEST*

QUEST with R = 0.68 m and a = 0.4 m was designed as a steady-state spherical tokamak (ST) using controlled Plasma Wall Interaction (PWI) [41]. A high temperature tungsten wall was chosen to control recycling.

#### 2.4.9.1 Steady-State 6900 s Discharges in a Spherical Tokamak with a Hot Wall

Tokamak reactors would have many advantages by being able to operate in steady-state. A very long pulse plasma lasting 6900 s [42] was achieved by EC current drive at 8.2 GHz and 40 kW. The discharge was terminated by the uncontrollability of the density. The hot wall plays an essential role in reducing the amount of wall-stored hydrogen and facilitates hydrogen recycling.

#### 2.4.9.2 Fully Non-inductive Current Ramp Up with Second Harmonic ECRH in a Spherical Tokamak

Tokamak design would be greatly simplified if an ohmic heating transformer coil producing inducing plasma current ramp-up were not required. A launcher system with two quasi-optical mirrors produced a sharply focussed incident microwave beam at the second electron cyclotron resonance [43, 44]. It was used to produce a record non-inductive plasma ramp-up to a plasma current of 86 kA, which was achieved with a focussed 230 kW beam at 28GHz.

## 2.4.10 *General Fusion's Spector and SLiC*

The General Fusion company developed a fusion reactor concept in which plasma injectors created a spherical tokamak which was then compressed by a surrounding liquid lithium metal wall. The concept was being tested on Spector and SLiC. The liquid lithium could both heat the plasma and be used for tritium breeding.

### 2.4.10.1 Spector

Spector versions 1–5 [45] enabled rapid development and testing of methods of plasma formation and compression by a moving metal wall. Most recently, a single-stage coaxial Marshall gun was used for forming a spherical tokamak plasma target, surrounded by a flux-conserving wall. A single-pulse fast Coaxial Helical Injection (CHI) was used with no additional heating or current drive. Plasma parameters included vessel inner radius 1 m, major and minor plasma radius 0.6–0.7 m and 0.3–0.4 m, elongation 1–1.6 and plasma current 0.3–0.5 MA. The plasma density and temperature achieved were $2–20 \times 10^{19}$ $m^{-3}$ and 100–500 eV at beta 2–8%.

### 2.4.10.2 SLiC (Spector Lithium Configuration)

The Spector machine was upgraded to SLiC [46], with a vessel radius of 1.5 m surrounded by liquid lithium walls driven by pistons capable of rapidly compressing the lithium and thereby the plasma and heating it. The range of compact plasma configurations possible included peak plasma current 300 kA, plasma density $5 \times 10^{19}$ $m^{-3}$ to $2 \times 10^{21}$ $m^{-3}$ and peak temperature of only 75 eV to 150 eV. They concluded that they were able to form a spherical tokamak plasma via fast CHI within a large area of liquid Li without significant problems.

### 2.4.10.3 Fusion Plant Construction Announced

In June 2021, it was announced [47] that General Fusion will build a nuclear fusion plant at Culham at the UK Atomic Energy Authority site, to show how their technology can be turned into a realistic source of clean energy. The project cost is projected to be $400 million and supposedly will pave the way for a commercial pilot plant. Construction is due to begin in 2022 and be completed by 2025.

## 2.5 Design Space Catalogue of Tokamak Operational Scenarios

This section contains a compendium of titles of important tokamak operational scenarios discussed in this monograph. Because scenarios involve many aspects of plasma and machine operation, the classification used is somewhat arbitrary. Each scenario is referenced to particular tokamaks, although it may have been demonstrated elsewhere. Not all tokamaks are included, nor all their results. The choice is reasonably comprehensive so as to serve as a design space basis. Within each category, the scenarios are listed in ascending order of publication date.

At the beginning of each section, a publication is discussed which either provides a review of the area, or illustrates how the combined knowledge in that area is being applied to the design of the proto-reactor ITER discussed in Chap. 5.

### *2.5.1 Vacuum Vessel and Component Conditioning*

Wall conditioning for ITER has been explored with experiments and modeling [48] in order to control fuel and impurity recycling, as well as tritium inventory. The conditioning experiments in JET with its ITER-Like Wall were used to study glow discharge conditioning on the in-vessel plasma-facing components. ITER foresees the use of low-temperature plasma-based techniques which are compatible with the permanently-on toroidal magnetic field from superconductors. These include ion or electron cyclotron wall conditioning for tritium removal between plasma pulses. Chosen scenarios are:

- High-Z tantalum-tungsten limiter, Doublet III [49]
- Wall cleaning and baking, discharge cleaning, Microtor [50] and Doublet III [51]
- Reactor relevant vessel wall high-Z material, cleaning and vertical plate divertor, Alcator C-Mod [52]
- Wall conditioning for ITER, JET [48]
- Vessel wall conditioning with ECRH, TCV [53]

### *2.5.2 Plasma Start-Up and Current Ramp*

A review of startup requirements for ITER and their simulation in DIII-D has been reported [54]. ITER startup presents unique challenges due to the low inductive toroidal electric field (0.3 V/m), power supply and poloidal field coil constraints, and (in some scenarios) plasma current ramp-up near the $n = 0$ vertical stability limit. A large-bore startup scenario with ECRH assist proved robust with electric fields as low as 0.21 V/m. Chosen scenarios are:

- Plasma initiation with electron cyclotron assisted breakdown, TCA [16]
- Plasma initiation and divertor control with shaping coils outside the toroidal field coil, Asdex-U [55]
- Plasma formation without an ohmic heating solenoid, JT-60U [56]
- Plasma breakdown with a carbon and ITER-like wall, JET [57, 58]
- Plasma ramp-up, ramp-down and operation scenarios for ITER, JET [59]
- Low voltage breakdown in a highly conducting vacuum vessel, TCV [60]
- Fully inductive current ramp up with second harmonic ECRH in a spherical tokamak, QUEST [43]

### 2.5.3 MHD Equilibrium and High Beta

The maximum normalized beta achieved in long-pulse tokamak discharges at low collisionality falls significantly below both that observed in short pulse discharges and that predicted by the ideal MHD theory [61]. Key issues were examined related to the long-pulse beta limits of ITER-size tokamaks. Plasma control methods that could increase the soft beta limit were examined, and several remedies were proposed. Chosen scenarios are:

- Toroidal beta limits from ohmic heating, TOSCA [9]
- In-situ pellet injection producing 2/1 vertically elongated plasma, Doublet III [4]
- Production of 3/1 elongated doublet-shaped plasmas, Doublet III [62]
- High toroidal beta at low toroidal field and limiter q of 1.7, Doublet III [63]
- High elongation (1.6) D-shaped plasmas, Doublet III [63]
- Beta limits with Be limiters and gettering and in H-modes, JET [64]
- High toroidal beta at low aspect ratio, DIII-D [65]
- High toroidal beta at extremely low aspect ratio, START [37]
- Limits of operating space for high plasma elongation, TCV [66]
- High-shared-flux Doublet configuration with separate lobe ECRH, TCV [60]
- Low aspect ratio plasmas with high vertical elongation, ST-40 [67]

### 2.5.4 Particle and Energy Confinement and H-Modes

The regimes of low and high confinement H-modes are summarized [31]. It was discovered that toroidally confined plasmas are able to self-organise in such a way that turbulence diminishes, resulting in improved confinement. A general understanding is that gradient driven drift-wave type turbulence develops to a level set by the spatial variation of neo-classical and self-driven plasma flows. Turbulence and flows have to be treated self-consistently. The ambipolar electric field causes sheared flows which interact with turbulence and regulate its level. Several methods have led to these regimes of improved confinement. Chosen scenarios are:

- Confinement studies in L and H-type Asdex discharges, Asdex [28–31]
- H-mode in expanded boundary with high stored energy and confinement time, Doublet III [68]
- Operation at very high toroidal field with pellet injection to reach the Lawson criteria, Alcator C [69]
- Hot-ion H-mode at 5 MA plasma current, JET [70]
- High fusion triple product $n_D(0)\tau_E T_i(0)$ in hot-ion H-mode, JET [71]
- Be limiter discharge performance at 7 MA plasma current, JET [72]
- High confinement ELM-free H-mode and impurity generation, JET [73]
- Very high confinement VH-mode with divertor and boronization during ELM-free H-mode, DIII-D [74]
- Double-barrier ELM-free H-mode with internal transport barrier inside edge-barrier with beam counter-injection and cryopumping, DIII-D [75]
- Highest fusion triple product in steady state ELMy H-mode, JT-60U [76]
- Ferritic insertion for reduction of toroidal magnetic field ripple, JT-60U [77]
- Variable toroidal field ripple experiments, JET [78]
- Electron internal transport barriers with EC current drive, TCV [79]
- Improved confinement (I-mode) ELM-free H-mode, Alcator C-Mod [80]
- Electron internal transport barrier (e-ITB), T-10 [25]
- Resolution of key physics issues for ITER and JT-60SA, JT-60U [81]
- EDA (Enhanced $D_\alpha$) H-mode, Alcator C-Mod [52]
- H-mode experiments in Dee shaped plasmas with an open divertor configuration with carbon tiles, COMPASS [21]
- H-mode regimes with ohmic heating in H, D and He, TCV [82]
- Improved confinement at high beta with an ITER-like wall, JET [83]
- Multi-machine ITER scenario to maximize edge pedestal height, TCV, JET, MAST [84]
- Operational scenarios in support to ITER, JET [85]
- Ohmic H-modes with electrostatic biasing of plasma edge, TCABR [18]
- ITER baseline (IBL) and IBL-A scenarios simulations, Asdex-U [86]
- Wide pedestal ITER-relevant QH mode, DIII-D [87, 88]
- Super H-mode, DIII-D [89]
- Formation of an internal transport barrier with NBI in advanced tokamak operation, HL-2A [34]
- Energy confinement in a 1 MA spherical tokamak, MAST [90]
- H-mode in spherical tokamaks, GLOBUS-M [39]

### 2.5.5 *Heating, Current Drive and Their Role in Controlling Instabilities and Disruptions*

An auxiliary heating power of >70 MW is envisaged at ITER in order to obtain the plasma temperatures and plasma profiles required to achieve Q > 10 for 400 s [91]. The systems are also expected to drive current, tailor the plasma profile, and control the plasma instabilities. While the neutral beam systems are expected to deliver 33 MW of heating power to the plasma and drive current, the ion cyclotron (IC) and the electron cyclotron (EC) systems deliver 20 MW each. IC covers a wide variety of heating and current drive schemes and EC heats the electrons providing local heating and current drive, which can be steered across the plasma cross section. Chosen scenarios are:

- Ohmic heating in tokamaks, T-3 [24]
- Toroidal field compression, TOSCA [8]
- Neutral beam heating to high ion temperature, PLT [19].
- Electron cyclotron heating at the second harmonic, TOSCA [10]
- Instability control with active m = 2 saddle coils, TOSCA [13]
- Fishbone instabilities, JET [92]
- Measurements of sawtooth crashes with neutron profile monitor, JET [93, 94]
- Neutral beam injection of 120 keV helium, JET [95]
- Ohmically heated plasma neutron production and ion thermal diffusivity, JET [96]
- Ion thermal conductivity from neutron measurements, JET [97]
- Lower hybrid current drive and Reactor Applications, JET [98]
- High elongation plasmas and ohmic H-modes, TCV [99]
- High energy gamma rays from runaway electrons, JET [100]
- Fully non-inductive steady-state plasmas with electron cyclotron current drive, TCV [101]
- Lower hybrid current drive (LHCD) and negative shear with internal transport barrier, JET [102]
- Partial current drive with 60% bootstrap current with high beta-normalized from Resistive Wall Mode (RWM) stabilization and electron cyclotron current drive (ECCD), DIII-D [103]
- Plasma heating and current drive by high energy negative ion neutral beam injection, JT-60U [104]
- High power electron cyclotron heating and bootstrap current drive, TCV [79]
- Advanced tokamak operation with full current drive and high $\beta_N$, JT-60U [105]
- Plasma heating with Alfven waves, TCABR [17]
- Instability control of Neoclassical Tearing Modes (NTM) with ECCD, DIII-D [106]
- Ion cyclotron heating, flow drive and impurity control with a Field Aligned (FA) Antenna, Alcator C-Mod [52]

- Lower Hybrid Current Drive (LHCD) at medium and high plasma densities, Alcator C-Mod [52]
- Disruption mitigation using massive gas injection (MGI), Alcator C-Mod [52]
- Non-inductive current drive operation with boronization, low electron density and NBCD and ECCD, Asdex-U [107]
- Neutral beam current drive and heating, TCV [108]
- High Field Side (HFS) Error Correction Coils (ECC), COMPASS [23]
- Disruption avoidance by real-time locked mode prevention with ECCD, TCV [109]
- Plasma disruption avoidance and mitigation using impurity doped shattered pellets, DIII-D [110, 111]
- Steady-state current drive advanced tokamak fusion reactor, DIII-D [112]
- Stabilization and suppression of post-disruption runaway electron beams, FTU [113]
- ELM-control RMP, LHCD, LBO-seeded impurities (Al, Fe, W) and impurity SMBI (Ar, Ne), HL-2A [34]
- Strategies to control Toroidal Alfven eigenmodes (TAE) driven by energetic ions, HL-2A [34]
- Stabilization of fishbone instabilities with ECRH, HL-2A [34]
- Elimination of disruption runaway electron current with current ramp down, TCV [114]
- High-$\beta_N$ fully non-inductive scenarios with combined EC and NBI, TCV [115]
- Heating by compression by liquid lithium walls, SLiC [46]

### *2.5.6 Divertor Radiation and Plasma-Wall Interaction*

The principal focus on divertors in ITER is on steady-state power fluxes in the d-t phases, obtained from simulations using plasma boundary codes, assuming the use of low-Z seeding impurities nitrogen (N) and neon (Ne) [116]. There are factors which may increase the peak steady state loads: divertor target shaping for component misalignment protection, the influence of fluid drifts, and the consequences of narrow scrape-off layer heat flux channels. Chosen scenarios are:

- Expanded boundary divertor and argon radiation, Doublet III [117]
- Divertor and SOL studies in the Mark I, II, IIA/AP, II Gas Box divertors, JET [118]
- Reactor-relevant power handling and operation of vertical target divertor, Alcator C-Mod [119]
- Exhaust control and detachment in snowflake and super-X divertors, TCV [108]
- High power plasma exhaust detachment with Ar and divertor x-point radiating zone, Asdex-U [107]
- ITER-grade heat flux to a tungsten limiter, T-10 [27]

- Small angle slot divertor with Resonant Magnetic Perturbation (RM) suppression of Edge Localized Mode (ELM) scenario, DIII-D [88]
- Radiating divertor combined with steady-state high $\beta_N$, DIII-D [88]
- Tin liquid limiter (TLL) power handling for divertors, FTU [113]
- Reduced heat flux with grassy ELM's at high triangularity, TCV and ASDEX-U [120]

### *2.5.7 Plasma Diagnostics and Machine Control*

Plasma diagnostics are based on a multitude of different physical processes and many different techniques are being employed for measuring the spatial profile and evolution of various plasma parameters [121]. With modern electronics and computers, they also form the basis for a real time control system based on a very rapid processing and analysis of their measurements. Chosen scenarios are:

- Essential plasma diagnostics, JET [122]
- Neutron spectrometry systems, JET [123]
- Neutron profile monitor, JET [93, 124, 125]
- Fast particle and neutron diagnostics, JET [126]
- Diagnostics for Experimental Thermonuclear Fusion Reactors, JET [127]
- Fully digital plasma control, TCV [128]
- Safe and flexible operations and plasma control, JET [129]
- Real time plasma control systems, TCV [130]
- Diagnostic enhancements for DT scientific campaigns, JET [131]

### *2.5.8 High Power Tritium Burning and Alpha Particle Heating*

A concerted physics and technology programme on JET was launched with a view to prepare the d-t campaign (DTE2) [151], which will prototype several scenarios for ITER. This programme includes the impact of isotopes in the operation and physics of d-t plasmas such as thermal and particle transport, high confinement mode access, Be and W erosion and fuel recovery and recycling, supported especially by neutron and fast particle diagnostics. Chosen scenarios are:

- Pure deuterium hot-ion H-mode target plasma for tritium experiments, JET [132]
- Neutral beam injection of 1%T in one PINI in hot-ion H-mode, JET [132]
- 1.7 MW of fusion power from neutral beam injection of 100% tritium in 2 PINIs into hot-ion H-mode, JET [132]
- Neutron emission profile measurements during PTE, JET [133]
- Particle and energy transport during PTE, JET [134]

- Ion cyclotron emission measurements during PTE, JET [135]
- Release of tritium from the carbon first wall of JET during PTE, JET [136]
- Supershots with maximum fusion power, TFTR [137]
- Gamma ray emission from high fusion power from RF accelerated Helium-3 ($^3$He) minority fuel ions, JET [138]
- Alpha particle physics, TFTR [139]
- Four years of d-t plasma operation, TFTR [140]
- Alpha particle heating of 1.3 MW in DTE1, JET [141]
- Bulk ion heating of deuterium and $Q \approx 0.22$ with ICRH in DTE1, JET [142, 143]
- Tritium retention, TFTR [144]
- Trace tritium and particle diffusion in DTE1, JET [145]
- Trace tritium and the H-mode density limit in DTE1, JET [146]
- D-T fusion power of 16.1 MW and $Q \approx 1$ in hot-ion H-modes with divertor in DTE1, JET [147]
- Internal transport barriers and high triple product in DTE1, JET [148]
- Bulk ion heating with $^3$He minority ICRH in DTE1, JET [143]
- Second harmonic tritium ICRH in DTE1, JET [143]
- Combined NBI and ICRH in ELM-free H-modes during DTE1 giving record stored energy and fusion power, JET [149]
- Tritium recycling, retention and clean-up in DTE1, JET [150]
- Scenarios extrapolated to DTE2 operation with tritium, JET [151]
- Extrapolated baseline scenario for ITER during DTE2, JET [151]
- Extrapolated hybrid scenario at high normalized beta for ITER during DTE2, JET [151]

### *2.5.9 Long Pulse and Steady-State Operation*

Superconducting tokamaks such as WEST aim at power/particle exhaust studies on an ITER-like actively cooled tungsten divertor and at bulk plasma performance in long-pulse discharges reaching 1000s [160]. It is equipped with 9 MW of ICRH power and 7 MW of LHCD. The upper divertor is made of actively cooled W coated cupper tiles. Chosen scenarios are:

- High power, 400 sec long pulse in circular high current plasma with LHCD and 1 gigajoule injected, Tore Supra [152]
- Long pulse 120 sec high confinement limiter operation with combined ICRH and LHCD, Tore Supra [152]
- Long pulse operation with LHCD and ECCD synergies, Tore Supra [153]
- Up to 5 h (18,000 s) fully non-inductive steady-state operation with LHCD, TRIAM-1M [154–156]
- Long pulse operation of 240 s with LHCD, HT-7 [157]
- Long pulse operation of 400 s with cooled belt limiter, HT-7 [158]

- Steady-state 6900 s discharges in a spherical tokamak with a hot wall, QUEST [42]
- Long pulse 30 s H-mode without ELM-crash using control of resonant and non-resonant modes, KSTAR [159]
- Plasma and impurity confinement in a long pulse, high power, tungsten divertor environment, WEST [160]
- Long pulse of 100 s with ECRH and LHCD in H-mode, EAST [161]
- Plasma–wall interaction control for H-mode operation over 100 s with an ITER-like tungsten divertor, EAST [162]

## 2.6 Tokamak Simulation Codes

### *2.6.1 Plasma Simulation and Design Codes*

An indispensable part of a systems approach is the use of simulation, analysis and design codes. Many physics and engineering codes have been developed to simulate various aspects. For plasma physics, there are codes to simulate MHD equilibrium, control of feedback from power supplies, stability, transport, plasma heating and current drive:

- TSC, dynamic equilibrium and transport [163, 164]
- TRANSP, transport modelling [165, 166]
- PTRANSP, integrated TRANSP and TSC modelling code [167]
- ONETWO, transport [168]
- POPCOM, scoping contour analysis [169]
- PEST-1, kink instability [170]
- ERATO, ballooning instability [171]
- BALMSC, ballooning instability [172]
- Coppi-Tang transport [173]
- GLF23, gyrokinetic transport [174]
- MMM95, drift mode transport [175]
- EPED1, pedestal stability [176]
- TORIC, ion cyclotron waves [177]
- NUBEAM, neutral bean injection [178]
- LSC, lower hybrid current drive [179]
- TORAY, electron cyclotron ray tracing [180]
- JSOLVER, MHD equilibrium evolution [181]

The range of plasma theory and simulation codes is typified by the PPPL Theory Department [182] website in the areas of:

- Research Codes
- Neutral Transport

- Magnetohydrodynamics
- Nonlinear Dynamics
- Basic Plasma Physics
- Turbulence & Transport
- Advanced Computing
- Energetic Particles
- Helio & Astro Plasmas.

For engineering aspects, simulations are made of coil stresses, power supply requirements, cooling needs, neutronics and tritium breeding, and a whole host of problems necessary for tokamak construction and operation. Digital computations can analyse plasma and machine diagnostics for use in real time control systems.

Computer Aided Design (CAD) software is used for detailed engineering of 3D models and is also used throughout the engineering process from conceptual design and layout of products, through strength and dynamic analysis of assemblies to definition of manufacturing methods of components. Software can also support Computer-aided engineering (CAE), finite element analysis and Computer-aided manufacturing (CAM) and computer numerical control (CNC) machines.

### 2.6.2 *Integrated Workflows of Simulation Codes*

The EUROfusion Project on Code Development for Integrated Modelling (WP-CD) provides integrated simulation workflows, incorporating core-edge-SOL/PFC coupling, first-principles models and control elements [183, 184], comprising:

- a high-resolution equilibrium and MHD stability chain, for core and pedestal;
- the fixed boundary core European Transport Simulator ETS, with various equilibrium modules and a full hierarchy of transport models, impurities, pellets, neutrals, sawteeth, Neoclassical Tearing Modes (NTM) modules, and full integration of Heating and Current Drive sources (Electron Cyclotron, Neutral Beam Injection, Ion Cyclotron, alpha), including synergies;
- a feedback controlled free boundary transport simulator;
- a Scrape-Off-Layer (SOL) turbulence workflow including a synthetic probe
- EC, IC and NBI codes;
- prototypes of self-consistent coupling between core and edge transport codes, coupling the ETS core transport code to the 2D edge transport code SOLPS;
- a prototype acyclic workflow for modelling the SOL and interaction with Plasma Facing Components (PFC) by coupling the 2D transport code SOPLS to the 3D Monte Carlo PWI and impurity transport code ERO.

### 2.6.3 *Systems Engineering Codes*

- SysML and OPM are two important systems architecture modelling language, discussed above. Other important systems codes include PROCESS [185], SYCOMORE [186], BLUEPRINT [187], MIRA [188] and POPCOM [169].

## 2.7 Component Design Options

Given all the tokamaks studied and their operational scenarios, it is possible to develop a component list for the tokamak design space. A guide is the interface matrix developed for ITER [5, 6] which is the basis for developing the elements, their form and function and interfaces in tokamak systems architecting. Design space options for most tokamaks are presented below with each system:

- plasma (see previous sections):
- coil conductor and cooling: copper (liquid or nitrogen cooled, inertial or continuous), superconducting NbTi (liquid or supercooled He, copper or Al matrix quench protection), $Nb_3Sn$ (various processes and matrix), high temperature superconductor yttrium-barium-copper-oxide
- coil feeders: copper or copper with superconductor interfaces
- toroidal field coils: continuously wound or demountable, multiple plates with joints to centre column, Bitter-type plate magnets for high field, superconductor or normal, coil number to minimize ripple and particle loss, ferritic inserts or correction coils for error fields or ripple
- coil structural support: prestressed, torque, strain, connected directly to coil, connected through individual coil cryostats, gravity supports
- transformer and ohmic heating coils: air core, iron core, saturated iron core, single or multiple outer limbs
- central solenoid: independent coils, epoxy bounded to toroidal field coils, inside or outside toroidal field coils, merged with or separate from poloidal shaping coils, omitted because of independent current drive
- poloidal field coils: few or many coils, closeness to plasma, inside or outside toroidal field coil, merged or not with central solenoid, high or low voltage and dI/dt, merged or not with divertor and rapid stabilization coils, independent or parallel wired, response to disruptions, complexity of control
- rapid stabilization coils: inside or outside vacuum vessel, series wired or independent, short or long pulse, cooling, disruption protection, response time, number of turns, vessel support and insulation system
- error correction and ELM control coils: separate or combined coils, inside or outside vacuum vessel and/or toroidal field coil, multipole structure, response time, steady for toroidal field errors, dynamic for instability control

- disruption control: sensing control system, vacuum vessel protection of other systems, rapid gas or impurity puff with argon or neon, multiple shattered pellet injection, runaway electron dissipation, halo current protection
- vacuum vessel main vessel: vacuum only on conducting shell function, stainless steel or inconel or low radiation activation material, solid or ribbed construction, resistivity, operating temperature, bakeable to 100°, 250°, 400 °C, installed whole or segments welded, insulating break or not, internal and external connection and support structures, actively cooled or not, insulating blanket and baking structure
- vacuum vessel port structures: number of ports, room between toroidal and poloidal coils, bakeable, actively cooled or not, size of ports, radial, tangential and anti-tangential access, personnel and remote access ports, removable, permanent or weldable in place, flexibility for modification
- vacuum pumping: rough pumps, turbopumps, cryopumps and their regeneration time, truly continuous operation or not
- vacuum vessel inner wall cladding: tile structure of carbon, steel, inconel, molybdenum, tungsten, coatings or gettering on surface, limiters of low, medium or high-Z materials, poloidal or toroidal limiters, protection against halo currents, method of installation and removal, reparability, active or passive cooling, insulated or not from vacuum vessel, mechanical resistance to eddy and halo currents, gettering with boron, silicon, titanium, and other evaporatable substances, high wall reflectivity to minimize synchrotron radiation losses
- conducting copper shell: thick copper shell with insulating gap to stabilize plasmas by induced image currents, stabilization by the vacuum vessel.
- glow discharge and wall gettering: glow discharge, RF discharge, Taylor plasma cleaning, alcohol cleaning, gettering with titanium, beryllium, boron or other low and medium Z metals
- auxiliary heating of plasma and current drive integration: either/both plasma heating and current drive from external electric field, inductive and bootstrap driven currents, NBI, ICRF, LH, ECRH, AWH, alpha-particles from fusion
- neutral beams: ion source of hydrogen, deuterium, tritium, $He^3$ or $He^4$, positively or negatively charged, one, two, or three atoms per hydrogen isotope module giving beams at 1/3, 1/2 or full acceleration energy, injection angle from co- to counter- to radial, injection energy and neutralization fraction, continuous or pulsed operation, cryopumping or normal pumping and regeneration time, optimized for heating or current drive or both, overall power efficiency, direct conversion on beam dumps.
- radio frequency heating: choice of launching antenna with faraday shield or waveguide, actively or passively cooled, choice of plasma density and edge characteristics for coupling and launching wave, ray tracing, plasma resonance, inboard or outboard or axis, type of resonance for heating, optimized for heating or current drive, comprising resonance and typical frequencies:

  - Alfven: choice of frequencies, 1–5 MHz
  - ion cyclotron: 30–120 MHz

  - lower hybrid, 1–8 GHz
  - electron cyclotron, 100–200 GHz

- neutron shields: optimized for d-d 2.5 MeV neutrons or d-t 14 MeV neutrons,, internal neutron shields near plasma for coils and systems protections, active or passively cooled, choice of material for minimum activation, radioactive isotopes, waste disposal considerations, minimal thickness optimization with lead, steel, inconel, tungsten, etc., need for integrated total stopping power with vacuum vessel and breeding blanket.
- bioshields: need for external personnel shielding special concrete mixture walls, optimized for multi-MeV x-rays from runaway electrons and neutrons.
- tritium breeding blankets: just breeding or also full neutron shield, active or passively cooled, circulating fluid for tritium removal or need to remove blanket for analysis, choice of lithium/beryllium/lead/etc. compound, breeding ratio, tritium inventory and retention, tritium diffusion barriers
- divertor coils: inside or outside vacuum vessel or toroidal field coil, types of x-points available (4 x-points, single null, double null, inboard or outboard of plasma axis, snowflake, expanded boundary, sweepable), cooling type, inertial or steady state, power requirements due to distance from plasma, integrated or not with shaping coils, disruption protection.
- divertor mechanical and plasma facing components: channels for plasma exhaust (JET Mark I and Mark II pumped divertor, slot divertor, slanted target plates, narrow or wide opening, scannable dump, active or passive cooling or steady state, erosion and replacement characteristics, moveable for testing scenarios, use of impurity gases to increase radiation losses, material choice and coatings (tungsten, carbon, inconel, copper).
- divertor pumping system: in-vessel cryopumps, radiation shielding, throughput of pumping systems, external cryopumps and regeneration for steady state reactors, short-pulse gettering.
- plasma fuelling systems: neutral atom beams, gas puffing, pellet injection, wall and divertor recycling control, initial gas filling pressure and in-situ pellets.
- remote handling, maintenance, repair and upgrades: interacts with overall machine design and flexibility for replacement and repairs, degree of radiation hardening, ability to operate and to get visual input in extremely high radiation environment, lifting capability, equivalency to capability of people inside non-radioactive vessel.
- heat shields: integral to cryogenic systems, reduces radiation and thermal conduction heating, cooling structures, choice of coolant, e.g. liquid nitrogen, choice of material for thermal conductivity and strength against eddy currents, support structures.
- cryostat: choice of individual coil cryostats or "bell jar" covering entire machine, needs strength against atmospheric pressure, neutron radiation resistance, penetrations for diagnostics, machine services and auxiliary heating, single piece or assembled composite.

- low temperature cooling systems: nitrogen cooling systems, helium refrigerators, supercooled or supercritical or gaseous helium and flow capability, capacity, redundancy in case of fault.
- water and high temperature cooling: total cooling power of water, radioactivation control and filtering, room temperature or boiling or pressurized high temperature water, other coolants for vessel and blanket, liquid metals or eutectics, CFC or equivalent for non-corrosive cooling.
- tritium processing: able to handle different sources and concentrations (from blanket breeding modules, neutral beam tritium injectors, cryopanel regeneration, leaks into larger volumes, processing rates, inventories, need for initial feedstock, separation rates, reprocessing divertor gas output with heavy impurities
- Plasma diagnostics (some examples of many):
  - magnetics: flux loops and magnetic field, Rogowski coils, high frequency pickup, equilibrium reconstruction, Hall probes
  - interferometry and reflectrometry (microwave, laser)
  - electron temperature (laser Thompson scattering, LIDAR, line broadening, soft x-ray spectrometer)
  - ion temperature and fast ion distribution (charge exchange with diagnostic beam, laser scattering, neutron rate)
  - radiation (spectroscopy, bolometers, soft-x-ray)
  - current profile (polarimeter, MHD equilibrium reconstruction)
  - edge and divertor fluctuation, flows and plasma diagnostics (magnetic probes, Langmuir probes, reflectometers)
  - divertor, limiter and first wall diagnostics (thermocouples, surface analysis)
- Plasma and machine control and feedback systems: pre-programmed, real time calculation and feedback control
  - MHD equilibrium and shape (fast reconstruction and feedback, analogue PID feedback plus digital pre-programming, fully digital control, feedback control with real time equilibrium calculation, high frequency stability control)
  - current profile (measurements plus current drive feedback)
  - density and temperature profile (heating and gas puffing feedback)
  - edge pedestal (ELM control)
  - magnetic coil currents (pre-programming plus feedback on plasma magnetics and parameters)
- Power supplies: electric grid, transformers, motor generator with flywheel (50–60 or 100–120 Hz), capacitors
  - toroidal field: silicon controlled rectifiers (SCR) or capacitor
  - ohmic and transformer: SCR, capacitor, vacuum breaker
  - poloidal field: SCR (3,6,12 phases), transistors, choppers
  - auxiliary heating: SCR or capacitors
  - plant: SCR from transformer.

- data collection and analysis(real time digital processing, large scale data acquisition for off line analysis)

## 2.8 Integration of the Systems Design Space

The discussion in Chap. 1 on systems approaches with detailed checklists, when combined with the design space above, provides a structured way of proceeding on the analysis of existing experiments or designing new ones. In deciding on goals and innovations, it is important to know the past, but not to be bound by it. Checklists and design spaces serve as a guide, not a restraint on creativity.

This chapter has provided a library of parts and tools for the design of systems architecture. The component systems and subsystems provide the basic elements for integration of the full machine. The descriptions of individual tokamaks in this and following chapters provide examples of how the form of the elements can be integrated and the scenarios show how form and function work together. They also provide a basis for developing goals based on the successes already achieved. The systems architecture can be further developed by looking at the different aspects or views, including the spatial view of form, the plasma view and the time view, while considering different time scales. Further details of the overall system are developed by looking at the full matrix of interfaces of all of the components. Using all these tools together from Chaps. 3, 4, 5 and 6, it is possible to begin either reverse or forward engineering of the machines to be considered in the following chapters.

## References

1. J. Wesson, *Tokamaks*, 4th edn. (Oxford University Press, Oxford, 2011)
2. E. Morse, *Nuclear Fusion* (Springer, Cham, 2018)
3. I. Jenkins et al., A SysML model of the tokamak subsystems involved in a DEMO pulse, report EUROFUSION WPPMI-CP(16) 15445, Proceedings of 29th Symposium on Fusion Technology (SOFT) (2016)
4. F.B. Marcus et al., In-situ hydrogen pellet experiments in Doublet III. Nucl. Fusion **21**(7), 859–870 (1981)
5. P. Barabaschi et al., System engineering and configuration management in ITER, presented by S. Chiocchio, Proceedings of 24th SOFT Conference, Warsaw (2006) https://slideplayer.com/slide/1514227/. Last accessed 17 June 2022
6. S. Chiocchio et al., Integrated systems engineering approach to manage the installation and start up of a fusion reactor. Semantics Scholar (2008) https://www.semanticscholar.org/paper/TO-MANAGE-THE-INSTALLATION-AND-START-UP-OF-A-FUSION-Chiocchio-Kuehn/11662dc0a6a04efce1eac4891a801a2f7924c3a5?p2df. Last accessed 17 June 2022
7. G. Grossettia et al., Systems engineering perspective to the integration of the heating and current drive system in the EU DEMO: analysis of requirements and functions. Fusion Eng. Des. **136A**, 53 (2018)

8. G. Cima et al., Shaping and compression experiments in a small, multipole tokamak, 7th EPS conference, Lausanne, 1975. Atom. Energy **40**(2), 227–228 (1976)
9. R. Birch et al., Attempts to optimise β in the TOSCA tokamak, 9th EPS meeting, Oxford. Nucl. Fusion **20**, 1 (1979)
10. M.W. Alcock et al., Local heating and current drive investigation on TOSCA, paper H-1, 10th EPS conference, Moscow, 1981. Nucl. Fusion **22**, 2 (1981)
11. M.W. Alcock et al., High power ECRH at the second harmonic in TOSCA, paper H-15, 10th EPS Conference, Moscow, 1981. Nucl. Fusion **22**, 2 (1981)
12. N.R. Ainsworth et al., High β and lower q studies on TOSCA using ECRH, paper E-26, 11th EPS conference, Aachen, 1983. Plasma Phys. Control. Fusion **26**, 1A (1984)
13. J. Ellis, A.W. Morris, Instability control with active saddle coils on TOSCA, paper O-04, 11th EPS conference, Aachen, 1983. Plasma Phys. Control. Fusion **26**, 1A (1984)
14. A.D. Cheetham et al., The TCA tokamak project report 1979, report LRP 162/80, CRPP, Ecole Polytechnique Federale de Lausanne, Lausanne (1980)
15. J. Stockel et al., Contribution of joint experiments on small tokamaks in the framework of IAEA coordinated research projects to mainstream fusion research, paper EX/P7–38, 26th IAEA Fusion Energy Conference, Kyoto (2016)
16. D.R. Whaley et al., X- and O-mode electron cyclotron heating breakdown and startup in TCA. Nucl. Fusion **32**(5), 757 (1992)
17. A.G. Elfimov et al., Results of localized Alfven wave heating in TCABR. Braz. J. Phys. **34**(4B), 1707 (2004)
18. G.G. Grenfell et al., H-mode access and the role of spectral shift with electrode biasing in the TCABR tokamak. Phys. Plasmas **25**(7), 072301 (2018)
19. H. Eubank et al., Neutral-beam-heating results from the Princeton large torus. Phys. Rev. Lett. **43**, 270 (1979)
20. R. Panek et al., Reinstallation of the COMPASS-D tokamak in IPP ASCR. Czech. J. Phys. **56**, B126 (2006)
21. R. Panek et al., Status of the COMPASS tokamak and characterization of the first H-mode. Plasma Phys. Control. Fusion **58**, 014015 (2015)
22. M. Hron et al., Overview of the COMPASS results, 28th IAEA Fusion Energy Conference, Nice, France (Virtual) 10–15 May 2021, Presentation OV/P-1, Paper IAEA–CN–286-1360, Submitted to Nuclear Fusion (2021)
23. R. Panek et al., ITER-relevant research on the COMPASS tokamak, 27th IAEA Fusion Energy Conference, Paper IAEA CN-258, Ahmedabad (2018)
24. L. Artsimovich et al., Plasma Physics and Controlled Nuclear Fusion Research, vol I, p. 157 (IAEA, Vienna, 1969)
25. V.A. Vershkov et al., Recent results of the T-10 tokamak. Nucl. Fusion **51**, 094019 (2011)
26. I. Zemtsov et al., Study of the ECR-heating influence on the anomalous transport of tungsten ions in T-10 plasma, 28th IAEA Fusion Energy Conference, Nice, France (Virtual) 10–15 May 2021, Presentation EX/P4–25, Paper IAEA–CN–286-1313 (2021)
27. L.N. Khimchenko et al., Ecton mechanism of energy load on ITER-grade tungsten limiter T-10 tokamak and forecast for ITER, 27th IAEA Fusion Energy Conference, Ahmedabad (2018)
28. F. Wagner et al., Regime of improved confinement and high beta in neutral-beam-heated divertor discharges of the ASDEX tokamak. Phys. Rev. Lett. **10**(19), 1408 (1982)
29. M. Keilhacker, Confinement studies in L and H-type Asdex discharges. Nucl. Fusion **26**(1A), 49 (1984)
30. F. Wagner, A quarter-century of H-mode studies. Plasma Phys. Control. Fusion **49**, B1 (2007)
31. F. Wagner, The history of research into improved confinement regimes. Eur. Phys. J. H. **43**, 523 (2018). https://doi.org/10.1140/epjh/e2016-70064-9
32. Y. Liu, Recent advances in the HL-2A tokamak experiments. Nucl. Fusion **45**, S239 (2005)
33. H.E. Hongda et al., Study of plasma MHD equilibrium in HL-2A tokamak. Plasma Sci. Technol. **8**(4), 397 (2006)

34. M. Xu et al., Overview of HL-2A recent experiments. Nucl. Fusion **59**(11), 112017 (2019)
35. X. Duan et al., Progress of HL-2A experiment and HL-2M program, 28th IAEA Fusion Energy Conference, Nice, France (Virtual) 10-15 May 2021, Presentation OV/4–3, Paper IAEA–CN–286-1200, Submitted to Nuclear Fusion (2021)
36. Y.-K.M. Peng, D.J. Strickler, Features of spherical torus plasmas. Nucl. Fusion **26**, 769 (1986)
37. A. Sykes et al., High-β performance of the START tokamak. Plasma Phys. Control. Fusion **38**, B247–B260 (1997)
38. E.J. Strait et al., Wall stabilization of high Beta tokamak discharges in DIII-D. Phys. Rev. Lett. **74**, 2483 (1995)
39. G.S. Kurskiev et al., Thermal energy confinement at the Globus-M spherical tokamak. Nucl. Fusion **59**, 066032 (2019)
40. Y. Petrov et al., Overview of Globus-M2 spherical tokamak results at the enhanced values of magnetic field and plasma current, 28th IAEA Fusion Energy Conference, Nice, France (Virtual) 10–15 May 2021, Presentation OV/4-5Rc, Paper IAEA–CN–286-0673, Submitted to Nuclear Fusion (2021)
41. K. Hanada et al., Physical design of MW-class steady-state spherical tokamak, QUEST, 22nd IEAE Fusion Energy Conference, Geneva, Paper IAEA-CN-165/ FT/P3–25 (2008)
42. K. Hanada et al., Investigation of hydrogen recycling in long-duration discharges and its modification with a hot wall in the spherical tokamak QUEST. Nucl. Fusion **57**, 126061 (2017)
43. H. Idei et al., Electron heating of over-dense plasma with dual-frequency electron cyclotron waves in fully non-inductive plasma ramp-up on the QUEST spherical tokamak. Nucl. Fusion **60**(1), 016030 (2020)
44. T. Onchi et al., Plasma current ramp-up with 28 Ghz second harmonic electron cyclotron wave in the Quest spherical tokamak, 28th IAEA Fusion Energy Conference, Nice, France (Virtual) 10–15 May 2021, Presentation EX/P7–15, Paper IAEA–CN–286-0844 (2021)
45. K. Epp et al., Recent progress in the plasma injector 3 spherical tokamak program, presentation CP11.00192, 60th Annual Meeting of the American Physical Society Division of Plasma Physics, Portland, Oregon, October 5–9 (2018)
46. S. Howard et al., Plasma-wall interaction on the SLiC spherical tokamak device with large-are, dynamic liquid lithium free surface, presentation GO13.8, American Physical Society Division of Plasma Physics, Virtual Meeting, November 10 (2020)
47. J. Dalton, Nuclear fusion plant backed by Jeff Bezos to be built in UK, Independent Newspaper, UK, 18 June (2021)
48. D. Douai et al., Wall conditioning for ITER: current experimental and modeling activities. J. Nucl. Mater. **463**, 150 (2015)
49. M.M. Sabado, F.B. Marcus et al., Doublet III limiter performance and implications for mechanical design and material selection for future limiters, Report GA-A15633, General Atomic Company, San Diego. Symposium on Engineering Problems of Fusion Research, 8th, San Francisco, Calif., November 13–16, 1979, Proceedings. Volume 3. (A81–18901 06–75) Piscataway, N.J., Institute of Electrical and Electronics Engineers, Inc., 1979, pp. 1361–1369. (1979)
50. L. Oren, R.J. Taylor, Trapping and removal of oxygen in tokamaks. Nucl. Fusion **17**, 1143 (1977)
51. F.B. Marcus, Impurity studies and discharge cleaning in Doublet III, Report GA-A15613, General Atomic Company, San Diego. Presented at the IAEA Technical Committee Meeting on Impurities in Tokamaks: Source, Effect, and Control. Alushta, U.S.S.R. October (1979)
52. M. Greenwald et al., 20 years of research on the Alcator C-mod tokamak. Phys. Plasmas **21**, 110501 (2014)
53. D. Douai et al., Development of helium electron cyclotron wall conditioning on TCV for the operation of JT-60SA. Nucl. Fusion **58**, 026018 (2018)

54. G.L. Jackson et al., Simulating the ITER plasma startup scenario in the DIII-D tokamak, Paper IT/P7–2, IAEA Fusion Energy Conference (2008)
55. U. Seidel et al., Discharge scenarios for the first operational phase of ASDEX upgrade, 1992 International Conference on Plasma Physics, Innsbruck. 29 June–3 July 1992, 16C, Part I, p. 451 (1992)
56. S. Shiralwa et al., Formation of advanced tokamak plasmas without the use of an Ohmicheating solenoid. Phys. Rev. Lett. **92**, 035001 (2004)
57. P.C. De Vries, Characterisation of plasma breakdown at JET with a carbon and ITER-like wall. Nucl. Fusion **53**(5), 3003 (2013)
58. D. Mueller, The physics of tokamak start-up. Phys. Plasmas. **20**, 058101 (2013)
59. A.C.C. Sips et al., Progress in preparing scenarios for operation of the international thermonuclear experimental reactor. Phys. Plasmas **22**, 021804 (2015)
60. B.P. Duval et al., Singlet breakdown optimization to a doublet plasma configuration on the TCV tokamak, Preprint: IAEA Fusion Energy Conference, Gandhinagar, India, EX/P1–6 (2018)
61. O. Sauter et al., Beta limits in long-pulse tokamak discharges. Phys. Plasmas **4**, 1654 (1997)
62. J.C. Wesley et al., Shaping and characteristics of Ohmically heated non-circular plasmas in doublet III, IAEA Fusion Energy Conference, Paper (IAEA-CN-38/A-3), Brussels (1980)
63. K.H. Burrell et al., Attainment of reactor level volume-averaged beta in Doublet III. Nucl. Fusion **23**(4), 536 (1983)
64. P. Smeulders, JET Team, High density regimes and beta limits in JET. Proceedings of the 13th International Conference on Plasma Physics and Controlled Nuclear Fusion Research, Washington D.C. 1990, vol. 1, pp. 219–227, IAEA, Vienna (1991)
65. E.J. Strait, Stability of high beta tokamak plasmas. Phys. Plasmas **1**, 1415 (1994)
66. F. Hofmann et al., Extension of the TCV operating space towards higher elongation and higher normalized current. Nucl. Fusion **42**(6) (2002)
67. M. Gryaznevich et al., *Theoretical and Experimental Studies of Confinement in High Field ST, EPTC 2019* (Ghent, Belgium, 2019)
68. N. Ohyabu et al., A regime of improved energy confinement in beam-heated expanded-boundary discharges in Doublet III. Nucl. Fusion **25**(1), 49 (1985)
69. M. Greenwald et al., Energy confinement of high-density pellet-fueled plasmas in the Alcator tokamak. Phys. Rev. Lett. **53**, 352 (1984)
70. M. Keilhacker, JET Team, The JET H-mode at high current and power levels. Proceedings of the 12th International Conference on Plasma Physics and Controlled Nuclear Fusion Research, Nice 1988, vol. 1, pp. 159–181, IAEA, Vienna (1989)
71. P.H. Rebut and the JET Team, Recent JET results and future prospects, Proceedings of the 13th International Conference on Plasma Physics and Controlled Nuclear Fusion Research, Washington, DC. 1990, vol. 1, pp. 27–52, IAEA, Vienna (1991)
72. P. Lomas, JET Team, Optimisation of performance in JET limiter plasmas. Proceedings of the 13th International Conference on Plasma Physics and Controlled Nuclear Fusion Research, Washington, DC. 1990, vol. 1, pp. 425–435, IAEA, Vienna (1991)
73. A. Tanga, JET Team (1991) High performance H-modes in JET, Proceedings of the 13th International Conference on Plasma Physics and Controlled Nuclear Fusion Research, Washington, DC. 1990, vol. 1, pp. 261–276, IAEA, Vienna (1991)
74. C.M. Greenfield et al., VH-mode discharges in the DIII-D tokamak. 1992 International Conference on Plasma Physics, Innsbruck 1992, European Physical Society, pp. I-11–I-14 (1992)
75. K.H. Burrell et al., Quiescent double barrier high-confinement mode plasmas in the DIII-D tokamak. Phys. Plasmas **8**, 2153 (2001)
76. A. Isayama et al., Achievement of high fusion triple product, steady-state sustainment and realtime NTM stabilization in high-βp ELMy H-mode discharges in JT-60U. Nucl. Fusion **43**(10), 1272 (2003)

77. K. Shinohara et al., Ferritic insertion for reduction of toroidal magnetic field ripple on JT-60U. Nucl. Fusion **47**(8), 997 (2007)
78. G. Saibene et al., Results of the variable toroidal field ripple experiments in JET (INIS-XA–08N0893). International Atomic Energy Agency (IAEA) (2008)
79. A. Fasoli et al., Overview of physics research on the TCV tokamak. Nucl. Fusion **49**, 104005 (2009)
80. A.E. Hubbard et al., Edge energy transport barrier and turbulence in the I-mode regime on Alcator C-mod. Phys. Plasmas **18**, 056115 (2011)
81. A. Isayama et al., Overview of JT-60U results towards the resolution of key physics and engineering issues in ITER and JT-60SA. Nucl. Fusion **51**, 094010 (2011)
82. R. Behn et al., Threshold power for the transition into H-mode for H, D and He plasmas in TCV. Plasma Phys. Control. Fusion **57**, 025007 (2015)
83. C.D. Challis et al., Improved confinement in JET high β plasmas with an ITER-like wall. Nucl. Fusion **55**, 053031 (2015)
84. I. Chapman et al., Joint experiments tailoring the plasma evolution to maximize pedestal performance, Nuclear Fusion IAEA Fusion Energy Conference, Kyoto (2016)
85. X. Litaudon et al., Overview of the JET results in support to ITER. Nucl. Fusion **57**, 102001 (2017)
86. H. Meyer et al., Overview of physics studies on ADDEX upgrade. Nucl. Fusion **59**, 112014 (2019)
87. K.H. Burrell et al., Discovery of stationary operation of quiescent H-mode plasmas with net-zero neutral beam injection torque and high energy confinement on DIII-D. Phys. Plasmas **23**, 056103 (2016)
88. C.C. Petty and the DIII_D Team, DIII-D research towards establishing the scientific basis for future fusion reactors. Nucl. Fusion **59**, 112002 (2019)
89. P.B. Snyder et al., High fusion performance in super H-mode experiments on Alcator C-mod and DIII-D. Nucl. Fusion **59**, 086017 (2019)
90. J.R. Harrison et al., Overview of new MAST physics in anticipation of first results from MAST upgrade. Nucl. Fusion **59**(11), 112011 (2019)
91. M.J. Singh et al., Status of heating and current drive systems planned for ITER. IEEE Trans. Plasma Sci. **44**(9), 1514 (2016)
92. M.F.F. Nave et al., Fishbone activity in JET. Nucl. Fusion **31**(4), 697 (1991)
93. F.B. Marcus et al., JET neutron emission profiles and fast ion redistribution during sawtooth crashes. Plasma Phys. Control. Fusion **33**(4), 277 (1991)
94. Marcus FB, JET Neutron Group, Effects of sawtooth crashes on beam ions and fusion product tritons in JET. Nucl. Fusion **34**, 687 (1994)
95. F.B. Marcus et al., JET experiments with 120 keV 3He and 4He neutral beam injection and neutron diagnostics applications. Plasma Phys. Control. Fusion **34**(8), 1371 (1992)
96. B. Esposito, F.B. Marcus, et al., Ohmic ion temperatures and thermal diffusivity profiles from the JET neutron emission profile monitor. Plasma Phys. Control. Fusion **35**, 1433 (1993)
97. Sasao, M, JET Neutron Group, Determination of the ion thermal diffusivity from neutron emission profiles in decay. Plasma Phys. Control. Fusion **36**, 1 (1994)
98. F.X. Soldner et al., Lower hybrid current drive in JET and reactor applications, Plasma Physics and Controlled Nuclear Fusion Research 1994, IAEA 15th Conference Proceedings, Seville, Spain, p. 423, Paper IAEA-CN-60/A3–2 (1994)
99. J.B. Lister et al., Variable configuration plasmas in TCV, paper IAEA-CN-60/A5–2, pp. 627–632, Plasma Physics and Controlled Nuclear Fusion Research 1994, Fifteenth Conference Proceedings, Seville, Spain, 26 September – 1 October 1994, IAEA, Vienna (1995)
100. B.B. Esposito et al., Runaway electron measurements in the JET tokamak. Plasma Phys. Control. Fusion **38**, 2035 (1996)

101. O. Sauter et al., Steady-state fully noninductive current driven by electron cyclotron waves in a magnetically confined plasma. Phys. Rev. Lett. **84**, 3322 (2000)
102. C. Gormezano et al., Overview of JET results in support of the ITER physics basis. Nucl. Fusion **41**(10), 1327 (2001)
103. M.R. Wade et al., Physics of High Bootstrap Fraction, high performance plasmas on the DIIID tokamak. 28th EPS Conference on Controlled Fusion and Plasma Physics. Funchal June 2001. ECA, vol. 25A, p. 1365 (2001)
104. T. Oikawa et al., Reactor relevant current drive and heating by N-NBI on JT-60U. Nucl. Fusion **41**(11), 1575 (2001)
105. N. Oyama et al., Overview of JT-60U results toward establishment of advanced tokamak operation. Nucl. Fusion **49**, 104007 (2009)
106. A.S. Welander et al., Advanced control of neoclassical tearing modes in DIII-D with real-time steering of the electron cyclotron current drive. Plasma Phys. Control. Fusion **55**(12), 124033 (2013)
107. A. Kallenbach et al., Overview of ASDEX upgrade results. Nucl. Fusion **57**, 102015 (2017)
108. S. Coda et al., Overview of the TCV tokamak program: scientific progress and facility upgrades. Nucl. Fusion **57**, 102011 (2017)
109. U.A. Sheikh et al., Disruption avoidance through the prevention of NTM destabilization in TCV. Nucl. Fusion **58**, 106026 (2018)
110. J.L. Herfindal et al., Injection of multiple shattered pellets for disruption mitigation in DIII-D. Nucl. Fusion **59**(10), 106034 (2019)
111. E.J. Strait et al., Progress in disruption prevention for ITER. Nucl. Fusion **59**(11), 112012 (2019)
112. R.J. Buttery et al., DIII-D research to prepare for steady state advanced tokamak power plants. J Fusion Energy **38**, 72–111 (2019)
113. G. Pucella et al., Overview of the FTU results. Nucl. Fusion **59**(11), 112015 (2019)
114. D. Carnevale et al., Runaway electron beam control. Plasma Phys. Control. Fusion **61**, 014036 (2019)
115. C. Piron et al., Extension of the operating space of high-βN fully noninductive scenarios on TCV using neutral beam injection. Nucl. Fusion **59**, 096012 (2019)
116. R.A. Pitts et al., Physics basis for the first ITER tungsten divertor. Nucl. Mater. Energy **20**, 100696 (2019). https://doi.org/10.1016/j.nme.2019.100696
117. M.A. Mahdavi et al., A review of the recent expanded boundary divertor experiments in the Doublet III device. J Nucl. Mater. **111 and 112**(November/December), 355–361 (1982)
118. R.D. Monk et al., Recent results from divertor and scrape-off layer studies at JET. Nucl. Fusion **39**(11Y), 1751 (2002)
119. B. Lipschultz et al., Divertor physics research on Alcator C-mod. Fusion Sci. Technol. **51**, 369 (2007)
120. B. Labit et al., Dependence on plasma shape and plasma fuelling for small edge-localized mode regimes in TCV and ASDEX upgrade. Nucl. Fusion **59**, 086020 (2019)
121. A.J.H. Donne, Introduction to plasma diagnostics. Trans. Fusion Sci. Technol. **49**, 349 (2006)
122. M. Keilhacker et al., Review of diagnostic systems 1986, JET report JET-IR(86)16 (1986)
123. N.P. Hawkes et al., A 2.5 MeV neutron spectrometry system with a tangential line of sight for the D-D phase at the JET tokamak. Nucl. Instrum. Methods in Physics Research Section A **335** (3), I 533 (1993)
124. J.M. Adams et al., The JET neutron emission profile monitor. Nucl. Instrum. Methods. Phys. Res. A **329**(1–2), 277 (1993)
125. O.N. Jarvis, J.M. Adams, F.B. Marcus, G.J. Sadler, Neutron profile measurements in the joint European torus. Fusion Eng Des. **34–35**, 59 (1997)
126. G.J. Sadler, JET neutron group, fast particle diagnostics at JET: status and plans. Nucl. Fusion **35**(12), 1609 (1995)

127. P.E. Stott et al., *Diagnostics for experimental thermonuclear fusion reactors* (Springer, 1998)
128. M. Ariola et al., A modern plasma controller tested on the TCV Tokamak. Fusion Technol. **36** (1999). https://doi.org/10.13182/FST99-A97
129. F. Sartori et al., JET operations and plasma control: a plasma control system that is safe and flexible in a manageable way, Proceedings – Symposium on Fusion Engineering (2009). https://doi.org/10.1109/FUSION.2009.5226522
130. H. Anand et al., Distributed digital real-time control system for the TCV tokamak and its applications. Nucl. Fusion **57**, 056005 (2017)
131. J. Figueiredo et al., JET diagnostic enhancements testing and commissioning in preparation for DT scientific campaigns. Rev. Sci. Instrum. **89**, 10K119 (2018)
132. P.H. Rebut et al., Fusion energy production from a deuterium-tritium plasma in the JET tokamak. Nucl. Fusion **32**, 187 (1992)
133. F.B. Marcus et al., Neutron emission profile measurements during the first tritium experiments at JET. Nucl. Fusion **33**(9), 1325 (1993)
134. B. Balet et al., Particle and energy transport during the first tritium experiments on JET. Nucl. Fusion **33**(9), 1345 (1993)
135. G.A. Cottrell et al., Ion cyclotron emission measurements during JET deuterium-tritium experiments. Nucl. Fusion **33**(9), 1365 (1993)
136. P. Andrew et al., Experiments on the release of tritium from the first wall of JET. Nucl. Fusion **33**(9), 1335 (1993)
137. M.G. Bell et al., Overview of DT results from TFTR. Nucl. Fusion **35**, 1429 (1995)
138. O.N. Jarvis et al., Gamma ray emission profile measurements from JET ICRF-heated discharges. Nucl. Fusion **36**(11), 1513 (1996)
139. S.J. Zweben et al., Alpha-particle physics in the tokamak fusion test reactor DT experiment. Plasma Phys. Control. Fusion **39**(5A), 275 (1997)
140. R.J. Hawryluk et al., Results from D-T experiments on TFTR and implications for achieving an ignited plasma. Philos. Trans. Math. Phys. Eng. Sci. **357**(1752), 443 (1999)
141. P.R. Thomas et al., Observation of alpha heating in JET DT plasmas. Phys. Rev. Lett. **80**(25), 5548 (1998)
142. D.F.H. Start et al., D-T fusion with ion cyclotron resonance heating in the JET tokamak. Phys. Rev. Lett. **80**(21), 4681 (1998)
143. D.F.H. Start et al., Bulk ion heating with ICRH in JET DT plasmas. Nucl. Fusion **39**, 321 (1999)
144. C.H. Skinner et al., Tritium experience in large tokamaks: application to ITER. Nucl. Fusion **39**(2), 271 (1999)
145. K.-D. Zastrow et al., Trace tritium transport in JET. Nucl. Fusion **39**(11Y), 1891 (1999)
146. G.F. Matthews et al., Trace tritium and the H-mode density limit. J. Nucl. Mater. **266–269**, 1134 (1999)
147. M. Keilhacker et al., High fusion performance from deuterium-tritium plasmas in JET. Nucl. Fusion **39**(2), 209 (1999)
148. M.L. Watkins et al., Physics of high performance JET plasmas in DT. Nucl. Fusion **39**(9Y), 1227 (1999)
149. F.G. Rimini et al., Combined heating experiments in ELM-free H-modes in JET. Nucl. Fusion **39**, 1591 (1999)
150. P. Andrew et al., Tritium retention and clean-up in JET, JET report JET-P(99)01 (1999)
151. E. Joffrin et al., Overview of the JET preparation for deuterium–tritium operation with the ITER like-wall. Nucl. Fusion **59**, 112021 (2019)
152. J. Jacquinot et al., Steady-state operation of tokamaks: key physics and technology developments on Tore supra, 20th IAEA fusion energy conference, Villamoura (Portugal), paper IAEA-CSP-25/CD; OV/2–2 (2004)
153. G. Giruzzi et al., Synergy of electron-cyclotron and lower-hybrid current drive in steady-state plasmas. Phys. Rev. Lett. **93**(25), 255002 (2004)

154. S. Itoh et al., Recent progresses on high performance steady-state plasmas in the superconducting tokamak TRIAM-1M, 17th fusion energy conference, Yokohama, Japan (IAEA-CN-69), paper OV2/3 (1998)
155. H. Zushi et al., Overview of steady-state tokamak plasma experiments in TRIAM-1M. Nucl. Fusion **43**(12), 1600 (2003)
156. K. Hanada et al., Power balance investigation in steady-state LHCD discharges on TRIAM-1M. Fusion Eng. Des. **81**, 2257 (2006)
157. B. Wan et al., Overview of the latest HT-7 experiments. Nucl. Fusion **45**(10), S132 (2005)
158. B. Wan et al., Recent experiments in the EAST and HT-7 superconducting tokamaks. Nucl. Fusion **49**(10), 104011 (2009)
159. H.K. Park et al., Overview of the KSTAR research progress and future plan toward ITER and K-DEMO, IAEA Fusion Energy Conference, Ahmedabad, India, paper OV/2–3 (2018)
160. C. Bourdelle et al., Ramping up RF power and increasing pulse length in the full tungsten environment of WEST, 46th EPS Conference on Plasma Physics, paper O5.101, Milan (2019)
161. B.N. Wan et al., Recent advances in EAST physics experiments in support of steady-state operation for ITER and CFETR. Nucl. Fusion **59**(11), 112003 (2019)
162. L. Wang et al., Advances in plasma–wall interaction control for H-mode operation over 100 s with ITER-like tungsten divertor on EAST. Nucl. Fusion **59**, 086036 (2019)
163. S.C. Jardin et al., Dynamic modelling of transport and positional control of tokamaks. J. Comput. Phys. **66**, 481 (1986)
164. S.C. Jardin et al., Tokamak simulation code modelling of NSTX, 27th EPS conference on controlled fusion and plasma physics, Budapest. ECA **24B**, 1549 (2000)
165. R.J. Hawryluk, An empirical approach to tokamak transport, in *Physics of plasmas close to thermonuclear conditions*, ed. by B. Coppi et al., vol. 1, (Commission of the European Communities, Brussels, 1980), p. 19
166. R.V. Budny, A standard DT supershot simulation. Nucl. Fusion **34**, 1247 (1994)
167. T. Rafiq et al., Fusion power production in international thermonuclear experimental reactor baseline H-mode scenarios. Phys. Plasmas **22**, 042511 (2015)
168. W. Pfeiffer, F.B. Marcus, et al., Giant sawtooth oscillations in the doublet III tokamak. Nucl. Fusion **25**(6), 655 (1985)
169. W.A. Houlberg, S.E. Attenberger, L.M. Hively, Contour analysis of fusion reactor plasma performance. Nucl. Fusion **22**(7), 935 (1982)
170. R.C. Grimm, J.M. Greene, J.L. Johnson, Computation of the magnetohydrodynamic spectrum in axisymmetric toroidal confinement systems. Meth. Comput. Phys. **16**, 253 (1976)
171. R. Gruber et al., Erato stability code. Comp Phys Commun. **21**(3), 323 (1981)
172. J.M. Greene, M.S. Chance, The second region of stability against ballooning modes. Nucl. Fusion **21**, 453 (1981)
173. W.M. Tang, Microinstability-based model for anomalous thermal confinement in tokamaks. Nucl. Fusion **26**, 1605 (1986)
174. J.E. Kinsey et al., Nonlinear gyrokinetic turbulence simulations of E×B shear quenching of transport. Phys. Plasmas **12**, 062302 (2005)
175. F.D. Halpern et al., Improved model for transport driven by drift modes in tokamaks. Phys. Plasmas **15**, 012304 (2008)
176. P.B. Snyder et al., Pedestal stability comparison and ITER pedestal prediction. Nucl. Fusion **49**, 085035 (2009)
177. M. Brambilla, *A Full Wave Code for Ion Cyclotron Waves in Toroidal Plasmas*, report IPP 5/66. (Max-Planck-Institut fur Plasmaphysik, Garching, 1996)
178. R.J. Goldston et al., New techniques for calculating heat and particle source rates due to neutral beam injection in axisymmetric tokamaks. J. Comput. Phys. **43**, 61 (1981)
179. D.W. Ignat et al., Dynamic modelling of lower hybrid current drive. Nucl. Fusion **34**, 837 (1994)

180. A.H. Kritz et al., Ray tracing study of electron cyclotron heating in toroidal geometry, Heating in toroidal plasmas Proceeding of the 3rd Joint Varenna-Grenoble International Symposium, Grenoble, France 2, 707 (1982)
181. J. DeLucia, S.C. Jardin, A.M.M. Todd, An iterative metric method for solving the inverse tokamak equilibrium problem. J. Comput. Phys. **37**, 183 (1980)
182. PPPL Theory Department, https://theory.pppl.gov/. Last accessed 17 June 2022
183. G.L. Falchetto et al., The European integrated tokamak modelling (ITM) effort: achievements and first physics results. Nucl. Fusion **54**, 043018 (2014)
184. G.L. Falchetto, et al., Multi-machine analysis of EU experiments using the Eurofusion integrated modelling (EU-IM) framework, P1.1081, 46th EPS Conference, Milan (2019)
185. M. Kovari et al., "PROCESS": a systems code for fusion power plants – part 2: engineering. Fusion Eng. Des. **104**, 9 (2016)
186. C. Reux et al., DEMO reactor design using the new modular system code SYCOMORE. Nucl. Fusion **55**, 073011 (2015)
187. M. Coleman, S. McIntosh, BLUEPRINT: a novel approach to fusion reactor design. Fusion Eng. Des. **139**, 26 (2019)
188. F. Franza et al., MIRA: a multiphysics approach to designing a fusion power plant, 28th IAEA Fusion Energy Conference, Nice, France (Virtual) 10–15 May 2021, Presentation TECH/2-3Rb, Paper IAEA–CN–286-0666, Submitted to Nuclear Fusion (2021)

# Part II
# High-Current Tokamaks

# Chapter 3
# Doublet III/DIII-D and 1–2 MA Tokamaks: Robustness and Adaptation

**Abstract** The 2 MegaAmpere (MA) plasma current tokamak Doublet III and its upgrade DIII-D are analysed with systems reverse engineering. The highly-flexible shaping-coils allowed a wide range of plasma configurations. The jointed toroidal-field coils supported a successful upgrade to a large diverted D-shaped plasma. The two machines were robust against failure modes and made important contributions to the advancement of fusion energy. The 1.4 MA Asdex-U used a more reactor relevant design with the poloidal field coils outside of the toroidal field coils. The design space variation of high toroidal field was explored by Alcator at 12 T, succeeded by Alcator C-Mod and complemented by FTU. Low aspect ratio spherical tokamaks reached 1 MA at only 0.5 Tesla toroidal field.

## 3.1 Systems Analysis Strategy for Doublet III/DIII-D

### *3.1.1 Overall Strategy*

Doublet III and its upgrade DIII-D are suitable for a case study of systems reverse engineering by using the methodology from Checklist 1.2 of Chap. 1. Although Doublet III began operation in 1978, the upgrade DIII-D which used many of the original components, is still highly productive four and a half decades later. Machine performance and metrics are evaluated, including robustness and fragility. This analysis is repeated for other tokamaks in less detail.

### *3.1.2 Application of Checklist 1.2*

The first part of Checklist 1.2 for reverse engineering concerns the identification of form and function and system architecture. The elements in their high level systems are:



F. B. Marcus, *Systems Approaches to Nuclear Fusion Reactors*, Springer Series in Plasma Science and Technology, https://doi.org/10.1007/978-3-031-17711-8_3

- toroidal field coils
- ohmic heating coil
- shaping and instability control coils
- vacuum vessel
- power supplies and coil connections
- plasma control and shaping
- neutral beam and electron cyclotron heating and current drive

### 3.1.3 Doublet III Form and Function Choices

Vertically elongated and shaped tokamaks are now widely recognized as having significant advantages for tokamak fusion reactors by allowing higher plasma current and stability improvements. The doublet concept was invented by Tihiro Ohkawa and was tested on smaller doublets and then used to design the Doublet IIA and the Doublet III tokamaks [1]. Figure 3.1 shows the Doublet III vacuum vessel, limiters, and coils for the toroidal, ohmic and shaping fields [2, 3].

The machine-defining parameters of Doublet III were major radius R = 1.43 m, minor radius a = 0.45 m, axial toroidal field $B_T = 2.6$ T, and plasma current $I_p$ up to 2.2 MA in the doublet configuration, and 1.0 MA in a D-shaped configuration in the upper half of the vacuum vessel. The surfaces of constant magnetic flux from actual plasma discharges were calculated by MHD equilibrium codes and are shown in Fig. 3.2 for singlet, droplet and doublet configurations.

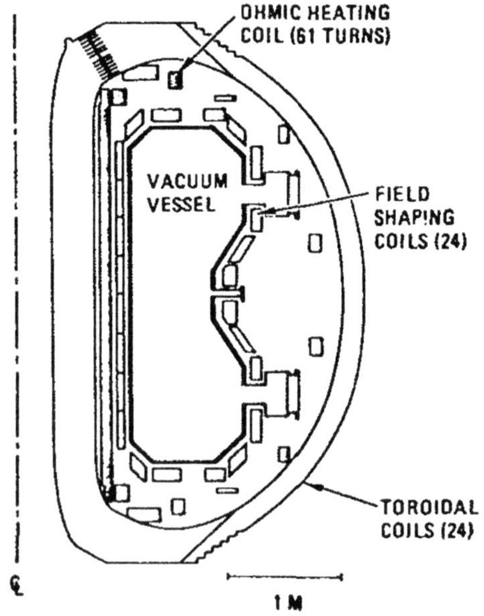


**Fig. 3.1** Doublet III vacuum vessel, limiters and magnetic coils. (From [3], Fig. 1, reused with permission from the IAEA)

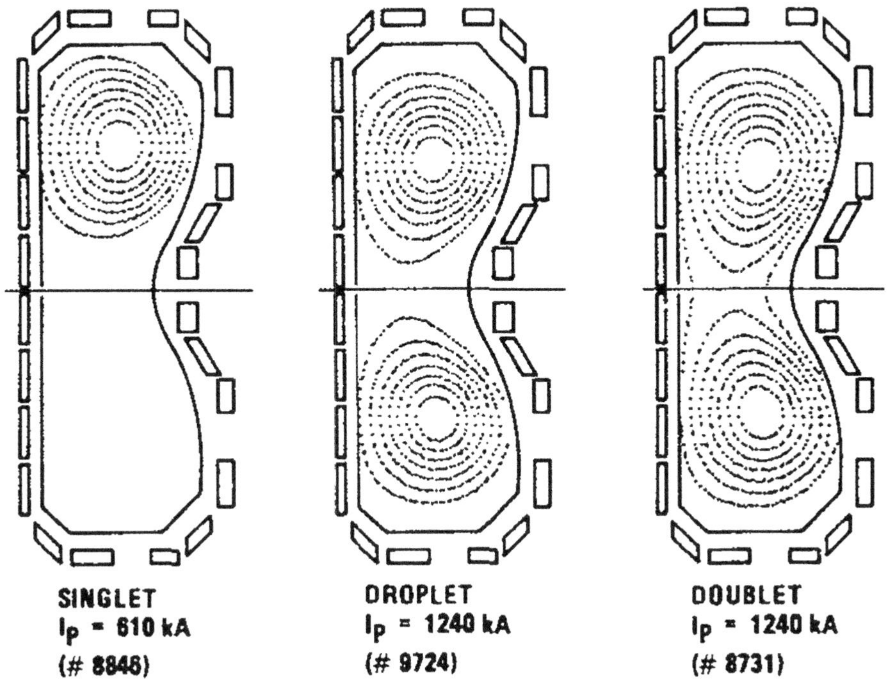


**Fig. 3.2** Doublet III with flux surfaces of singlet, droplet, and doublets. (From [3], Fig. 2, reused with permission from the IAEA)

The form of the elements and the view of their spatial relationships consisted of a tokamak central column bound with epoxy containing the inner plates of the toroidal field coil, the ohmic heating transformer coil and inner shaping coils. The vacuum vessel which encircled the central column was composed of a double layer of thin high-resistivity inconel sheeting, with corrugations connected by welding to form a strong construction. Outside the vessel are multiple shaping coils, all inside the outer toroidal field coils. The shaping coils are controlled by a few highly flexible power supplies, with some shaping coils connected in parallel. The functional interaction matrix for such a poloidal coil system is shown in Table 2.1 of Chap. 2, with some variations.

This chapter examines the poloidal field shaping coil subsystem's ability to flexibly create and shape plasmas for high performance. The operational and theoretical teams developed a wide range of functional models to interpret results, e.g. MHD-equilibrium and transport codes.

#### 3.1.3.1 Toroidal Field Coils

Doublet III and DIII-D employed design choices for the toroidal field coils that produced extreme robustness and the means for protecting against fragilities. These coils had joints and were demountable, hence any faulty construction could be replaced, repaired or upgraded The demountable joint at the inner top of the toroidal field coil allowed installation of the vacuum vessel, flux return coils, field-shaping coils, and supporting structure without any joints in these components. The overall design and assembly sequence of Doublet III and DIII-D was therefore critically dependent on proper functioning of the demountable joints. The selected interlocking-finger design was manufactured using standard machining tolerances and procedures and could be assembled and disassembled with relative ease and rapidity.

The central column core contained the inner part of the toroidal field coil, ohmic heating coil and the inner shaping coils. It was a single structure constructed in place and bonded together with epoxy resin. The machine was assembled and operating in 1978, 6 months after core completion.

The 24 toroidal field coils were D-shaped to reduce mechanical stress. The large number of coils provided low magnetic ripple at the plasma outer edge, which was important for plasma confinement. The coil shape ultimately allowed a D-shaped vacuum vessel replacement that efficiently filled the coil volume.

Each conductor could carry up to 195 kA, grouped in 24 bundles of six turns. Stress analysis showed that the assembled toroidal field coil was structurally adequate for 100,000 cycles of operation at 4 T toroidal field. Due to limitations from the prestressed mechanical structure, the toroidal field was operated up to only 2.6 T at the plasma axis, leaving a wide margin of safety in the coils themselves.

#### 3.1.3.2 Poloidal Field Coils and Vacuum Vessel

The plasma shaping coils were connected in parallel groups to provide plasma stability via image currents induced in the coils by the plasma, supplemented by several power supplies to achieve the desired doublet plasma in the full doublet vacuum vessel or D-shaped plasmas in the doublet or D-shaped vessels. The chopper power supplies provided multi-kilohertz high-frequency for feedback control against vertical plasma instabilities.

#### 3.1.3.3 Vacuum Vessel

The vacuum vessel was composed of high strength and high electrical resistivity Inconel 625 and was an all-welded, thin-wall, continuous chamber, with a corrugated sandwich construction connecting inner and outer walls, giving a current penetration time scale of 1 ms. This short time allowed a large starting voltage of 100 V/turn to fully penetrate to the plasma region, producing a very rapid current rise. The vessel allowed baking to 400 °C to achieve excellent vacuum conditions.

The vacuum vessel access ports allowed for vacuum pumping, diagnostics and auxiliary heating, initially high powered neutral beams. A personnel access port was provided for construction and manual cleaning inside the vessel.

#### 3.1.3.4 Power Supplies and Coil Connections

The plasma shaping circuitry [2] was optimized for doublet production, but its high flexibility allowed a full range of configurations. The plasma shape was controlled by six power supplies. Control of the magnetic flux linking these coils allowed the cross-section and position of the plasma to be varied. These coils were arranged in several parallel-connected groups, each with a power supply.

Two principal shaping functions were required to obtain a doublet discharge. The vertical field to maintain radial equilibrium was provided by a combination of an inductor in series with the inner shaping coils and the vertical power supplies with an 8 ms response time in series with the outer coils. The flux required to form the doublet waist was provided by the doublet supply, also 8 ms, assisted by a capacitor discharge circuit during the initial plasma current rise.

Fine tuning of the plasma shape was provided by two "chopper" supplies [4] with 0.2 ms response that controlled the flux at the top and bottom of the plasma. The chopper power supplies were complex and involved silicon controlled rectifiers, inductances, condensers and high speed switches. Their use on the upper and lower coils provided high frequency response to maintain plasma vertical stability.

Finally, two resistance compensation supplies operating at 16 ms minimized differences in the shaping flux on the inner coils produced by the coil resistance.

#### 3.1.3.5 Plasma Control and Shaping

The measures used for plasma equilibrium control [5] included control signals used for the flux surface position regulation They were derived from measurements of the poloidal field and flux obtained with an array of flux loops and magnetic field measuring coils located immediately outside the vacuum vessel. The close proximity of the sensors to the plasma surface allowed the position of the plasma surface to be accurately computed with simple analogue circuitry. Control of the position of the outer plasma surface to $\pm 0.5$ cm was routine.

#### 3.1.3.6 JAERI Collaboration and Auxiliary Heating

A major international collaboration was negotiated between General Atomic (GA) and the Japan Atomic Energy Research Institute (JAERI). JAERI agreed to furnish a large motor generator and neutral beam injection upgrades. The motor generator allowed longer plasma pulses and high power auxiliary heating, including electron cyclotron heating when it became available.

In return, JAERI was granted half the operating time on Doublet III. JAERI sent a team of a dozen physicists, and GA provided operations and diagnostics support for both JAERI and GA physics teams. Each team performed experiments on alternate weeks, operating the machine with 16-hour shifts for 7 days a week. The understanding was that JAERI would explore D-shaped singlet plasmas in the upper half of the vacuum vessel, and GA would explore doublet plasmas.

The first results [6] from the JAERI team demonstrated singlet plasmas with 1.0 MA plasma current, vertical elongation of 1.6 and density up to $1.0 \times 10^{20}$ $m^{-3}$. More advanced results [7] on energy confinement of beam-heated divertor and limiter discharges were also obtained.

Since neutral beam heating was divided equally between upper and lower half of the machine, D-shaped plasmas received half of the total power available. Each neutral beam injector box [8] had two ion sources that accelerated hydrogen ions to 80 kV resulting in a total injected power of 3.6 MW per box. The beams were composed of 1.4, 1.0 and 1.2 MW at 80, 40 and 26.7 kV respectively from the neutralization of $H^+$, $H_2{}^+$ and $H_3{}^+$ ions initially accelerated. The injection angle was nearly perpendicular to the toroidal field lines at 76°. The use of auxiliary heating and current drive is described in more detail in the following scenarios.

## 3.2 Doublet III Experiments and Scenarios

Continuing the procedure from Checklist 1.2, machine experiments and operational scenarios are examined here to determine the constructional, operational and experimental outcomes. The experimenters and theorists had developed models to interpret the results.

### *3.2.1 MHD Equilibrium*

Shaped and vertically elongated plasmas can have important advantages for high plasma current, stability and confinement time. The JAERI group produced diverted and non-diverted D-shaped plasmas, the latter with elongations of up to 1.6. The GA group produced droplet plasmas (two separated teardrops) with elongations up to 1.5, doublet plasmas with an internal separatrix and elongations up to 3.2, and an expanded boundary configuration. Record values of plasma current were achieved: 1.0 MA in the D and 2.2 MA in the doublet at a toroidal field of 2.4 T. D-shaped plasmas with q(a) = 2.6 and divertor plasmas with q(a) = 2 were obtained.

### *3.2.2 Plasma Energy Confinement in Different MHD Equilibria*

The energy confinement time of a plasma is a key metric of success. Energy confinement times were determined for a variety of plasmas using high atomic number (Z) primary limiters of 90% tantalum – 10% tungsten and secondary limiters of molybdenum [9]. With near-circular (b/a = 1.1–1.3) and doublet (b/a = 3) cross section plasmas, energy confinement measurements were made during the steady-state portions of ohmically heated discharges. The electron energy confinement was found to be equal at a given density in both the near-circular and doublet plasmas. This result in comparing doublet to D-shaped or near circular plasmas was a key element in deciding at first to concentrate solely on near circular plasmas in half of the vacuum vessel. The energy confinement of ohmically-heated plasmas were further studied [10].

Research was ended on the doublet configuration because the doublets created were acting like two separate plasmas connected only by a small amount of the total magnetic flux surrounding both cores. This ultimately led to the upgrade of Doublet III with a new vacuum vessel to obtain maximum performance with a D-shaped plasma.

## 3.2.3 *Doublet III Scenarios*

### 3.2.3.1 High-Z Limiters and Wall Conditioning

Plasma behaviour is very dependent on vessel cleanliness, plasma-wall interactions and density control. A major study of limiter material design [11] used primary and backup limiters. Initial materials were a 90% tantalum and 10% tungsten compound for the primary limiter and molybdenum for the backup limiters, chosen for their thermal and structural properties. Tantalum was preferred due to high temperature transition problems with tungsten. In later experiments, the nickel-based alloy Inconel X-750 was selected for a medium-Z limiter material. It was demonstrated that good plasma results could be obtained.

Plasma discharges with effective plasma charge $Z_{eff}$ values near unity, i.e. nearly all protons, and low concentrations of medium and high-Z metallic impurities were obtained [12] by a combination of wall cleaning with alcohol, high temperature vessel baking, low-power hydrogen discharge cleaning, gas puffing, precise plasma shape and position control, and high-Z limiters. The high-Z limiters showed only small areas of melting or arcing.

The cleaning experience in Doublet III was relevant to the next generation of tokamaks. Further studies [13] explored the role of MHD instabilities on impurities near the plasma axis. Soft x-ray data implied that peaked density discharges had an impurity profile that was also peaked, leading to a hollow current density profile and instability with current redistribution.

#### 3.2.3.2 High Elongation Single Axis Plasmas

The plasma hydrogen density in plasmas is usually controlled by gas puffing, neutral beam injection or injection of solid hydrogen pellets at high velocity. Solid hydrogen pellet injection can increase plasma density and also affect the temperature, hence the conductivity and the plasma current profile, in turn influencing plasma shaping. An alternative method [14] injected a 2 mm diameter cylindrical pellet of solid hydrogen at low velocity to the centre of the upper half of Doublet III before plasma initiation.

During plasma initiation, the pellet ablated in 6–9 ms, increasing the initial plasma density by 2–3 times. The profiles of electron and current density of discharges initiated with a pellet are very flat compared to normally initiated discharges. Lower electron temperatures and impurity levels persist for hundreds of milliseconds. The resultant D-shaped plasmas were vertically stable and had elongations of 1.8 on axis and 2.0 on the outermost flux surface as shown in Fig. 3.3.

Other ways of controlling the plasma current profile were explored theoretically by using electron cyclotron resonance heating to drive plasma current in a particular localized gradient region of the plasma [15]. The current produced by this mechanism flows in the same direction as the Fisch-Boozer [16] current and is equivalent in magnitude.

#### 3.2.3.3 Expanded Boundary Divertor

The magnetic field lines in a divertor region that is outside the last closed flux surface receives the heat flowing out of the main plasma. Radiating away that power can significantly reduce the direct heat load to the divertor plates. The flexibility of the Doublet III coil system allowed the external coil connections to be rewired to create a diverted magnetic configuration that had an expanded boundary [17]. The magnetic flux outside the x-point was guided into the lower half of the machine, greatly expanding the plasma flow volume. Experiments on the expanded boundary divertor in Doublet III revealed a medium density operating regime in which the divertor density rises with the cube of the main plasma density.

Injection of argon into the discharge resulted in a concentration of argon which was two orders of magnitude higher in the divertor than in the main discharge. This trapping effect suggested that the argon could be useful for cooling the plasma boundary, without being deleterious to the plasma as a whole.

#### 3.2.3.4 H-mode in Expanded Boundary

A plasma in H-mode that can radiate power entering the divertor should be an ideal operating mode. The knowledge built up during Doublet III operation and the high flexibility of the system allowed the combination of an expanded boundary

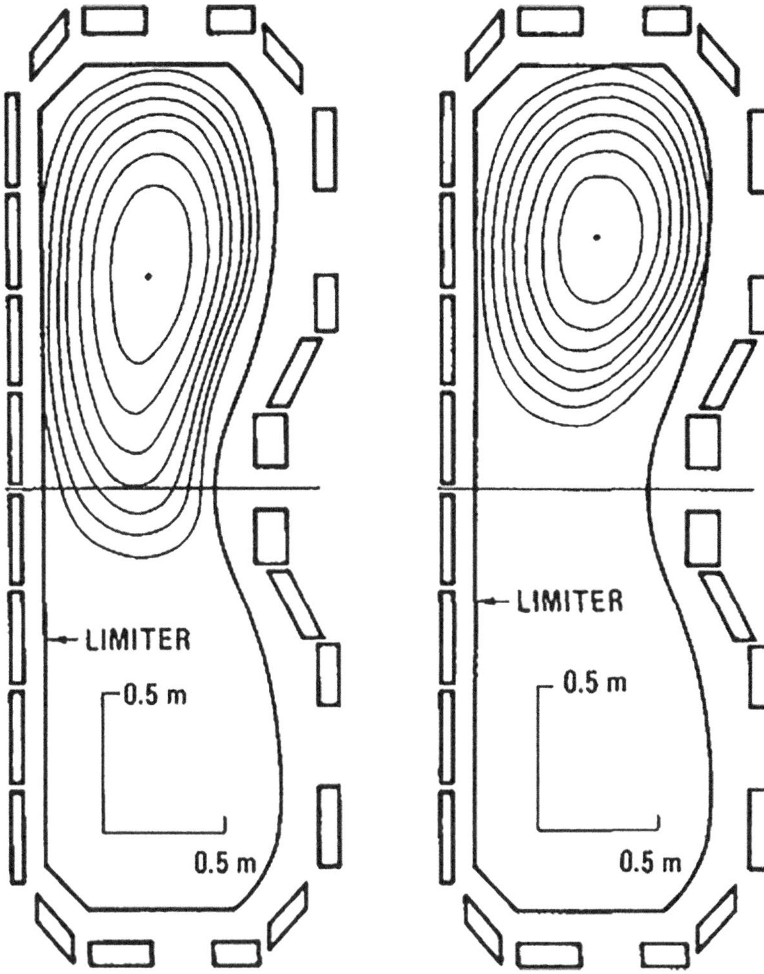


**Fig. 3.3** Plasma configurations after in-situ pellet (surfaces of constant magnetic flux) at 45 ms after discharge initiation, height-to-width ratio = 2.00 and without pellet, height-to-width ratio = 1.37. (Reused with permission from [14], Fig. 9)

configuration and intense neutral beam heating to achieve the H-mode [18, 19]. This improvement was obtained for a wide range of plasma parameters $290 < I_p$ (kA) $< 800$, $0.8 < B_T(T) < 2.5$, $2.0 < n_e$ $(10^{19}\ m^{-3}) < 10.0$, and $1 < P_B$ (MW) $< 4.5$. The energy confinement time at high plasma current reached 80 ms for injected beam power 3.5 MW. The $n\tau$ parameter reached $1 \times 10^{19}\ m^{-3}$ s at $P_B = 2.0$ MW, accompanied by a record energy content of 280 kJ.

#### 3.2.3.5 High Toroidal Beta

One of the main aims of tokamak experiments is to maximize toroidal beta ($\beta_T$), in order to minimize the strength of magnetic fields required to achieve fusion. At a low toroidal field of 0.62 T [20], volume-averaged toroidal beta of up to 4.5% was obtained, using a neutral-beam injection power of 3.4 MW into a 345 kA plasma that was low q = 1.7 and vertically elongated = 1.38 in the upper half of Doublet III. This level of beta was above the minimum level required for a tokamak reactor.

#### 3.2.3.6 Giant Sawtooth Instability

Sawtooth instabilities result when electron temperatures are peaked on axis, so the plasma resistivity is low. The plasma current then tends to peak there, leading to q less than one on axis. This causes an instability that redistributes the current. Sawtooth instabilities were studied [21] in high NBI power, low-q discharges. Large-amplitude sawtooth oscillations were measured in many diagnostics, including soft-X-ray emission, central electron temperature, neutron production rate, radiated power, D-alpha emission, fast-neutral flux and one-turn voltage. Giant sawteeth tend to appear at low $q_a$ and high toroidal beta. Many of their features were simulated with the ONETWO transport code to which a sawtooth mixing model and several diagnostic models were added. Substantial redistribution of particles and energy was shown to occur over the plasma interior. Global energy confinement was only moderately degraded by the sawteeth.

## 3.3 DIII-D Design and Results

### *3.3.1 Upgrade of Doublet III to DIII-D*

Having reached the maximum levels of performance in Doublet III, it was decided to upgrade Doublet III to the DIII-D configuration, originally called Big Dee because of the new vacuum vessel [22–24]. The installation of the large D-shaped vessel in Doublet III was made possible by the large low-ripple volume of the 24 demountable toroidal field. Ripple at the plasma edge was less than 1%. A central requirement for the design of the vessel was to maximize access to the plasma for auxiliary heating (NBI and RF), diagnostics, developmental components, material testing, and pumping. The Dee vessel was structurally designed along the same lines as that of Doublet III, an Inconel 625, all-welded, continuous chamber in a corrugated sandwich construction. The project also involved replacement of outer shaping coils and addition of support structure, limiters, vessel armour and power supplies. The cross-section of the device is shown in Fig. 3.4, with old components shaded, highlighting the unshaded upgraded coils and vacuum vessel that are new.

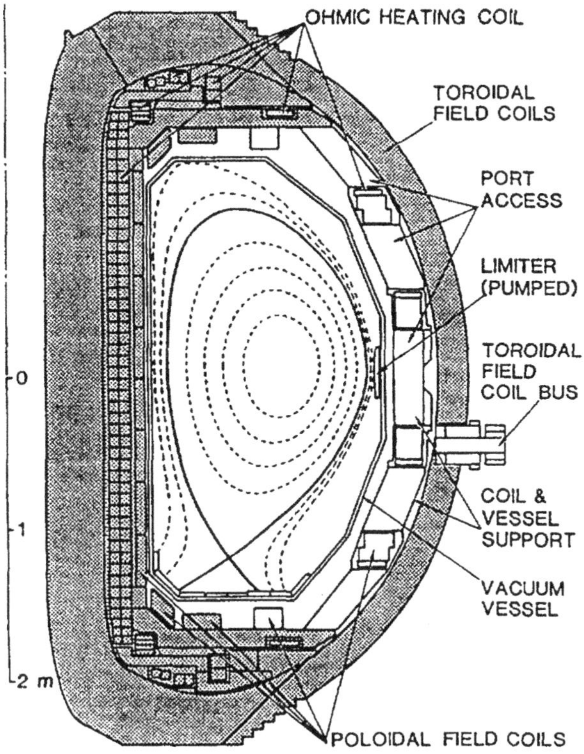


**Fig. 3.4** Big Dee/DIII-D cross section. (From [23], Fig. 1)

Parameters were major and minor radius R = 1.67 m and a = 0.67 m, elongation up to 2/1, toroidal magnetic field $B_T = 2.2$ T, flux swing 7.5 Volt-Second, plasma current $I_p = 3$ MA. Additional heating was provided by neutral beams all on the midplane, ECRH and ICRH. Either limiter or divertor plasmas could be produced.

The DIII-D plasma had a low aspect ratio R/a of 2.5, similar to that of the JET tokamak. Its high elongation was designed to obtain high-current non-circular discharges at moderate toroidal field. This relatively low aspect ratio approached that of spherical tokamaks (see below) with inner wall neutron shielding. The plasma shaping and control systems were extensions of those used on Doublet III and provided precise control of elongation and triangularity. The outer flux surfaces were diverted into armour at the bottom of the device.

The system producing high voltages for plasma breakdown was the same as for Doublet III [25]. Reaching 2.5 MA plasma current was thought to require a peak one turn voltage of up to 250 V (16 kV across coil terminals) and a magnetic flux swing of 5 Volt-seconds. The interruption of the transformer coil current was accomplished by an array of vacuum circuit breakers and a counter pulse network.

### 3.3.2 DIII-D Scenarios

The upgrade to DIII-D proceeded excellently and led to decades of successful plasma operation.

#### 3.3.2.1 DIII-D Initial Results and VH-Mode

A main goal of tokamak operation is to achieve plasma regimes with very high (VH-mode) energy confinement. In initial operation [26], very high confinement was observed in divertor discharges. The VH-mode followed boronization of the DIII-D vessel. It exhibited total energy confinement a factor of 2.5 to 3.5 greater than that predicted by the "ITER L-mode scaling". Confinement of thermal energy alone sometimes exceeded twice that of the JET/DIII-D H-mode scaling.

VH-mode was observed during a long phase of the discharges that was free of the ELM instability. At the beginning of the ELM-free period, the plasma appeared to be in H-mode, with confinement near that predicted by the JET/DIII-D scaling, remaining nearly constant for up to hundreds of milliseconds. Afterwards, the behaviour sharply increased over that of H-mode as the confinement began to increase. This increase continued until the occurrence of a beta-related global MHD event, which rapidly decreased the plasma stored energy.

#### 3.3.2.2 Double Barrier H-mode

Internal transport barriers (ITB) can be formed inside a plasma with sharp gradients, similarly to pedestals at the plasma edge in H-mode. They could greatly increase plasma confinement time. A detailed study of ITB in DIII-D led to significant improvements in understanding of plasma performance [27, 28]. Standard H-modes come at a significant cost because ELMs appear. DIII-D demonstrated a quiescent H-mode regime that is ELM-free and yet has good density and impurity control. An internal transport barrier could be maintained for long periods in a quiescent double barrier regime. The key factors in creating the quiescent H-mode operation were NBI in the direction opposite to the plasma current (counter injection) plus cryopumping to reduce the density. Density and impurity control were possible because of an edge harmonic oscillation which enhanced the edge particle transport while leaving the energy transport unaffected.

#### 3.3.2.3 Partial Current Drive with Bootstrap, NBI and ECRH

A tokamak operating in truly steady state with plasma current driven internally from pressure-gradient driven (bootstrap) currents and/or by auxiliary heating would have many advantages as a reactor. DIII-D explored advances towards a steady-state plasma current in a tokamak utilizing the bootstrap current, NBI, ECRH and Electron Cyclotron Current Drive (ECCD) [29]. High performance plasmas that simultaneously combined high normalized performance with $\beta_n \sim 4$, $H_{89} \sim 3.0$, and $f_{BS} \sim 60\%$, with $q_{min} > 1.5$ were sustained for several energy confinement times. The $\beta$ limit in these plasmas high with $\beta_n \sim 6 \times l_i$ (internal inductance) as a result of active feedback stabilization of resistive wall modes. Density control compatible with high efficiency ECCD was demonstrated.

#### 3.3.2.4 Instability Control of Neoclassical Tearing Modes with ECCD

Instability control is a major goal of tokamak operation in order to maintain the plasma and to avoid energy loss. ECCD was used to control instabilities in the plasma [30]. The system for controlling neoclassical tearing mode (NTMs) instabilities suppressed the NTM when it became unstable, by activating the ECCD and bringing it back to stable. This technique was made possible by several advancements:

- ECCD was very accurately aligned to the NTM.
- The alignment was maintained even when the NTM was almost gone, made possible by real-time equilibrium reconstructions including measurements of the motional Stark effect and a refraction estimator.
- Real-time steerable mirrors need fast and accurate actuators for the alignment adjustments.
- Early NTM detection was made possible by a real-time mode analysis that filters noise to minimize false positives.

#### 3.3.2.5 Plasma Disruption Avoidance and Pellet Mitigation

Since disruptions cause the rapid loss of plasma current and energy, their mitigation and control are essential for fusion reactors [31]. The causes of disruptions are many and complex [32, 33]. The result of a disruption can be vertical motion of the plasma which drives currents in the vacuum vessel and in-vessel components, a rapid dump of energy in small areas, the generation of runaway electron beams, large induced currents in the vacuum vessel, poloidal coils and structures. This provides a strong incentive for preventing disruptions or detecting them as early as possible and taking action to mitigate the damage.

Experiments in disruption avoidance relevant to ITER using passive and active control were tested in DIII-D [34]. A model-based feed-forward plus feed-back algorithm was used to control both neutral beams and gyrotrons to achieve desired levels of plasma energy and current density profile in discharges for non-inductive current-drive operation.

In cases where disruptions actually began to occur, multiple shattered pellet injection was used at different locations [35]. Investigations were made of variations in numbers of pellets and their amounts of neon to optimize radiation loss.

#### 3.3.2.6 Small Angle Slot (SAS) Divertor with Resonant Magnetic Perturbation (RMP) Suppression of Edge Localized Mode (ELM)

Optimization of the divertor plate geometry is a major design challenge for optimal distribution of the plasma heat loading and control of gas recycling. A SAS divertor [31] was installed to evaluate the role of divertor closure. Core-boundary studies showed that the SAS achieved detachment at lower density. It achieved extended heat loading across the target plate, which is essential for controlling heat flux and erosion.

The SAS divertor also allowed a large electron temperature gradient across the edge pedestal. Using a rotating $n = 2$ Resonant Magnetic Perturbation (RMP) combined with a stationary $n = 3$ RMP, the coil current for ELM suppression is reduced. For ELM suppression by RMPs in an ITER similar shape ($q_{95} = 3.5$), a clear and consistent threshold in the impurity rotation was observed, at which point ELM suppression was lost, fortunately at a level less than that expected in ITER.

#### 3.3.2.7 Wide Pedestal ITER-Relevant QH-mode Scenario

In the wide pedestal regime [36] that exhibited improved performance and no ELMs, labelled as QH, the start-up counter torque was eliminated so that the entire discharge used zero injected torque. The observed improvement in both core and pedestal confinement suggested that fast ions from neutral beam injection were not a requirement for wide-pedestal QH-mode [31]. The operating space was expanded to include a lower-single-null configuration with ITER triangularity. The maximum volume average beta obtained without ELMs was $\beta = 2.1\%$ ($\beta_N = 2.3$).

#### 3.3.2.8 Radiating Divertor Combined with Steady-State High $\beta_N$

In a reactor, it will be necessary to combine high plasma pressure operation with a divertor that can handle output power in steady-state. The high- $\beta_N$ ($\leqslant 3.8$) hybrid scenario was extended to the high-density levels necessary for radiating divertor operation [31], achieving about 40% divertor heat flux reduction using either argon or neon, combined with NBI and off-axis ECRH. The pedestal density was increased by over 50%. A mixture of co- and counter-injected neutral beams was necessary to achieve the zero plasma torque required for high performance.

#### 3.3.2.9 Super H-mode

A Super H-mode regime was discovered and extended to high plasma current (2.0 MA) and density to achieve very high pedestal pressures (~30 kPa) and stored energy (3.2 MJ) with $H_{98y2} \approx 1.6$–$2.4$ [37]. This regime was predicted to enable pedestal height and fusion performance substantially higher than standard H-Mode operation. This regime existed due to a bifurcation of the pedestal pressure as a function of density. Experiments on Alcator C-Mod and DIII-D achieved access to this improved regime. DIII-D Super H experiments demonstrated strong performance, including the highest stored energy in the present configuration of DIII-D (W ~ 2.2–3.2 MJ), while utilizing only about half of the available heating power ($P_{heat}$ ~ 7–12 MW). These DIII-D experiments have obtained the highest value of peak d-d fusion gain, with a d-t equivalent $Q_{DT,equiv} \sim 0.5$, achieved on a medium scale ($R_{major} < 2$ m) tokamak. Sustained high performance operation ($\beta_N \sim 2.9$, $H_{98} \sim 1.6$) was achieved utilizing $n = 3$ external magnetic perturbation coils for density and impurity control. Pedestal and global confinement was maintained in the presence of deuterium and nitrogen gas puffing, which enabled a more radiative divertor condition, hence reducing the heating at the target plates. Super H-Mode access was predicted to allow ITER to achieve its goals ($Q = 10$) at $I_p < 15$ MA.

#### 3.3.2.10 Steady-State Current-Drive Advanced Tokamak Fusion Reactor

Advanced tokamak operation involves means of reaching steady state conditions at high performance. The possibility of a steady-state advanced fusion reactor was explored [38] which involved high pressure operation, favourable stability and transport properties and a highly self-driven ('bootstrap') plasma current without disruptions. These scenarios involved high $\beta$ access, off-axis current drive techniques, profile control, and stability and ELM control. Radiative techniques were explored to develop improved compatibility with divertor requirements and simultaneous access to high performance pedestals. A current drive density twice as high was obtained using a top launch compared to an outside launch [39].

#### 3.3.2.11 High Beta of 12.5% at Low Aspect Ratio

A high toroidal beta of 12.5% [40] was obtained in the low aspect ratio DIII-D. These discharges demonstrated that a resistive wall can stabilize MHD modes. In discharges with broad current profiles, beta exceeded the ideal MHD stability limit, but remained below the limit calculated for a perfectly conducting wall.

## 3.4 DIII-D Systems Analysis of Emergent Properties and Tradeoffs

### *3.4.1 Analysis of Emergent Properties*

The following sections use the remainder of Checklist 1.2 to examine, identify and evaluate metrics and performance. In what follows, various emergent properties of the system are examined, including those that contribute to the overall properties of robustness, fragility and performance.

### *3.4.2 Systems Control*

At the overall systems level of the entire machine, the control systems have evolved from a mixture of digital pre-programming and analogue control to fully digitally controlled systems. All of the components of the Doublet III /DIII-D system worked together to produce very high performing plasmas. At the construction and upgrade levels, the machine components were constructed on time and on cost to give an integrated and working tokamak.

At the subsystem level of plasma control, the overall magnetic detection of the plasma boundary, signal analysis and feedback from the shaping coils successfully produced all of the design plasma configurations. It was flexible enough to explore an even broader range of plasma configurations than was originally envisaged. Over the decades, as electronic technology advanced rapidly, more sophisticated controllers were installed, resulting in fully digital systems that can calculate plasma equilibria in real time and add signals from diagnostics.

### *3.4.3 Fault-Tolerance*

A high level of similar components and the ease of dismantling the system for repair and replacement gave a high tolerance against any faults. A fault in an external coil could easily be replaced. A particularly crucial and vulnerable element was the central column of magnets, which was designed to operate at a toroidal field on the plasma axis of 4 T. The maximum ever attempted was 2.6 T. For machine operation, the shaping coils and vacuum vessel were specially designed to have strong tolerance against plasma disruptions. Internal access for repair workers provided for the replacement of components, so redundant protection of fault-tolerance was provided against unforeseen accidents. The ease of potentially replacing external coils also made an upgrade to a larger vacuum vessel and larger shaping coils relatively straightforward.

### 3.4.4 Modularity

Modularity provides isolation of a perturbation from the rest of the system. When the plasma makes a rapid motion during a disruption, it would not damage the external magnetic field coils. When the plasma was in contact with a limiter that defined the plasma, there were several limiters at different locations that could also protect the surrounding vacuum vessel from the plasma. A disruption potentially affecting the shaping coils could be controlled by protection mechanisms.

### 3.4.5 Decoupling

Decoupling isolates low-level noise and fluctuations from functional structures. The plasma was surrounded by a conducting metal vacuum vessel which produces compensating image currents in the metal to compensate plasma motion. They filter and counter high-frequency, low-level fluctuations in the plasma. For medium frequency perturbations, image currents in the plasma shaping coils assumed this role, because the vessel resistivity was chosen to allow rapid penetration of the plasma breakdown voltage of 100–250 V. The mixture chosen of close shaping coils and rapid power supplies allowed the overall system to be successful.

### 3.4.6 Resistance

The design of Doublet III and the subsequent DIII-D modification was characterized by a large initial margin of overdesign on the assumption of possible maximum performance. Therefore, toroidal magnetic field, ohmic and shaping coils were designed for a maximum toroidal field of 4 T and plasma currents up to 5 MA. The vacuum vessel was designed to withstand plasma disruptions at this maximum value. For plasma control, a multi-layered concept for resistance to plasma fluctuations was put in place, starting with limiters, then the vacuum vessel image currents, then the shaping coil passively induced currents, and finally a multi-layered plasma position and shape control system that smoothed out the low frequency perturbations.

### 3.4.7 Avoidance

In order to avoid instabilities of the doublet shape of plasma, and later on D-shaped plasmas, the parallel connected shaping coils and vacuum vessel were designed to the same shape as the desired plasma, to obtain the maximum in passive stability from image currents and to make feedback shaping as effective as possible. When it proved difficult to optimize the performance of the doublet plasma, at first more stable circular plasmas were investigated, and then a whole new D-shaped vacuum vessel was installed to optimize the performance and stability of the new shape. Very extensive theoretical calculations were carried out for equilibrium and stability designed to find the optimum plasma shapes.

### 3.4.8 Resource Demands

The initial funding for Doublet III was obtained at a time when ample funds were available in the USA for new tokamaks, promising results having been obtained on smaller doublet experiments. Additional funding was obtained in the early years by a collaboration agreement with JAERI. The flexibility and high performance of DIII-D encouraged further collaboration and machine upgrades.

The goal of a 4 T upgrade required a powerful motor generator, which provided spare capacity that allowed long-pulse high-power experiments limited only by coil temperature rise.

### 3.4.9 Robustness

Systems have tradeoffs between robustness (against perturbations) and competing requirements of fragility, performance and resource demands. Doublet III and DIII-D proved that they possessed extreme robustness.

The original Doublet III design allowed both construction and operation robustness. When the Doublet shaped plasma was found to be non-optimal, the overall system was able to change by installing a new vacuum vessel. The high flexibility of the coil systems and power supplies allowed the production of circular, D-shaped and expanded boundary divertor plasmas in the original doublet vacuum vessel, and D-shaped and diverted plasmas in the big Dee vessel. A very wide range of experiments were undertaken that were not envisaged when the machine was originally designed. The large access ports in the vacuum vessel allowed for the installation of many diagnostics and heating and current drive systems, which greatly increased the operating range of the machine.

### 3.4.10 Fragility

In Doublet III and DIII-D, the maximum fragility in the machine design was probably in the central core of the machine bound with epoxy resin. If a short circuit or a cooling water leak had occurred, the entire central column would have had to be replaced after disassembling the machine. However, this fragility was avoided by means of extremely large safety margins and care in fabrication. No major fault has occurred. During plasma operation, disruptions were the main fragility, and mitigated against.

### 3.4.11 Performance and Metrics

DIII-D was able to achieve plasma performance beyond the original objectives of the first design. By using D-shaped plasmas optimized by external heating, plasma diagnostics and theoretical and experimental analysis, the DIII-D plasmas produced near fusion reactor conditions.

However, the machines never achieved the 4 T and 5 MA original design objectives, which might have led to higher plasma performance. During various construction phases, the decision was made not to attempt to strengthen the machine structure for safety reasons related to maximum allowable stress in the coils.

However, the actual operating toroidal field and plasma current limits allowed DIII-D to carry out a full programme of tokamak physics and to reach high levels of plasma performance. The machine was never designed to operate in tritium; hence it was not necessary to try to operate at full fusion reactor conditions.

In terms of performance, the DIII-D Super-H experiments demonstrated the highest stored energy (W ~ 2.2–3.2 MJ), while utilizing only about half of the available heating power ($P_{heat}$ ~ 7–12 MW). The highest value of peak fusion gain, $Q_{DT,equiv}$ ~ 0.5, was achieved. Super H-Mode access was predicted to allow ITER to achieve its goal of Q = 10. Toroidal beta values up to 12.5% were achieved.

The Doublet III – DIII-D programme can therefore be considered a success.

## 3.5 Asdex-Upgrade

### 3.5.1 Asdex-Upgrade Design

Asdex-Upgrade reported its first results [41, 42] in 1992, and has made major contributions to tokamak and fusion research. Asdex-Upgrade was a follow-on machine from Asdex where the H-mode was discovered.

Asdex Upgrade's parameters [43, 44] are R = 1.65 m, a = 0.5 m, $\kappa \leq 1.8$, $B_T \leq 3.2$ T, $I_p \leq 1.4$ MA, pulse length = 10 s, plasma heating = 27 MW, operating gas H or D. Even for deuterium operation, the machine was surrounded with a 2 m thick concrete wall. A schematic of Asdex Upgrade is shown in Fig. 3.5.

The toroidal magnetic field was generated by 16 large water cooled and continuously wound copper magnet coils wrapped around the plasma vessel. The poloidal coils were placed mostly outside the toroidal field coil and produced the plasma current, shaping and positioning of the plasma and the divertor x-point fields. The

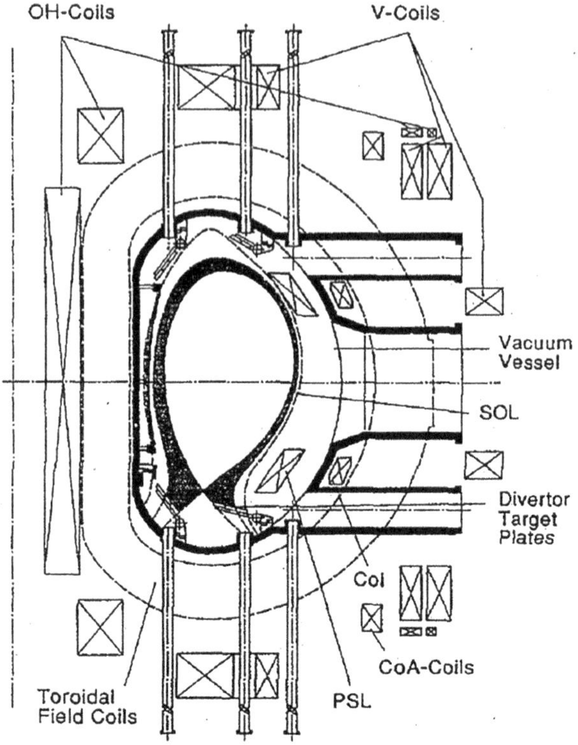


**Fig. 3.5** A schematic of the Asdex Upgrade tokamak. (From [42], Fig. 1, reused with permission from the IAEA)

only coils inside the vessel were for high frequency response to vertical position instability.

Three different plasma heating methods are installed:

- neutral particle injection with a power of maximum 20 MW
- radio-frequency heating with 6 MW and frequency tuneable between 30 and 120 MHz.
- electron cyclotron microwave heating with 8 MW at frequencies of 105 and 140 GHz.

The plasma was investigated with 40 diagnostics. Up to 4 gigabytes of raw data were recorded per discharge. The plasma shapes were each calculated within less than a millisecond for feedback control. The electrical energy was provided by large flywheel generators. With its flexible coil system, the Asdex Upgrade tokamak was able to develop several reactor-relevant scenarios. To model the future ITER tokamak, the steel vacuum vessel inside tiles were altered to tungsten.

## 3.5.2 *Asdex-Upgrade Scenarios*

### 3.5.2.1 Plasma Initiation and Divertor Control with Shaping Coils Outside the Toroidal Field Coil

The reactor-relevant feature of having the poloidal coils outside the toroidal field coils and thus far away from the plasma required relatively high power and large currents during the plasma start-up phase. Equilibrium calculations showed that the plasma started as a small circle on the inner wall limiter and then grew to a full-sized diverted plasma.

The starting voltage of greater than 25 V per turn induced large currents in the vacuum vessel and shaping coils, making plasma start-up difficult. A lower starting voltage of about 15 V per turn was better for initiating the plasma and more appropriate for machines that would have superconducting coils.

### 3.5.2.2 NBI and ECCD

Auxiliary heating methods such as NBI and ECRH also can produce plasma current, creating the possibility of steady state operation. Current drive experiments with low collisionality [43] used freshly boronized walls to reduce the tungsten influx at high edge temperature conditions. Discharges aimed at non-inductive operation and high normalized pressure used an optimized shape and reduced gas puffing. The safety factor q on axis was kept above 1.5 to avoid sawteeth and NTM. Full ELM suppression and non-inductive operation up to a plasma current of Ip = 0.8 MA was obtained at low plasma density. About 35% of the plasma current was driven by neutral beams. ECCD was used to obtain an optimized current profile. About 50% of the plasma current was due to bootstrap current.

### 3.5.2.3 High Power Plasma Exhaust Detachment with Argon and Divertor x-Point Radiating Zone

If the divertor region plasma can radiate away most of the power it receives from the plasma, it can effectively detach from divertor plates so they receive a greatly reduced heat load. Argon is a suitable impurity for inducing radiative cooling in the plasma outer core and pedestal region [43]. The radiation inside the X-point was related to divertor detachment, line-averaged and pedestal densities and ELM size. The X-point radiating regime with strong detachment is an interesting reactor scenario.

#### 3.5.2.4 ITER Baseline (IBL) and (IBL-A) Simulations

A key goal is to simulate the ITER Base-Line (IBL) scenario [44]. Experiments at $I_p = 1.2$ MA ($q_{95} = 3$) and $I_p = 1.0$ MA ($q_{95} = 3.6$) (IBL-Alternative) were carried out with $N_2$ seeding and reduced ELM size, by using pellet pacing and Resonant Magnetic Perturbations (RMP). Both high density scenarios fell short of the required confinement by about 20%. Control techniques had a limited impact on the ELM frequency, though a slight density pump-out was observed with the application of RMPs. High pressure in the divertor and lower average shear led to small ELMs.

The performance of the IBL-A scenario was also tested using only ECRH and ICRH to dominantly heat the electrons and also reduce the torque input. No marked changes in performance were observed. Operating at lower density led to a higher confinement even in the presence of RMPs.

#### 3.5.2.5 Preparing the Physical Basis of ITER Operations and DEMO Scenarios

Where machines could not fully simulate ITER or DEMO plasmas, it was necessary to develop codes that could simulate operation in those machines. The Asdex Upgrade Flight Simulator incorporated plasma models validated on Asdex Upgrade, with profiles provided by the ASTRA model [45].

### *3.5.3 Robustness and Performance Analysis*

Asdex Upgrade was very successful, having demonstrated that a fusion reactor-like configuration with a set of poloidal coils outside continuously wound toroidal field coils could create stable elongated plasmas with a divertor. Experiments were carried out on a wide range of plasmas and made major contributions to fusion reactor development. The very large horizontal port access meant that high power auxiliary heating systems could be coupled efficiently to the plasma. The original goals of the machine to explore a wide range of divertor scenarios and to act as a prototype for several features of a fusion reactor were fulfilled.

When Asdex Upgrade first operated, it was discovered that the originally designed large plasma breakdown voltage was too high for successful operation due to induced eddy currents in surrounding conductors and the vacuum vessel. Fortunately, the machine was sufficiently robust that the starting voltage for plasma start-up could be as low as 15 V/turn.

## 3.6 Alcator C and C-Mod

### 3.6.1 Alcator and Alcator C-Mod Design

ALCATOR C-MOD [46] was the latest version in a series of high toroidal field tokamaks built at MIT and is shown in Fig. 3.6. High field technology was developed at MIT's Francis Bitter National Magnet Laboratory. The first Alcator [47] tokamak had design parameters of $R = 0.54$ m, $a = 0.12$ m, and high toroidal field of 12 T on axis. The toroidal field coils used liquid-nitrogen cooling in Bitter-type plates made of composite-copper-steel. They produced 15.8 T at the inside edge of the plasma. The design limiter q of the plasma was 2.5, and the goal of the machine was to explore the limits of ohmic heating. Alcator used a traditional thick-copper conducting shell (with an insulating gap) located just outside the vacuum vessel to stabilize the plasma via image currents. As the successor to Alcator (A), B and C, the most recent version Alcator C-Mod dispensed with this shell and had a thick

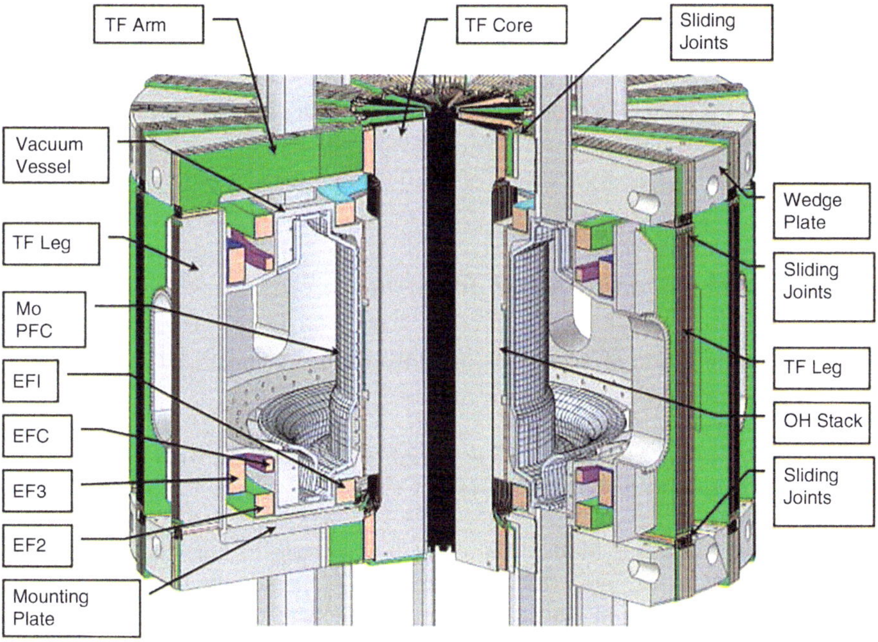


**Fig. 3.6** Cutaway view of Alcator C-Mod. The TF and PF magnets, vacuum vessel, and inner wall hardware are shown. The EF coils are installed in up-down symmetric pairs with labels indicated on the lower coils only. Not shown are the retaining cylinder and domes and the EF4 magnet that attaches to the cylinder. (Reprinted by permission of Taylor & Francis Ltd., http://www.tandfonline.com on behalf of copyright © American Nuclear Society, http://www.ans.org, Fig. 2 From J. Irby et al., Alcator C-Mod Design, Engineering, and Disruption Research, Fusion Science and Technology, 51:3, 460–475, DOI: 10.13182/FST07-A1433 (2007))

conducting vacuum vessel. The degree to which the vacuum vessel serves as a conducting shell is an important design option.

Alcator C-Mod was designed as a much larger machine in order to allow significant port access for auxiliary heating and to accommodate a vertically-elongated plasma and a divertor. The design parameters were R = 0.68 m, a = 0.21 m, maximum vertical elongation 1.85, toroidal field = 8.11 T (9 T design), plasma current 2.02 MA (3 MA design), RF heating power = 8 MW, lower hybrid source power = 3 MW. The maximum magnetic field at the coils was 17 T. The ohmic coil was wound directly onto the toroidal field central column.

The toroidal field coils contained joints to permit the poloidal field coils surrounding the vacuum vessel to be inside the toroidal field coils. They were also cooled by liquid-nitrogen. The entire machine was inside a cryostat. Wedge plates were inserted between the toroidal field coils in order to support the high field stresses. The plasma current was driven by an air core ohmic heating coil. Poloidal field coils were located well above and below the machine midplane in order to allow horizontal port access.

The plasma-shaping poloidal field coils were initially operated by a hybrid digital-analogue controller provided through a collaboration agreement with the TCV group. It was later replaced with an all-digital controller. As in TCV, the vacuum vessel was highly conductive with no toroidal break, with a resistance of 40 μΩ and an L/R time of 20 ms.

Thanks to the high ratio of auxiliary and ohmic heating power to plasma size, the parallel heat fluxes into the divertor reached 400 $MW/m^2$, comparable to that expected in ITER.

## 3.6.2 *Alcator C and C-Mod Scenarios*

Results from operation of Alcator C-Mod tokamak are summarized [48, 49].

### 3.6.2.1 Operating Alcator-C at Very High Toroidal Field with Pellet Injection to Reach the Lawson Criteria

Before discussion the C-Mod scenarios, it is important to note that the previous Alcator-C machine had made important progress. It reached the Lawson product of $10^{20}$ $m^{-3}s$ of the electron density and the confinement time [50] by operating at $B_T$ = 10 T and $I_p$ = 750 kA. With high velocity pellet injection, central plasma densities of $2 \times 10^{21}$ $m^{-3}$ and confinement times of up to 50 ms were obtained. The central electron temperature from ohmic heating was 1.5 keV.

#### 3.6.2.2 C-Mod Reactor-Relevant Vessel Wall High-Z Material, Cleaning and Vertical Plate Divertor

The vacuum vessel of tokamaks needs to be conditioned to remove impurities that could enter the plasma and reduce performance. C-Mod internal components and conditioning were designed to simulate fusion reactor conditions, and included a vertical target lower divertor with refractory metals on all plasma-facing surfaces [48]. The machine also ran with an upper x-point on a flat target divertor with a toroidal cryopump. The wall used 7000 tiles made of molybdenum alloy TZM (99.5% Mo, 0.5% Ti, and 0.08% Zr). The small tile size was designed to limit the forces due to eddy currents induced during disruptions. A belt of tungsten tiles was installed in the highest heat flux areas.

For machine conditioning, the thick, low-resistance vacuum vessel made pulsed "Taylor" discharge cleaning impossible. Surfaces were prepared for operation via Electron Cyclotron Discharge Cleaning [51]. The toroidal field was operated near 0.09 T and slowly swept so that the discharge intercepted all of the internal structures. After baking, discharge cleaning and initial operation, the plasma facing surfaces were covered with boron by discharge cleaning with deuterated diborane in a 90% He background.

#### 3.6.2.3 Reactor-Relevant Power Handling and Operation of Vertical Target Divertor

The divertor chamber in a fusion reactor will need to handle very high power loads. The C-mod vertical target divertor used a shallow angle between the magnetic field and an extended divertor leg [52] to spread out the heat load. In this geometry, neutrals arising from recombination at the divertor strike point were directed toward the divertor channel, enhancing reionization and providing a natural baffle. Neutrals created in the divertor were isolated from the main chamber by the divertor plasma itself. The use of high-Z materials resulted in low sputtering, low erosion, and low hydrogen isotope retention at high temperatures. The concept was adopted for ITER.

#### 3.6.2.4 EDA (Enhanced $D_\alpha$) H-mode

The detailed nature of plasma operation, and especially plasma boundary conditions, can greatly affect plasma confinement properties. The EDA mode [48] was the most common form of H-mode on C-Mod. The method for accessing this mode of operation was to place the strike point deep in the divertor slot. While sufficient to bring the discharge to a stationary state, the ELMs were always "small" in the sense that the reduction in particle and energy inventory was well under 1% per ELM.

#### 3.6.2.5 Improved Confinement (I-mode) ELM-Free H-mode

An Improved confinement (I-mode) [53] H-mode which is ELM-free was obtained by operating at high-power conditions where the L to H mode threshold was elevated. The I-mode combined H-mode-like energy confinement with the L-mode particle and impurity confinement and was generally ELM-free. The change in global confinement and the drop in core thermal diffusivity were mirrored by a drop in core fluctuations. Most commonly, the I-mode was accessed by running with the ion drift from the toroidal magnetic field gradient in the direction unfavourable for H-mode operation. I-modes had low collisionality and improved confinement. Weak and energetically insignificant ELMs were observed in some I-modes. Impurity transport in the I-mode was essentially at the L-mode levels.

Overall, the I-mode had the advantages of the H-mode without its drawbacks. The density was controlled by gas puffing and the density profiles were mildly peaked as in the L-modes or the low-collisionality H-modes. Strong fuelling also helped avoid an I-H transition. The I-mode in ITER might allow Q = 10.

#### 3.6.2.6 ICRH, Flow Drive and Impurity Control with a Field Aligned (FA) Antenna

In very high field tokamaks, due to space restrictions and plasma resonance conditions, ion cyclotron waves from an antenna inside the vacuum vessel is often the preferred auxiliary heating method. ICRH [48] was the main heating method by accelerating minority ions which preferentially transfer power to electrons. Many instabilities were stabilized by plasma flow which could be generated with ICRH. Plasma flow was demonstrated in the mode conversion regime with flow levels well above that expected from intrinsic rotation due to heating alone. The driven rotation was proportional to power per plasma particle.

Impurity generation has long been recognized as an important problem with an antenna made of high-Z materials. These impurities were reduced by a Field Aligned Antenna by minimizing the component of the RF electric field parallel to the tokamak magnetic field. The antenna box, straps, and Faraday screens (to shield the plasma from antenna electric fields) were all aligned with the total magnetic field. The molybdenum source from the antenna was significantly lower and the electrical properties of the antenna were excellent.

#### 3.6.2.7 LHCD at Medium and High Plasma Densities

Lower hybrid waves can often produce the most efficient means of plasma current drive. LHCD was extended to medium and high plasma densities [48]. At moderate densities, up to $0.6 \times 10^{20}\ m^{-3}$, the LH system produced 100% of the plasma current (0.5 MA). For simulating ITER "steady-state" scenarios, C-Mod provided a test-bed for studies of current profile control, stability, and transport. The population of non-thermal electrons predicted by LH theory was measured with a multi-chord hard x-ray diagnostic and found to build up and decay at a rate consistent with models. A Motional Stark Effect (MSE) diagnostic used to constrain a magnetic equilibrium reconstruction demonstrated that the current was driven well off-axis. The current profile could be modified to create a sawtooth-free reversed-shear regime.

At high densities, $n_e > 10^{20}\ m^{-3}$, but below the limit for wave accessibility, LHCD efficiency dropped faster than expected. The problem was partly identified as power deposition in the scrape-off layer [49]. A path was found to reduce the edge density and thereby reduce these effects at $n_e = 1.5 \times 10^{20}\ m^{-3}$.

#### 3.6.2.8 Disruption Mitigation Using Massive Gas Injection (MGI)

Plasma disruptions were a major danger for in-vessel components, the vacuum vessel itself and external coils. The danger was greater if accompanied by a vertical instability while significant current was still flowing in the plasma. Asymmetric disruptions in C-Mod were found to generate especially dangerous halo currents flowing between conducting materials inside the vacuum vessel. Disruption mitigation had two goals: to convert plasma kinetic and magnetic energy to uniformly distributed radiation, and to cause the plasma current to decay before the plasma suffered from a vertical instability. The most successful mitigation was massive gas injection [48], by injecting a mixture of a noble gas mixture through a fast valve connected to a high-pressure plenum. Up to $10^{23}$ atoms could be injected. The mixture travelled rapidly into the plasma and radiated in a high ionization state. The optimum mixture was found to be 10% argon in helium. Multiple gas jets helped symmetrise radiation during the pre-thermal quench. In the thermal quench itself, the MHD effects were seen to dominate, and multiple injection points were not necessarily helpful.

### *3.6.3 Robustness and Performance Analysis*

The Alcator series of tokamaks, and especially C-Mod, demonstrated the ability to operate at high toroidal fields. The structural requirements associated with high field also limit access to the plasma, excluding the possibility of tangential neutral beam injection. The Alcator series was extremely successful in operating at high field and could closely model some of the conditions expected in a fusion reactor.

## 3.7 FTU

### *3.7.1 FTU Design*

The Frascati Tokamak Upgrade (FTU) [54] is a compact, high field tokamak designed on the same principles as the Alcator series. It began operation in 1990 and achieved its goals of plasma current 1.6 MA at 8 T and average electron density greater than $4 \times 10^{20}$ $m^{-3}$. The plasma was circular with major and minor radius equal to 0.935 and 0.30 m, initially defined by a nickel-based Inconel limiter.

More recently [55], the FTU stainless steel vacuum chamber with a thickness of 2 mm was covered at the high field side by a toroidal limiter made of 2 cm thick molybdenum tiles and outboard by a molybdenum poloidal limiter. The auxiliary plasma heating is provided by a 140 GHz electron cyclotron system with 1.5 MW and an 8.0 GHz lower hybrid system at 2.0 MW. The walls were conditioned with boron coating.

### *3.7.2 FTU Scenarios*

#### 3.7.2.1 Stabilization and Suppression of Post-disruption Runaway Electron Beams

Plasma disruptions can lead to very high induced voltages inside the decaying plasma and produce beams of "runaway" electrons at multi-MeV energies. Stabilization and suppression of post-disruption runaway electron beams [55] was achieved by a control architecture that allowed detection of the current quench The runaway electron beam current was controlled by a ramp-down while the beam was kept away from the vessel. The beam energy reduction was confirmed by analysing hard x-ray monitors and runaway electron imaging spectroscopy.

#### 3.7.2.2 Tin Liquid Limiter (TLL) Power Handling for Divertors

A liquid plasma-facing surface has excellent heat transfer properties and is safe from damage. Liquid tin was a candidate for a plasma-facing component material, with a large operating window (300–1300 C) before vaporization. It has low or negligible activation and low hydrogen isotope retention. The tin liquid limiter (TLL) [55] was based on a capillary porous system consisting of a molybdenum tube wrapped around strips made of tungsten felt filled with tin. A tin concentration of $5 \times 10^{-4}$ $n_e$ was measured in the plasma. The plasma confinement time was practically unchanged and no degradation of performance was observed. Liquid lithium limiters were also explored [56].

#### 3.7.2.3 TM Stabilization by Pellet Injection or ECH

In sawtooth-free low-density plasmas, magnetic islands formed by TM instabilities around the q = 2 surface were saturated at large amplitudes. Stabilization was produced by pellet injection in the presence of a rotating magnetic island [57].

TM stabilization was also produced by ECRH. Power deposition was varied in real time around the q = 2 surface, with a "sweeping" strategy, allowing an acceleration of the stabilisation.

### *3.7.3 Robustness and Performance Analysis*

Like the Alcator series of tokamaks, FTU operated successfully at high toroidal magnetic fields and was able to explore a wide range of plasma regimes. With the Alcator experiments, it demonstrated the advantages of operating at a high toroidal field, especially in achieving high density plasmas.

## 3.8 Spherical Tokamaks MAST, MAST-Upgrade and STEP

### *3.8.1 MAST Design and Results*

The Mega-Ampere Spherical Tokamak MAST [58] was a follow-on from START, and was constructed with a very low R/a aspect ratio, shown in Fig. 3.7. Such machines have a natural vertical elongation and relatively high plasma current even with a low toroidal field. Its parameters were R = 0.85 m, a = 0.65 m, R/a = 1.3, vertical elongation 2.2, $B_T = 0.51$ T, plasma volume 10 $m^3$, and plasma currents up to 1 MA.

All the poloidal field coils were placed inside the vacuum vessel. The toroidal field coil consisted of a rod going through a small diameter central column in the toroidal vacuum vessel, with returning portions of the rod also outside the vacuum vessel. Hence, the maximum field at the central rod is much higher than at the plasma axis.

Operationally, two start-up schemes were demonstrated: merging-compression and the more conventional solenoid induction. The former provided over 400 kA of plasma current with no demand on solenoid flux. Boronization with trimethylboron enabled good vacuum conditions. Plasma currents of over 1MA were obtained together with central plasma temperatures (ohmic) of order 1 keV. The Hugill and Greenwald electron density limits were significantly exceeded. H-mode operation was achieved at low additional NBI power. Particle and energy confinement showed an immediate increase at the L-H transition.

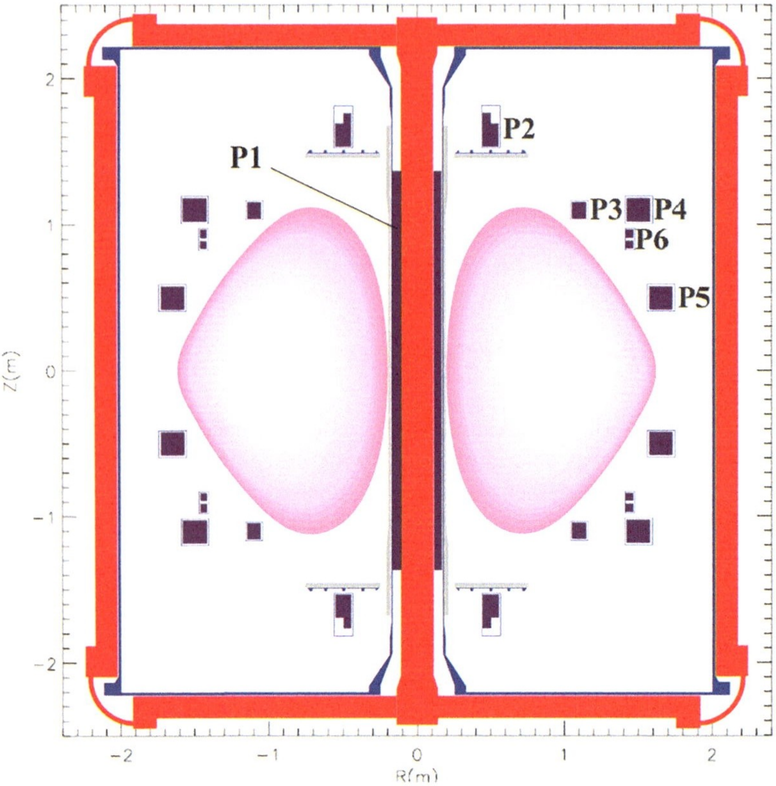


**Fig. 3.7** Plan view of MAST, showing the PF coils. (Reused with permission from [58], Fig. 1)

## *3.8.2 Energy Confinement in MAST*

It is important to determine the mechanisms for energy loss in spherical tokamaks. In the MAST H-mode [59], the turbulence at the pedestal top was shown to be consistent with electron temperature gradient modes. A reconnection process occurred during ELMs and filaments determined the power profile at the divertor. Resonant magnetic perturbations were able to mitigate ELMs. The mitigation of intrinsic error fields was shown to be important for plasma performance.

Fast ion physics was studied in detail [60]. Measured d-d fusion products from a neutron camera and a charged fusion product detector were 40% lower than predictions from TRANSP/NUBEAM, highlighting possible fast ion losses due to deficiencies in the guiding centre approximation for particle motion. Modelling in the presence of resonant magnetic perturbations (RMPs) reproduced observations.

### 3.8.3 Robustness and Performance Analysis

The operation of MAST was so successful that MAST-Upgrade [61] was designed. MAST had produced 1 MA of plasma current and produced H-mode operation without the need for manor repairs.

### 3.8.4 MAST Upgrade First Results

The objectives of MAST-Upgrade [61] were

- Test reactor concepts (in particular exhaust solutions via a flexible divertor allowing Super-X and other extended leg configurations).
- Add to the knowledge base for ITER (by addressing important plasma physics questions and developing predictive models to help optimize ITER).
- Explore the feasibility of a fusion Component Test Facility.

MAST-Upgrade began operation [62] with plasma currents of 0.8 MA. Compared to MAST, the new capabilities included:

- 19 new poloidal field coils for substantially improved shaping of the core plasma and divertor;
- New solenoid with increased inductive flux;
- Increased toroidal field;
- On- and off-axis neutral beam heating and current drive;
- Closed, tightly baffled divertor, up-down symmetric divertor chambers with Super-X capability;
- Improved suite of diagnostics in the divertors.

### 3.8.5 STEP

The UK government was impressed with MAST and MAST-U, and committed funds for designing a Spherical Tokamak for Energy Production (STEP) [63, 64]. It would be a fusion reactor producing 100 MW(e) of net power and is projected to cost £2 billion.

## 3.9 Spherical Tokamaks NSTX and NSTX-U

The National Spherical Torus Experiment (NSTX) at Princeton Plasma Physics Laboratory had the machine parameters of $R = 0.85$ m, $a = 0.68$ m, $B_T = 0.3$ T, $I_P = 1.4$ MA and auxiliary heating 11 MW.

NSTX was replaced by NSTX Upgrade (NSTX-U) [65] with a planned toroidal field up to 1.0 T and plasma current of 2 MA. During initial operation, diverted plasmas with $I_P = 0.8$ MA, $B_t = 0.6$ T and a pulse length of 1 s were obtained. H-mode plasmas were achieved with 2.5 MW of NBI. The maximum parameters were achieved at 4 MW and 1 MA with a stored energy of 200 kJ, $\beta_N = 4$, $\kappa = 2.2$, $\tau = 50$ ms and 50% increase in pulse length due to n = 1 error field correction coils.

NSTX-U was stopped in late 2016 soon after operation began, due to a failure of a poloidal coil inside the vacuum vessel, requiring complete dismantling of the machine and a major redesign. A restart in 2022 was planned.

## 3.10 Low Aspect Ratio with High Vertical Elongation ST-40

The high field spherical tokamak ST-40 [66] was built by Tokamak Energy Ltd. and had the design parameters $R_0 = 0.4$–$0.6$ m, A = 1.7–2.0, $I_{pl} = 2$MA, $B_t = 3$ T and vertical elongation $\kappa = 2.5$. Ultra-capacitors were used to power liquid-nitrogen cooled copper magnets. Plates to stabilize vertical instabilities were installed. NBI power of 2 MW was available, allowing for a pulse length of ~3 s. Most of the centre post area was used for TF coils, the rest for a minimal central solenoid to help maintain the plasma current. For start-up, merging compression was used to first form two plasma rings around the coils and then combine them, creating a hot plasma which was compressed in major radius. Preliminary results [67] showed that ST-40 achieved 1.5 T toroidal field and 0.4 MA plasma current at an elongation of 2.5. Up to 4 MW of auxiliary NB and EC heating was injected into a plasma with a pulse length of 200 ms.

## 3.11 Discussion of the Robustness and Performance of 1–2 MA Tokamaks

The tokamaks analysed here were mostly very successful in their operational campaigns, with the notable exception of NSTX-U. Such failures highlight the need for modular and easily repaired systems, especially for fusion reactors. Some component systems were overdesigned, which left important safety margins. Some designs also allowed upgrades, especially of diagnostics and auxiliary heating power.

## References


1. R.L. Freeman et al., Doublet research at General Atomic Company. Paper IAEA-CN-37/C-2 in Proceedings of Plasma Physics and Controlled Nuclear Fusion Research 1978, Vol. I, IAEA, Vienna (1979)
2. T. Ohkawa, Shaping and characteristics of doublet plasmas in Doublet III. Report GA-A15671, General Atomics, San Diego. Proceedings of the 9th European Conference on Controlled Fusion and Plasma Physics, September 1979, Oxford. Paper CONF-790912-12 (1979)
3. J.C. Wesley et al. Shaping and characteristics of ohmically heated non-circular plasmas in Doublet III, Plasma Physics and Controlled Nuclear Fusion Research 1980, vol. 1, Eighth Conference Proceedings Held in Brussels, Belgium, 1–10 July 1980. International Atomic Energy Agency (IAEA), Vienna, pp. 35–42 (1981)
4. P.J. Rock, J.C. Wesley, 1.5 Megawatt D.C. chopper power supplies for plasma shape control on Doublet III. Report GA-A15658, General Atomic Company, San Diego. 8th Symposium on Engineering Problems of Fusion Research, November 1979, San Francisco, Paper CONF-791102-75 (1979)
5. R. Stambaugh et al., Plasma equilibrium control in Doublet III. Report GA-A16049, General Atomic Company, San Diego. 11th Symposium on Fusion Technology, September 15–20, 1980 at Oxford. Paper CONF-80058-14 (1980)
6. A. Kitsunezaki et al., First results of Dee experiments in Doublet III, 8th IAEA Fusion Energy Conference, Brussels, Belgium, 1–10 July 1980, paper IAEA–CN–38/N-3. Nucl. Fusion Suppl. **1**, 405 (1981)
7. M. Nagami et al., Energy confinement of beam-heated divertor and limiter discharges in Doublet III. Nucl Fusion **24**(2), 183 (1984)
8. J.H. Kamperschroer, B.K. Bhadra, Neutral beam injection for the Doublet III device. IEEE Trans. Plasma Sci. **PS-7**(3), 151 (1979)
9. F.B. Marcus et al., Energy confinement in Doublet III with High-Z limiters. General Atomic Company Report GA-A15771, San Diego. Submitted to the Proceedings of the American Physical Society Meeting 1979, Boston (1980). https://inis.iaea.org/search/search.aspx?orig_q=RN:11550390. Last accessed 17 June 2022
10. S. Ejima et al., Scaling of energy confinement with minor radius, current and density in Doublet III ohmically heated plasmas. Nucl. Fusion **22**(12), 1627–1649 (1982)
11. M.M. Sabado, F.B. Marcus et al., Doublet III limiter performance and implications for mechanical design and material selection for future limiters, Report GA-A15633, General Atomic Company, San Diego. Symposium on engineering problems of fusion research, 8th, San Francisco, CA, November 13–16, 1979, Proceedings. Vol. 3. (A81-18901 06-75). Institute of Electrical and Electronics Engineers, Inc., Piscataway, 1979, pp. 1361–1369, (1979)
12. F.B. Marcus, Impurity studies and discharge cleaning in Doublet III, Report GA-A15613, General Atomic Company, San Diego. Presented at the IAEA Technical Committee Meeting on Impurities in Tokamaks: Source, Effect, and Control. Alushta, U.S.S.R. October (1979)
13. Doublet III Groups, Status of impurity studies in Doublet III. J. Nucl. Mater. **93 & 94**, 259–266 (1980) North Holland Publishing Company
14. F.B. Marcus et al., In-situ hydrogen pellet experiments in Doublet III. Nucl. Fusion **21**(7), 859–870 (1981)
15. P.B. Parks, F.B. Marcus, Current production by cyclotron heating of electrons passing through an axial magnetic field gradient. Nucl. Fusion **21**(9), 1208–1211 (1981)
16. N.J. Fisch, A.H. Boozer, Creating an asymmetric plasma resistivity with waves. Phys. Rev. Lett. **45**, 720 (1980)
17. M.A. Mahdavi et al., A review of the recent expanded boundary divertor experiments in the Doublet III device. J. Nucl. Mater. **111 and 112**(November/December), 355–361 (1982)
18. K.H. Burrell, Doublet III Groups. A regime of improved energy and particle confinement in neutral beam heated divertor discharges in Doublet III. 11th European Conference on Controlled Fusion and Plasma Physics, Aachen, 5–9 September 1983, Part I, pp. 11–14. (1983)

19. N. Ohyabu et al., A regime of improved energy confinement in beam-heated expanded-boundary discharges in Doublet III. Nucl. Fusion **25**(1), 49 (1985)
20. K.H. Burrell et al., Attainment of reactor level volume-averaged beta in Doublet III. Nucl. Fusion **23**(4), 536 (1983)
21. W. Pfeiffer, F.B. Marcus, et al., Giant sawtooth oscillations in the Doublet III tokamak. Nucl. Fusion **25**(6), 655 (1985)
22. J.L. Luxon et al. The Big Dee upgrade of Doublet III. 11th European Conference on Controlled Fusion and Plasma Physics, Aachen, 5–9 September 1983, Part I, pp. 99–102 (1983)
23. J. Luxon, Initial results from the DIII-D tokamak, 11th IAEA Fusion Energy Conference, Kyoto, Japan, 13–20 November 1986, Paper IAEA-CN-47/A-HI-3, Nuclear Fusion Supplement, p. 159 (1987)
24. L.G. Davis, Design of a Dee Vacuum Vessel for Doublet III, 29th National Symposium of the American Vacuum Society, November 1982, Baltimore, Maryland, Paper CONF-821111-25. J. Vac. Sci. Technol. A. **1**, 1319 (1983)
25. R.W. Callis et al. Function and operation of the Doublet III E-coil vacuum breaker system. 8th Symposium on Engineering Problems of Fusion Research, San Francisco, California, Paper CONF-791102-76 (1979)
26. C.M. Greenfield et al., VH-mode discharges in the DIII-D Tokamak. 1992 International Conference on Plasma Physics, Innsbruck 1992, European Physical Society, pp. I-11–I-14 (1992)
27. K.H. Burrell et al., Quiescent double barrier high-confinement mode plasmas in the DIII-D tokamak. Phys. Plasmas **8**, 2153 (2001)
28. E.J. Doyle, Internal transport barriers in the DIII-D tokamak. 12th International Toki Conference and 3rd General Scientific Assembly of the Asia Plasma Fusion Association, Toki, Japan, December 11–14, 2001
29. M.R. Wade et al., Physics of high bootstrap fraction, high performance plasmas on the DIII-D tokamak. 28th EPS Conference on Controlled Fusion and Plasma Physics. Funchal June 2001. ECA Vol. 25A 1365 (2001)
30. A.S. Welander et al., Advanced control of neoclassical tearing modes in DIII-D with real-time steering of the electron cyclotron current drive. Plasma Phys. Control. Fusion **55**(12), 124033 (2013)
31. Petty CC, The DIII_D Team, DIII-D research towards establishing the scientific basis for future fusion reactors. Nucl. Fusion **59**, 112002 (2019)
32. J. Wesson, *Tokamaks*, 4th edn. (Oxford University Press, Oxford, 2011)
33. E. Morse, *Nuclear Fusion* (Springer, Cham, 2018)
34. E.J. Strait et al., Progress in disruption prevention for ITER. Nucl. Fusion **59**(11), 112012 (2019)
35. J.L. Herfindal et al., Injection of multiple shattered pellets for disruption mitigation in DIII-D. Nucl. Fusion **59**(10), 106034 (2019)
36. K.H. Burrell et al., Discovery of stationary operation of quiescent H-mode plasmas with net-zero neutral beam injection torque and high energy confinement on DIII-D. Phys. Plasmas **23**, 056103 (2016)
37. P.B. Snyder et al., High fusion performance in Super H-mode experiments on Alcator C-Mod and DIII-D. Nucl. Fusion **59**, 086017 (2019)
38. R.J. Buttery et al., DIII-D research to prepare for steady state advanced tokamak power plants. J. Fusion Energy **38**, 72–111 (2019)
39. M.E. Fenstermacher, DIII-D research advancing the physics basis for optimizing the tokamak approach to fusion energy. 28th IAEA Fusion Energy Conference, Nice, France (Virtual) 10–15 May 2021, Presentation OV/1-3, Paper IAEA–CN–286-0686, Submitted to Nuclear Fusion (2021)
40. E.J. Strait et al., Wall stabilization of high beta tokamak discharges in DIII-D. Phys. Rev. Lett. **74**, 2483 (1995)

41. U. Seidel et al., Discharge scenarios for the first operational phase of ASDEX Upgrade. 1992 International Conference on Plasma Physics, Innsbruck. 29 June to 3 July 1992, Volume 16C, Part I, p. 451 (1992)
42. W. Koppendorfer et al., Results of the first operational phase of ASDEX upgrade, Plasma Physics and Controlled Nuclear Fusion Research 1992, vol. 1, Fourteenth Conference Proceedings Held in Wurzburg, Germany, 30 September–7 October 1992, International Atomic Energy Agency (IAEA), Vienna, pp. 127–40 (1993)
43. A. Kallenbach et al., Overview of ASDEX upgrade results. Nucl. Fusion **57**, 102015 (2017)
44. H. Meyer et al., Overview of physics studies on ASDEX upgrade. Nucl. Fusion **59**, 112014 (2019)
45. U. Stroth, Progress from ASDEX-upgrade experiments in preparing the physical basis of ITER operation and DEMO scenario development, et al. 28th IAEA Fusion Energy Conference, Nice, France (Virtual) 10–15 May 2021, Presentation OV/2-2, paper IAEA–CN–286-1046, Submitted to Nuclear Fusion (2021)
46. J. Irby et al., Alcator C-Mod design, engineering, and disruption research. Fusion Sci. Technol. **51**, 460 (2007)
47. L.M. Lidsky, Alcator Design Group. Design of alcator – the MIT high field torus. Proceedings of the 4th European Conference on Controlled Fusion and Plasma Physics, Rome, p. 20 (1970)
48. M. Greenwald et al., 20 years of research on the Alcator C-mod tokamak. Phys. Plasmas **21**, 110501 (2014)
49. E.S. Marmar et al., Overview of research results from the Alcator C-Mod tokamak. 27th IAEA Fusion Conference, IAEA, Gandhinagar, India, Paper IAEA-CN-258-690 (2018)
50. M. Greenwald et al., Energy confinement of high-density pellet-fueled plasmas in the Alcator Tokamak. Phys. Rev. Lett. **53**, 352 (1984)
51. R.T. Nachtrieb et al., Electron cyclotron discharge cleaning (ECDC) experiments on Alcator C-Mod. J. Nucl. Mater. **266–269**, 896 (1999)
52. B. Lipschultz et al., Divertor physics research on Alcator C-Mod. Fusion Sci. Technol. **51**, 369 (2007)
53. A.E. Hubbard et al., Edge energy transport barrier and turbulence in the I-mode regime on Alcator C-Mod. Phys. Plasmas **18**, 056115 (2011)
54. F. Alladio et al., Energy confinement of FTU ohmic plasmas. Plasma Phys. Control. Fusion **34**, 13 (1992)
55. G. Pucella et al., Overview of the FTU results. Nucl. Fusion **59**(11), 112015 (2019)
56. G. Pucella et al., Overview of the FTU results. Nucl. Fusion **57**(10), 102004 (2017)
57. G. Pucella, Overview of the FTU results. 28th IAEA Fusion Energy Conference, Nice, France (Virtual) 10–15 May 2021, Presentation OV/P-2, paper IAEA–CN–286-0884, Submitted to Nuclear Fusion (2021)
58. A. Sykes et al., First results from MAST. Nucl. Fusion **41**(10), 1423 (2001)
59. A. Kirk et al., Overview of recent physics results from MAST. Nucl. Fusion **57**(10), 102007 (2017)
60. J.R. Harrison et al., Overview of new MAST physics in anticipation of first results from MAST Upgrade. Nucl. Fusion **59**(11), 112011 (2019)
61. J. Milnes et al., MAST-upgrade – construction status. Fusion Eng. Des. **96–97**, 42–47 (2015)
62. J. Harrison Overview of first physics results from MAST upgrade. 28th IAEA Fusion Energy Conference, Nice, France (Virtual) 10–15 May 2021, Presentation EX/P6-39, Paper IAEA–CN–286-1538 (2021)
63. Vaughn A. UK takes step towards world's first nuclear fusion power station. New Scientist, Issue 3312, 2 December (2020)

64. STEP website. UK Atomic Energy Authority, Culham Centre Fusion Energy, STEP (Spherical Tokamak for Energy Production). https://ccfe.ukaea.uk/research/step/. Last accessed 17 June 2022
65. J. Menard, Overview of first results from NSTX-U and analysis highlights from NSTX. IAEA Fusion Energy Conference, Kyoto, Paper OV-5-2 (2016)
66. M. Gryaznevich et al., Overview and status of construction of ST40. Fusion Eng. Des. **123**, 177 (2017)
67. M. Gryaznevich et al., Theoretical and Experimental Studies of Confinement in High Field ST, EPTC 2019, Ghent, Belgium (2019)

# Chapter 4
# TCV: A Case Study in Systems Forward Engineering of a MA Tokamak

**Abstract** The TCV (Tokamak à Configuration Variable) is used as a forward engineering example of machine design, construction and plasma physics experiments. System robustness was achieved by highly flexible and modular construction, separate power supplies for each shaping coil and upgradeable control systems. The design was based on integrated MHD equilibrium and stability calculations together with overall systems modelling of plasma evolution, shaping coil current, voltage and power supplies. A key design element was a thick conducting vacuum vessel that stabilized high- and medium- frequency plasma instabilities. The design exploited the use of high vertical elongation and shaping capabilities to achieve MA plasma current at low toroidal field and low cost along with extreme shaping flexibility and divertor operation.

## 4.1 Systems Forward Engineering Strategy for TCV

### *4.1.1 Case Study Strategy*

This chapter examines TCV (Tokamak à Configuration Variable) as a case study in systems forward engineering for designing, funding, building, commissioning and operating a new fusion experimental machine. This study uses systems approaches to optimize design and robustness and counter fragilities. Checklist 1.3A from Chap. 1 gives general systems principles for forward engineering and Checklist 1.3B outlines detailed steps to follow using these principles. The actual design process of TCV followed such methodologies closely.

The method uses systems thinking to set goals, determine systems elements, relations and levels, develop systems architecture from several variants, finalize design and complete construction. Experimental results are summarized in order to examine robustness and fragilities and to learn lessons. The scenarios developed during TCV operation were added to the systems design space in Chap. 2.



F. B. Marcus, *Systems Approaches to Nuclear Fusion Reactors*, Springer Series in Plasma Science and Technology, https://doi.org/10.1007/978-3-031-17711-8_4

### 4.1.2 TCV Capabilities

For perspective on TCV's current importance, an overview article [1] was published 27 years after the first plasma by 270 authors from 56 institutions plus the EUROfusion MST1 (Medium Sized Tokamak 1) Team. TCV is a versatile device with unequalled shaping capabilities due to 16 independently controlled shaping coils and flexible heating systems for plasma currents designed to exceed 1 MA, although operation is usually less than 500 kA. Its research programme ranged from conventional topologies and scenarios in support of ITER to advanced tokamak operation and alternative divertor configurations. TCV also served as a university research facility at the Swiss Plasma Center (SPC), formerly Centre de Recherche en Physique des Plasmas (CRPP) of the Swiss École Polytechnique Fédérale de Lausanne (EPFL, also Federal Institute of Technology at Lausanne). TCV engineering parameters are given in Table 4.1.

The TCV vacuum vessel and several of the plasma poloidal cross sections produced in TCV [2] are shown in Fig. 4.1. The machine was also designed to allow extensive port access into the vacuum vessel inserted between the outer coils to allow installation of additional heating [3], as shown in Fig. 4.2. Access was also available at the top and bottom of the vacuum vessel, including large portholes for personnel access during construction. Other ports for heating and diagnostic access are also shown, as well as the carbon tiles shielding the inside of the vessel. The inner ohmic and shaping coil stacks and the outer 8 shaping coils are coloured green, with the access ports for co- and counter- tangential neutral beam injection are coloured purple. Also visible is the inner central column of the toroidal field coils, the 16 outer returns, finger joints and coil support structure.

**Table 4.1** TCV engineering parameters

| TCV engineering parameters | Values |
|---|---|
| Major/minor radius of vacuum vessel Centre | 0.88 m / 0.24–0.25 |
| Plasma half-height/maximum elongation | 0.72 m / 3/1 |
| Plasma current (design value)/actual | (1.2) / 1.06 MA |
| Toroidal Field at vessel Centre | 1.41 tesla (1.5 available) |
| Maximum flux swing/loop voltage | 3.4 V-s / 12 V/turn |
| Vessel internal full width/height | 0.56 m / 1.54 m |
| Plasma-Wall separation side/top | Typically 0.04 m sides with 0.05 m at maximum elongation |
| Vessel thickness/resistance/time constant | 20 mm / 55 μΩ / 8 msec |
| 16 shaping supplies/ motor generator frequency | 4quad - 12pulse / 120-96 Hz |
| Auxiliary heating | 4.5 MW EC, 2 MW NBI |

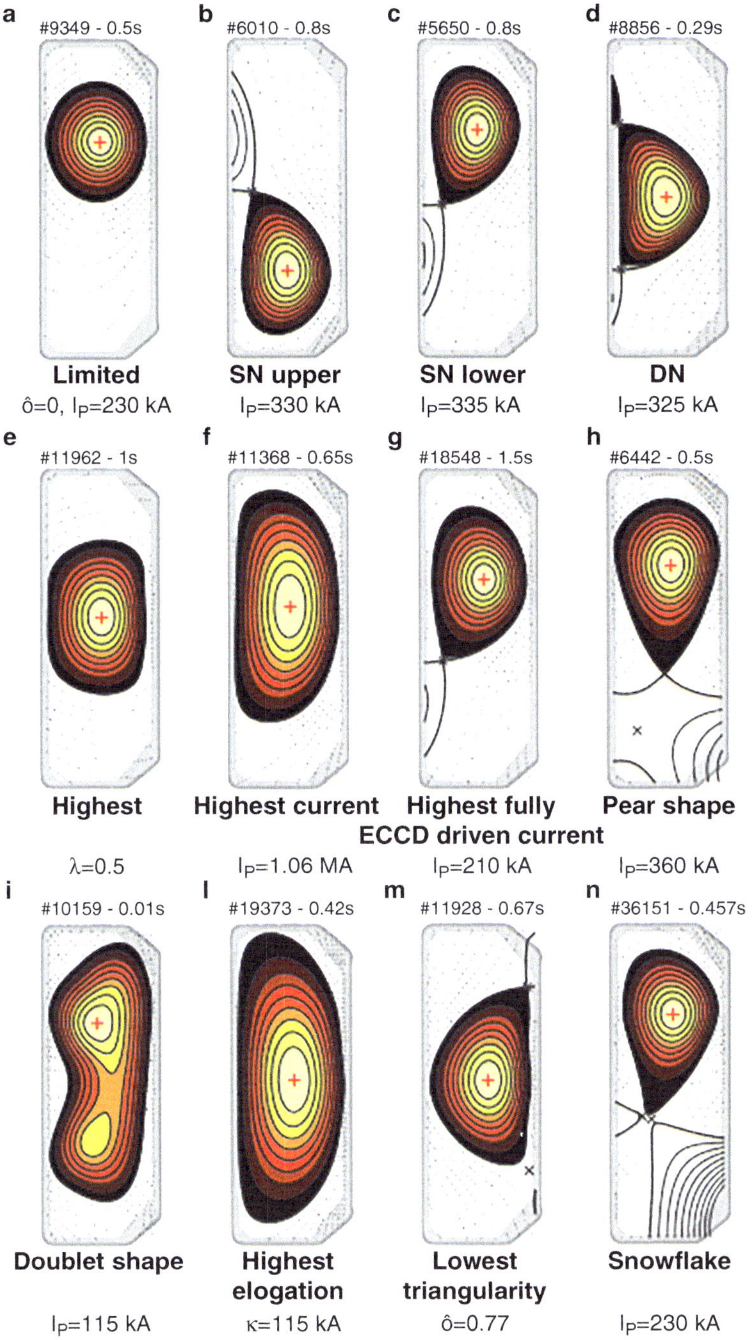


**Fig. 4.1** Plasma shapes achieved on TCV: (**a**) standard limited configuration, (**b**) diverted upper single-null, (**c**) diverted lower single-null, (**d**) diverted double-null, (**e**) highest squareness, (**f**) highest plasma current, (**g**) highest fully ECCD driven plasma current, (**h**) pear shaped plasma, (**i**) doublet shaped configuration (see Chap. 3), (l) highest plasma elongation, (m) lowest plasma triangularity and (n) snowflake divertor configuration. (From [2], Fig. 2.4.1)

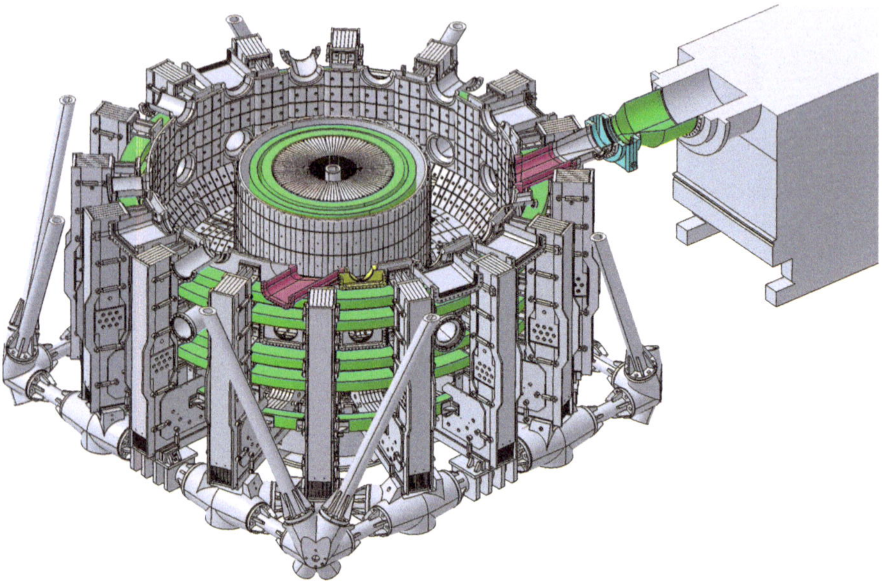

**Fig. 4.2** TCV 3-D schematic showing lower half cutaway and neutral beam injection. (From [3], Fig. 3)

## 4.2 Systems Approaches to the Design of TCV

### *4.2.1 Suitability of TCV as a Case Study in Systems Approaches*

TCV was chosen for a detailed case study because it was a highly agile and robust system which has made major contributions to fusion research. It reached the class of MA plasma current tokamaks. Construction was nearly on time with a budget about a third of comparable machines. It had many innovations compared to its predecessor tokamak TCA.

TCV was designed at a propitious time in 1983. Switzerland was about to join the European Atomic Community (Euratom), giving access to Euratom preferential funding for new fusion-directed experiments. At the same time, the university was moving to a new campus just outside Lausanne, with funds available for new offices and experimental buildings.

### *4.2.2 Systems Forward Engineering – Getting Started*

In 1983, the CRPP director Francis Troyon called Ferdinand Hofmann and this author into his office and asked us to design the most advanced and ambitious tokamak possible, in terms of plasma shaping, current and pressure, and within the budgets available by exploring an upgrade or a new machine.

### 4.2.3 Predecessor Tokamak TCA Design

At the time of the TCV design, the CRPP was operating the TCA tokamak with 400 kW of Alfvén wave plasma heating. The parameters of the circular plasma were $R = 0.615$ m, $a = 0.18$ m, $B_T = 1.52$ T, maximum $I_p \leq 170$ kA at $q_L \geq 2.2$ [4]. To establish preliminary goals for TCV, the first step was to review the original TCA construction goals, which were:

- Investigate the possibility of plasma heating in a circular plasma tokamak by dissipation of Alfvén waves, which had been investigated on a linear pinch.
- Provide a tokamak to study laser collective scattering, plasma wall interaction and bolometric measurements.
- Allow the possibility of upgrading the vacuum vessel to study non-circular shaped plasmas similar to JET and to investigate other heating schemes.

To accomplish these goals, the tokamak was supposed to satisfy the following:

- The electron temperature should exceed 500 eV for soft x-ray measurements.
- The discharge duration should be long enough to establish quasi-steady state to observe the effects of auxiliary heating.
- The machine should have a low aspect ratio $R/a \approx 3$ to reach high $\beta_T$ values and test stability limits for non-circular plasmas.
- The toroidal field coils must be demountable and elongated in the vertical direction such that the vacuum chamber can be replaced in case of a Phase II upgrade.

  These criteria led to the TCA tokamak with the dimensions given and an operating plasma current of 135 kA at $q = 3$.

### 4.2.4 TCA Technical Choices

The toroidal field coils were based on those of the ISX (Impurity Studies Experiment) at Oak Ridge National Laboratory. The coils were demountable so that the inner components of the machine could be accessed vertically for installation or removal. They would heat inertially during a pulse and were water cooled to allow full power pulses every 5 min. They were powered from transformers directly connected to the Swiss electricity grid, and used thryristors to rectify the electrical power to the coils.

The water-cooled air-core ohmic coil was designed to provide 0.5 Volt-seconds of flux, powered from capacitor banks coupled to ignitrons, vacuum switches and resistors. These were used to reverse charge the coil, induce 100 volts/turn at the plasma and apply a pre-programmed low voltage to achieve an approximate plasma current flattop.

The eight shaping coils were interconnected so as to produce a vertical field for plasma position and a horizontal field for vertical stability. A quadrupole field was applied only at plasma initiation and an oscillating field was available for discharge cleaning. These coils were also powered by capacitor banks.

The vacuum vessel had 5 mm thick sides and 12 mm thick top and bottom and was made of stainless steel. It was fabricated in two halves that were toroidally separated by an insulating gap with Viton rings to allow the ohmic electric field to penetrate the vessel. However, the rings outgassed and limited the baking temperatures to slightly over 100 °C, thereby compromising machine cleanliness, particularly to water vapour.

## *4.2.5 TCV First Goals and Design Options*

The TCV team initially consisted of the two design and project managers plus an engineering draftsman. The rest of the CRPP staff were available for consultation. The original project focus was to develop an upgrade of the existing TCA. A first preliminary design was available in October 1983.

### 4.2.5.1 First TCV Goals

The first primary TCV goals were to:

- Incorporate a divertor in TCA to improve Alfvén heating by impurity control and H-mode.
- Study Alfvén wave heating and current drive in elongated and circular plasmas.

The first secondary TCV goals were to:

- Study electron cyclotron heating at the second harmonic with CRPP developed gyrotrons in the 100 GHz range.
- Study plasma shaping, high beta and divertor physics.
- Study current drive with ECH and Alfvén waves in a longer pulse machine.

These goals led to a first design option of using a larger vacuum vessel in the same style as TCA to profit from the vertically elongated toroidal field coils.

### 4.2.5.2 The First TCV Design Option

The first TCV design parameters were R = 0.62 m, a = 0.20 m, $B_T$ = 1.5 T, vertical elongation = 1.75 and plasma current up to 400 kA.

However, as the design evolved and other options were considered, it was felt that the existing TCA coils provided only limited scope for upgrade, and therefore more ambitious options and goals were considered. The emphasis of the project began to concentrate more on the high beta aspects. By January 1984, the project goals had evolved considerably.

#### 4.2.5.3 Revised TCV Goals

Goals for a new machine became:

- Demonstrate Alfvén wave heating of a non-circular plasma to reactor levels.
- Maximize plasma pressure and beta by vertical elongation of plasmas with broad current profiles.
- Demonstrate impurity control at high plasma pressure with expanded boundary divertor and pumped limiter.
- Study Alfvén heating in a long pulse.
- Study 2nd harmonic electron cyclotron heating and current drive with advanced CRPP quasi-optical gyrotrons.

These goals were subject to **constraints** which, in fact, were altered as the design proceeded:

- The upgraded machine should be of a similar size to fit in the existing building.
- The aspect ratio should be larger than R/a = 3 to have enough volt seconds and avoid very high toroidal field on the coils.
- The toroidal field should be 2.2 T for second harmonic ECRH.

These ambitious goals depended on certain **assumptions**:

- Alfvén wave heating was to be demonstrated in the existing TCA.
- Beta limits were determined by Troyon scaling at high elongation.
- Plasmas with elongations up to 3/1 could be produced.
- Engineering solutions could be found to the problems of Alfvén antennas.

These goal alterations led to a **revised TCV design option** with parameters R = 0.62 m, a = 0.20 m $B_T$ = 2.2 T, vertical elongation 3/1, plasma current = 2 MA and a poloidal divertor with baffles. In support of this design option, a number of MHD equilibria were calculated.

These design options and goals were presented to the CRPP scientific council which decided to proceed along the lines of the more ambitious option and to prepare a more detailed design for consideration for Euratom preferential funding.

## 4.3 Review and Prioritize Goals and Evaluate Variants for TCV

### *4.3.1 Systems Review Process in Early Stages and Prioritizing Goals*

The process for reviewing and prioritizing goals and evaluating variants for TCV followed the procedure of Checklist 1.3B. The early design revealed the need to prioritize between some of the earlier machine goals. The most important goal seemed to be the demonstration of high beta plasmas at very high plasma current and high vertical elongation, which would allow exploration of the limits of the Troyon scaling.

In addition, very promising results had been demonstrated on other tokamaks at medium elongation, as is shown from early scenarios in the tokamak design space of Chap. 2.

### *4.3.2 Scenarios for TCV from the Tokamak Design Space and Revised Priorities*

Relevant scenarios from the tokamak design space were identified.

- Doublet III had operated a D-shaped plasma in its upper half and had reached 5% beta with neutral beam injection. It had demonstrated that a 2.2/1 elongated plasma was stable with wide current profile from in-situ pellet start-up.
- TOSCA had shown that high beta could be achieved both with ohmic heating and electron cyclotron heating.
- High beta equilibria at very high elongation (8/1) had been transiently observed in the Lausanne Belt Pinch [5].
- Doublet III and Asdex had shown the importance of divertor plasmas and the H-mode in improving confinement.

Theoretical calculation with a free boundary MHD equilibrium code had shown that plasmas with 3/1 and 4/1 elongations could be produced. It was concluded that plasma shaping and high plasma current should be the highest priority goals in the design. Starting plasma operation with just ohmic heating in a highly elongated plasma would already provide an interesting and challenging experimental programme. This prioritization was expressed in the name of the new machine, TCV (Tokamak à Configuration Variable).

The use of Alfvén heating to reach high pressure had not yet been demonstrated. As was already evident in 1984 and confirmed in the 1992 experiments of TCABR, Alfvén waves did indeed couple to the plasma and produce heating, but was always accompanied by a rapid density rise, so that plasma parameters did not greatly exceed those obtained from Ohmic heating and gas puffing.

With the highly variable plasma configurations planned for TCV, it was hard to determine the optimal location for the Alfvén antennas. Therefore, it was decided to keep Alfvén heating as an important option, but to avoid making it the main focus of the machine. In fact, Alfvén heating was never attempted in TCV.

The third important strand was electron cyclotron heating. The CRPP had made good progress in designing quasi-optical high-power gyrotrons for plasma heating and current drive at harmonics of the electron cyclotron heating, as for example demonstrated on TOSCA. Therefore, electron cyclotron heating and current drive could also be considered as a primary goal on TCV. Neutral beam heating of plasma had also been demonstrated to be effective, so vacuum vessel ports were designed for tangential accent when funding for beams became available, although the actual installation required some vessel modifications.

### *4.3.3 The Intermediate Design Research Areas for TCV*

As stated in the 1985 Phase I proposal for Euratom Preferential Funding [6], the intermediate design research areas were to study:

- scenarios for the creation of elongated plasmas;
- the possibility of maintaining positional stability and optimization of the control system;
- the dependence of the maximum current on elongation;
- the scaling laws for the confinement time and the density limit at high elongation;
- the regime in which the plasma eventually settles into a steady state (monotonic or non-monotonic q-profile, minimum internal q, island structure);
- the dependence of the beta limit on current.

### *4.3.4 Tokamak Systems Basic Parameters for Further Design*

Having settled on the intermediate research goals for TCV, the next steps from Checklist 1.3B were to scope the highest system level of major variants, develop models, estimate performance, and produce a first report.

The primary systems to be considered were the plasma, vacuum vessel, plasma facing components, divertor, toroidal field coils, poloidal field and ohmic field coils, instability coils, electrical power supplies, diagnostics, machine and plasma control, gas pumping and injection, auxiliary plasma heating, coil cooling and machine structure.

Of these systems, the most important consideration was to scope the highly interactive systems of the plasma, vacuum vessel and shaping coils as described in Table 2.1 in Chap. 2 and determine the maximum performance of plasma current and vertical elongation.

The next step from Checklist 1.3B was to produce definitive proposals and engineering design for Phase I proposal, by choosing one or more most promising variants and begin detailed design of level 1 and level 2 subsystems of key areas. A Phase I proposal could be produced with a specific design involving appropriate physics and engineering modelling plus cost and time estimates, which involved detailed systems architecture.

## *4.3.5 TCV Systems Architecture*

The fundamental choices of high level system form for TCV were dictated by the overall goal of reaching highly vertically elongated plasmas, and to retain extreme flexibility in plasma shaping and control. This led to the choice of having a highly conductive vacuum vessel near the plasma to provide maximum wall stabilization against vertical and plasma instabilities. A continuous vacuum vessel without an insulating break would provide improved plasma stability, improved protection of external coils against disruptions, and the possibility of baking the vessel up to 350 ° C, allowing much improved vacuum and thus plasma purity. Similarly, it was important to have many independently controlled shaping coils just outside the vacuum vessel, to provide maximum shaping flexibility for given coil currents. Close-fitting coils would also minimize volt seconds required by reducing the plasma's external inductance and, hence, transformer volt-second requirements. Therefore, demountable toroidal field coils would be necessary and could be outside the shaping coils.

### 4.3.5.1 Maximum Elongation

The essential parameter of TCV was the maximum plasma elongation that was chosen to be 4/1 in a racetrack configuration. Equilibrium calculations showed that above an elongation of 2/1, hollow q profiles could develop even for peaked current profiles due to the elongation of the central flux surfaces. Varying the elongation from 1 to 4 allowed a clear identification of the dependence of the plasma parameters on elongation.

A basic scenario for obtaining this 4/1 elongation was chosen so as to start with a circular plasma with low $q \approx 2$ and then to elongate it in a controlled manner. A relatively long pulse of hundreds of msec was planned so as to explore resistive phenomena and current profile diffusion and to allow the plasma to reach quasi-steady state conditions. The effects of wall stabilization were also modelled.

#### 4.3.5.2 Machine Size, Toroidal Field and Plasma Current

The choice of three parameters, toroidal field, minor radius and aspect ratio, determined the size and the characteristics of the machine. The toroidal field was chosen to be 1.5 Tesla, the same as in TCA, which was low enough to use the TCA-proven water-cooled copper jointed-coil technology and high enough to give good quality near-circular plasmas. The field would allow for second and third harmonic electron cyclotron heating at CRPP gyrotron frequencies which allowed wave transmission at reasonably high plasma densities.

The horizontal minor radius of 0.18–0.20 m was also chosen to be the same as in TCA. The plasma minor radius determined the machine size, port access and shaping coil spacing. The resulting coil spacing allowed standard diagnostic ports (0.15 m inner diameter). A smaller plasma would have required an excessively compact construction and also reduced diagnostic access and volt-second capability. A larger machine was considered too expensive.

#### 4.3.5.3 Poloidal Coils

The plasma current for a 4.1 elongated tokamak would be about 1.6 MA. This high plasma current required far more ohmic coil volt seconds than in TCA and a vertical elongation too high to fit in its existing coils. These calculations confirmed the decision that a TCA upgrade was definitively rejected.

An increased major radius of R = 0.80 m led to an aspect ratio A = 4/1 and had been increased from the TCA radius of 0.61 m in order to allow for a large increase in available volt-seconds, up to a swing of −1.2 to +1.2 Volt-sec. This capability provided many options. For a maximum 4/1 elongated plasma with q = 2, then 2.4 Volt- sec would have provided 1.5 Volt-sec for the inductive requirement and 0.9 Volt-sec for both plasma build-up resistive loss and a flat-top current lasting hundreds of msec. A lower elongation plasma would be obtained with a small fraction of the available volt-seconds and allow much longer pulses.

#### 4.3.5.4 Vacuum Vessel

The planned rectangular vacuum vessel was 0.48 m wide inside and 1.54 m high, allowing 0.40 m full-width plasmas with 0.04 m gaps on each side. A gap of 0.04 m on the sides and at the top was a compromise between good plasma-wall isolation and stabilization from the conducting vessel wall. The box height would have accommodated plasmas with elongations up to 4/1 of a 0.36 m width plasma, with top/bottom gaps of 0.05 m.

The two vertical shaping coil stacks had uniformly spaced coils at small major radius, and 4 groups of 2 at large major radius, allowing 3 large vertical gaps 0.22 m high. This arrangement allowed good diagnostic access from top, bottom and outside and would allow the use of transferred TCA diagnostic and vacuum systems, consistent with shaping flexibility and a minimum of total shaping current. For highly elongated plasmas, the sum of the absolute values of the coil currents was less than 1.5 times the plasma current for racetracks.

In exploring options for shaping coil positions, it was found that the shaping coils could be considerably further away from the vacuum vessel and still produce the same equilibria, with a tolerable increase in required total current. Even so, to obtain the best plasma stabilization and to minimize power requirements, the coils were positioned as close to the plasma as possible. The shaping coils were planned to be individually controlled by feedback signals from flux loops and magnetic probe coils on or inside the vessel.

#### 4.3.5.5 Budget and Buildings

The total estimated budget that would be available for TCV, including preferential support from Euratom, was 17.5 million Swiss Francs (about USA $8 million) at 1984 prices and exchange rates, with 1/3 of the cost each for the load assembly, power supplies, and motor generator. A larger plasma would result in too expensive a machine, both in machine costs, power supplies and a motor generator, since power supply requirements would far exceed that available from the grid to the EPFL.

Since TCA had been in a building on loan from another EPFL department which now wanted it returned, a whole new set of buildings would also be required. Concrete shielding would be needed against x-rays from runaway electrons and possible neutrons from d-d reactions. Much more space would be required for a motor generator and power supplies. The EPFL was willing to provide these buildings as part of its budget to construct a new campus.

### *4.3.6 Scoping the Systems for Creating High Current and Highly Elongated Plasmas*

Specifying the systems architecture of TCV involved developing models to scope various subsystems. The most important was the electromagnetic interactions of the shaping coils, the vacuum vessel and the plasma, including its build-up phase from small, circular plasma. These interactions were summarized at a system interaction level in Chap. 2 by Table 2.1, which displays the function matrix for tokamak poloidal field system - electromagnetic view.

The scientific feasibility of the TCV concept of creating elongated plasmas used the Tokamak Simulation Code (then called Startup) to model these interactions, showing that plasma could be evolved continuously from a near circular shape to a 4/1 vertically elongated racetrack [7], as shown in Fig. 4.3.

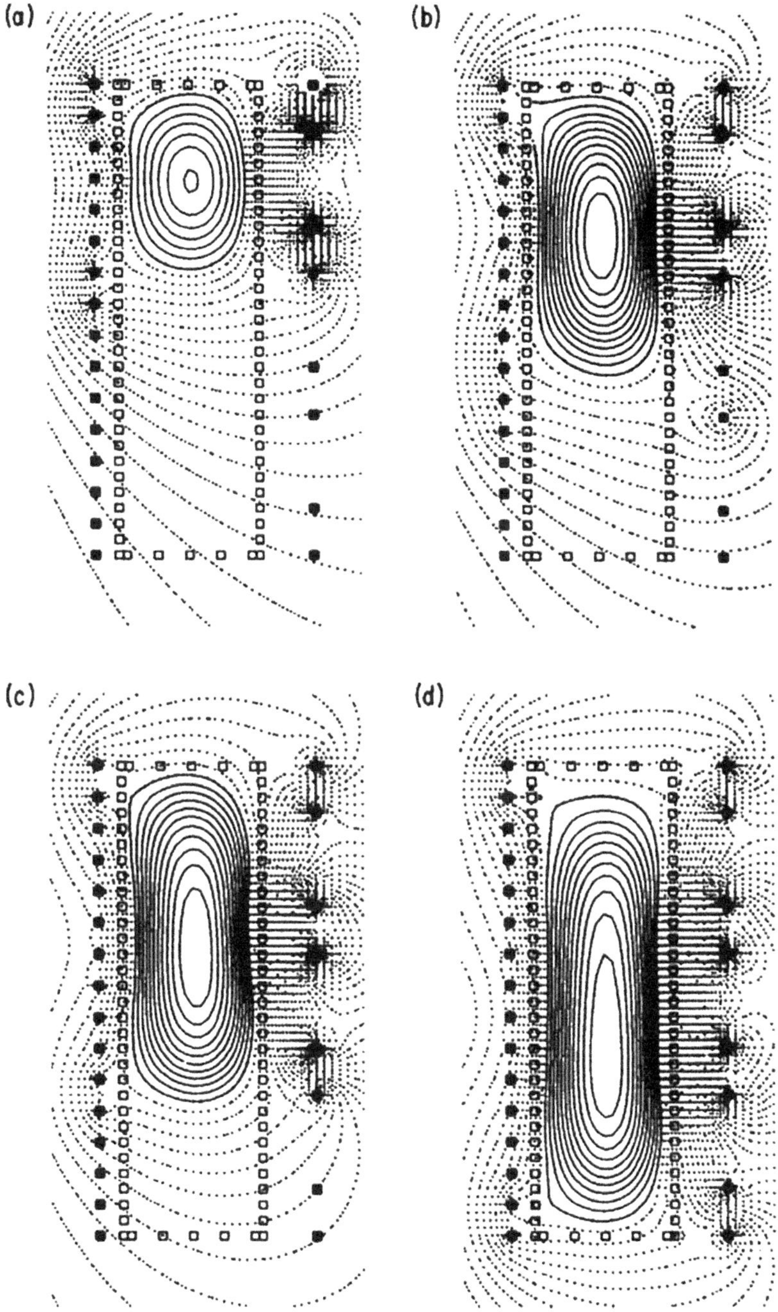


**Fig. 4.3** Controlled evolution of plasma shape in TCV. (From [7], Fig. 2)

All intermediate stages (a, b, c) and the final state (d) were found to be stable to axisymmetric MHD modes. The stabilization was provided by the vacuum vessel walls on the ideal time scale and by an orthogonal field superposition driven by active feedback system controlling the shaping coil currents on the resistive time scale. Flux surfaces inside the plasma are shown as solid lines and outside the plasma as dots.

In equilibrium (a), there are currents and surrounding flux only in coils in the upper half, whereas in subsequent equilibria, currents in the lower half are activated also. All coil currents are changing as the shape is changed and the plasma current increased. The pre-programmed currents were determined as described in the next section.

The virtual control system used a combination of pre-programmed currents, passive conductors, and an orthogonal feedback system. The equilibria all had $q = 2$ at the limiter and broad, but not hollow, current profiles consistent with the MHD evolution equations assuming classical Spitzer resistivity.

The stability of individual plasma shapes was also explored with another method [8] which computed global axisymmetric instabilities of tokamak plasmas in a conducting shell. The method used a free-boundary equilibrium code to find displacement vectors compatible with ideal MHD, which was then applied to the computation of vertical instability growth rates of highly elongated race-track shaped plasmas. Results obtained for up/down symmetric configurations were compared with ERATO [9] calculations. The gain in stability produced by axisymmetric positioning of a plasma in a given shell was evaluated. Growth rates of axisymmetric modes of elongated tokamak plasmas in the presence of a conducting shell were computed [10].

The effects of plasma-wall distance, current profile and elongation were analysed. For highly elongated plasmas, it was found that the side walls were more important than the top and bottom walls for stabilizing these modes. Furthermore, it was shown that decreasing the width of the radial current profile has a much greater effect on vertical stability than increasing the inside width of the conducting shell by the same amount.

## *4.3.7 Power supply Requirements for Shaping and Controlling Plasma*

### 4.3.7.1 MHD Equilibrium Calculations

The shaping coil currents and power supply requirements to achieve those currents [11] using MHD equilibria were calculated with a free boundary code which requires shaping coil currents as input. To estimate these currents, the shape of the outermost flux surface was specified in a separate computation to be an elongation $\kappa = 4/1$ racetrack, and a guess at the final plasma current profile was added. A linear matrix equation was inverted to obtain the shaping currents, by using mutual inductances to relate uniformly-spaced constant poloidal flux values on the desired flux surface to the shaping currents plus plasma current. The shaping coils were in vertical columns at $R_0 \pm d$, where $d/a_{plasma} = 1.67$. To obtain stationary equilibria with elongation $\kappa > 2$, it was necessary to superimpose a feedback-controlled quadrupole-field which maintained the elongation of the outermost flux surface. Without this feedback, plasmas with $\kappa > 2$ either shrank or diverged into two halves. Once currents were known, detailed stress calculations were made on coils and structures.

#### 4.3.7.2 Plasma Shaping and Control

An extension of these methods was used [12] to study ways of controlling and shaping elongated plasmas. In TCV, with 16 independent shaping coils, the ohmic coils, a continuous thick current-carrying vacuum vessel and with fast internal coils, there were many more control system possibilities than in most tokamaks, which have hard-wired moment control.

Two possible methods were considered. The first used methods developed for finding and optimising equilibria in the FBT code, which allowed the computation of arbitrarily shaped tokamak equilibria with external or internal separatrices and multiple magnetic axes. It was shown that bifurcations can be controlled by an iterative procedure using shape feedback. Three applications were discussed:

- the problem of shape accuracy;
- the calculation of a start-up evolution for a highly elongated tokamak;
- methods to control x-points on a separatrix.

The control method was adapted to be compatible with the TCV simulation models which multiplied input measurements by a time-varying matrix.

In the second method, complete circuit equations, including self and mutual inductances and resistances, were written for the plasma, vessel, shaping and ohmic coils, in terms of current increments during the following time step. The vessel current increments were calculated. A cost function Q was minimized so as to minimize flux errors and current magnitudes according to weighting functions. Shaping and vessel currents were determined and then shaping and ohmic voltages required could be determined.

#### 4.3.7.3 Stabilization of Vertical Instability at High Growth Rate

For high growth-rate, highly-elongated plasmas, extra stabilization was possible by putting rapid reaction inside the outer corners of the vacuum vessel. The operation of these coils was evaluated [13]. The control and evolution of these plasmas were simulated with TSC, which showed that pressure perturbations such as sawteeth and externally programmed displacements created initial velocity perturbations in addition to equilibrium changes. It was demonstrated that these perturbations may be stabilized by low power, rapid response coils inside the passively stabilizing vacuum vessel, together with slower shaping coils outside the vessel. This method works because coils inside the vessel do not suffer a flux penetration delay.

Vertical disruption induced voltages and forces on the rapid coils and the vessel were investigated to establish coil and vessel design requirements. A model was proposed and used to estimate the additional vertical force due to poloidal currents, which was comparable to the force from toroidal currents.

This concept of small, fast coils inside the vessel developed for TCV is now in use in many larger tokamaks.

#### 4.3.7.4 Plasma Disruptions

A major danger to tokamak systems is a disruption. The TSC code was used to predict [14] induced voltages, currents, and forces due to such disruptions in TCV:

- Inside the vessel, up to 1200 volts/turn at the plasma centre would be generated during a disruption with the given current disruption rates.
- Induced voltages outside the vessel near the shaping coils might reach only 85 volts/turn due to shielding currents induced in the vessel.
- With non-symmetric disruptions, the maximum values of induced voltages and currents are similar, but spatial and temporal distributions change.
- The total vertical force on the vacuum vessel would be about 17 tons.

The very high internal plasma voltage during a disruption could result in runaway electrons at tens of MeV. These could hit the plasma limiters and produce high energy x-rays giving a dose calculated to be 300 REM in a person at a distance of 8 m, thereby requiring significant concrete shielding around TCV [15].

### *4.3.8 Phase I Proposal and Euratom Review*

The next step in the checklist was to review the proposal at team, laboratory and project level, and submit it for funding. The Phase I proposal [6] was submitted in mid-1985 to an ad-hoc review group which produced a formal report that broadly supported the proposal. The Euratom Programme Committee gave the CRPP the approval to proceed to a final Phase II detailed design for preferential funding. The following points had to be examined for the Phase II presentation:

- The analysis of the stability with respect to axisymmetric perturbations as a function of various current distributions should be provided.
- Further theoretical work on MHD stability should be carried out as far as possible, including ballooning modes.
- Flexibility should be maintained to allow a pragmatic approach to additional heating schemes and profile control in order to meet problems as they arise.

- The Ad-hoc group requested further investigation of wall conditioning, density control, limiter design, the effects on plasma stability of error fields from diagnostic ports, and the stability of configurations other than racetracks.

The CRPP of course welcomed this decision and proceeded to the Phase II detailed design, emphasizing immediate analysis of the points raised.

## 4.3.9 *Phase II Proposal to Euratom*

Checklist 1.3b describes the activities for preparing the Phase II proposal, which include incorporating review suggestions and proceeding to a full scale design.

### 4.3.9.1 MHD Stability Calculations and Revised Design

Immediately after the external review, efforts were intensified to determine the stability of plasmas with elongation in the range of 2/1 up to 4/1 to vertical, ballooning and kink instabilities. Using the methodology of ERATO [9, 10], the conditions for axisymmetric stability during the evolution from a circular plasma to a highly elongated racetrack were investigated. After comparisons of the plasma instabilities, the maximum beta for a racetrack occurred for an elongation of 3/1, and declined above that. Values of beta up to 5% were stable to ballooning modes at an elongation of 2.5 and an aspect ratio of 4.4. Plasmas up to an elongation of 3/1 possessed axisymmetric stability for a range of current profiles.

Taking into account these new findings, and also the recommendations to maintain maximum flexibility, it was therefore decided to alter the TCV design to a 3/1 elongated plasma. The vacuum vessel shape and field coils were altered to this final design configuration at minimum changes in overall size and cost. The tokamak central core and height were unchanged, but the major radius and vessel width were increased. With the same plasma wall separation of 4 cm, defined by limiters or wall tiles, the minor radius of the plasma became 0.24 m, and the major radius became 0.88 m, resulting in an aspect ratio of 3.6.

Plasma vertical elongations of up to 4 were still found to be possible for racetracks by reducing the plasma width at maximum height, with stability at maximum elongation requiring a flat current profile.

### 4.3.9.2 Implementing Other Review Recommendation

The first stage of planned experiments would be to determine the optimum shape and performance obtainable with ohmic heating. The vacuum vessel ports and access and space around the tokamak were designed to allow for future upgrades. Vacuum vessel cleaning, plasma density control and limiter design would rely on best practices from other tokamaks as outlined in the tokamak design space.

The accompanying power supplies and motor generator were redesigned for the new TCV load assembly with flexibility for future auxiliary heating. This process involved continuous interaction and feedback from the analysis models. A detailed design was carried out involving:

- re-engineering of all systems including materials and stress calculations for all mechanical components;
- electrical analysis for all control and measurement systems;
- programming and analysis for all supporting control systems.

## 4.4 TCV Top Level Robustness Properties and Tradeoffs

Checklist 1.3B further proposes an assessment of robustness and fragilities and to take suitable precautions.

### *4.4.1 Robustness*

The machine was designed to be able to operate reliably over decades without any major faults occurring. Standard machine maintenance was facilitated by personnel access to the inside of the vacuum vessel. Other major standard maintenance activities such as complete disassembly and rebuild of the motor generator could be planned to coincide with other upgrades.

The most important contribution to robustness was the novel and simple design of a thick stainless-steel vacuum vessel by:

- ensuring its own integrity
- protecting the rest of the tokamak against plasma disruptions
- stabilizing high frequency plasma oscillations
- allowing the use of standard low-frequency thyristor power supply technology
- dispensing with an insulating gap that generated impurities and allowing high temperature baking

The result of this increase of robustness was a far more reliable machine, which could be constructed at much lower cost than for a high resistivity vessel.

## 4.4.2 *Fragility*

As with all highly robust systems, there were points of fragility in the design, but many backups were planned to prevent their occurrence.

### 4.4.2.1 Plasma breakdown and Current Rise with Continuous Vacuum Vessel

The fragility due to the vacuum vessel design was the reliance on creating the plasma with a low one-turn breakdown and ionization voltage of less than 12 volts per turn. If the gas in the TCV vacuum vessel had failed to ionize and form a plasma at this voltage or if the current could not be ramped up, the whole machine would have been a failure. All power systems and coils had been designed to deliver low voltages. Other machines at the time used around 100 volts per turn.

Several mitigations were developed against this risk. The solid vessel with few structural welds would allow the vessel to be baked to 350 degrees C, which allowed a very low level of impurities in the plasma, greatly reducing the need for a high breakdown voltage to overcome impurity radiation. Careful calculations of hydrogen-isotope atomic and molecular physics rate equations indicated the required E/p (electric field/filling gas pressure) range which would be available to obtain breakdown. Also, high power first harmonic electron cyclotron heating was an option for ionizing the filling gas and heat the plasma. Vacuum switch breakers could have been added to the ohmic coil if necessary, although the vacuum vessel would have attenuated higher starting voltages. Finally, other impurity reduction methods before plasma breakdown were planned to be employed, including glow discharge cleaning, as discussed in the tokamak design space.

Simulations with TSC which calculated current penetration in realistic plasmas using transport models showed that current rise could be successful at low applied voltage to the plasma.

The shaping systems could also produce optimized multipole vacuum fields such as octopoles with null central fields to aid breakdown.

As further preparation, experiments on low voltage start-up and assisted electron cyclotron breakdown [16] were performed on the still operating TCA tokamak, using 39 GHz at the first harmonic. The most efficient breakdown on axis was achieved using a high-field launched O-mode. Future experiments, after TCV was built and operating, would confirm that plasma breakdown and ionization and current rise were easily obtained at the machine design loop voltage thanks to excellent machine cleanliness. Even though this critical fragility was ultimately proven not to be dangerous, it was appropriate strategy that many mitigations were in place.

#### 4.4.2.2 Vertical instability of Highly Elongated Plasmas

Another fragility was the possibility of vertical instability of highly elongated or highly shaped plasmas. This would be overcome in the design of the machine by using extensive MHD equilibrium and stability calculations to reduce the probability of such instabilities. The TCV design also included low current small coils inside the vacuum vessel at its outer corners, with a multi-kilohertz low-power supply to suppress any vertical instability. The use of these coils was simulated using TSC and their operating parameters were confirmed.

#### 4.4.2.3 Need for Auxiliary Heating

An important fragility in terms of achieving TCV goals was the possibility that ohmic heating alone could not test plasma pressure beta limits.

This fragility was overcome by designing TCV to be compatible with:

- electron cyclotron heating at the second and/or third harmonic using CRPP high power gyrotrons being developed;
- neutral beam injection heating by designing tangential access ports and relying on NBI development occurring elsewhere. Actual installation of NBI required major in-vessel modifications for maximum heating access.

Both were successfully installed, allowing strong plasma heating and high beta.

Future operation confirmed that the innovations in design on TCV did involve fragilities, but a careful analysis and mitigation of these fragilities in advance allowed the highly successful operation of TCV.

## 4.5 TCV Final Design for Construction

### *4.5.1 Phase II Committee Review and Approval*

The proposal was submitted for Phase II approval in 1986 and was reviewed by another Euratom Ad-hoc committee with different members in September 1986. They considered that the Phase II proposal had adequately addressed the questions that were raised in the Phase I review by re-designing TCV to have a maximum current of 1.2 MA and maximum vertical elongation of 3/1. The minor and major radius had increased, and the flux swing had been increased to 3.4 volt-seconds from 2.5 to allow a more progressive build-up of shape versus time and to allow longer pulse plasmas.

However, they felt that auxiliary heating had not been addressed in enough detail, and that there should have been more transport simulations, more attention to installation problems and more concern about plasma heat removal in case of auxiliary heating. Therefore, there was concern that the beta limit could not be reached with ohmic heating only. They considered that the proposal design was laudably flexible in shaping options and control, including separatrix configurations. The committee felt that construction within the revised budget of 20 ½ million Swiss Francs (1986 prices) was feasible, but that the time schedule would require more time than estimated.

In conclusion, they felt that the team was especially suited; the proposed engineering design was technically feasible and should be achieved in a very cost effective manner. Finally, the Ad-hoc group recommended that TCV be awarded Preferential Support Phase II funding.

### 4.5.2 Systems Level Considerations for Final Design and Construction

Checklist 1.3b recommends several steps for the final engineering design and construction after final funding approval. After receiving Phase II approval for TCV, the final design and construction phase began. Upgrade proposals for high power gyrotrons were immediately prepared.

### 4.5.3 TCV System Level Final Design for Construction

The built TCV tokamak [17] and the design, construction and initial operational status were summarized [18]. The following construction choices were:

- sixteen independently controlled shaping coils;
- maximum design plasma current of 1200 kA;
- a thick-walled (average 1.8 cm stainless steel), continuous, all welded vacuum vessel with and inside width of 56 cm which allows a nominal plasma radius of 24 cm and wall gap of 4 cm, bakeable to 350 ° C;
- an ohmic heating flux swing of 3.4 Vsec;
- demountable toroidal field coils generating a nominal 1.43 T at $R_o = 0.875$ m. Further calculations concluded that operation up to 1.5 T is safe for a limited number of discharges;
- supplementary fast control windings inside the vessel outer corners;
- large area carbon primary and backup tile limiters with carbonization.

With this highly flexible configuration, a very wide range of plasma shapes were possible at maximum plasma elongation: outer triangle, inner triangle, bulging outwards, pinched in centre, inner and outer wall divertor. There were other shapes available also at lower elongation, as shown in Fig. 4.1.

### 4.5.4 *Implementation*

TCV components were built and assembled with a combination of in-house and external resources. External call-for-tenders were issued, contractors selected, and manufacturing and quality control procedures decided. The machine assembly and initial operations procedure were planned in advance.

TCV was built at a cost of at least three times less than machines of comparable plasma current. Major components were manufactured outside the CRPP, and were open to international bidding under WTO rules. The vacuum vessel design permitted the major items such as the motor generator and the thyristor power supplies to be standard technology and well suited for competitive bidding.

## 4.6 TCV Subsystem-Elements Descriptions and Robustness Analysis

This section goes from the "black box" level of subsystem description to the "grey box" or "white box" level of detail for the TCV elements actually built. A systems analysis of robustness and fragility is performed for each major subsystem, with more detail than in the overall top level systems analysis.

### 4.6.1 *Toroidal Field Coils, Power Supplies and Systems Analysis*

#### 4.6.1.1 Toroidal Field Coils

The TCV toroidal field coils were constructed from 23 mm thick copper plates to obtain 16 coils of 6 turns each. The maximum current was 65.3 kA giving a toroidal field of nearly 1.5 T at R = 0.88 m. Electrical insulation between turns was provided by sheets of fibreglass-epoxy. In the central section of the coil, the copper plates were wedged to form a vault structure. This section was epoxy vacuum impregnated, along with the ohmic and plasma inner shaping poloidal coils. It constituted a rigid assembly which resisted torsion about the vertical axis.

The horizontal limbs and the outer vertical limbs were not impregnated; they were strengthened by a number of bolts holding the 6 turns together. This basic design had been used in other tokamaks and was proven technology. Copper tubes for water cooling were inserted into grooves which were cut in the edges of the copper plates.

The stresses in the coil were modelled by a 3-D finite-element grid with 979 nodes. The current density in the copper was found by solving Laplace's equation for the electric potential. These stresses were predominantly compressional and uniform throughout the central column, reaching 8.2 MPa. Thermal expansion caused additional stresses. Additional structures were added externally to the coils. All stresses calculated were within acceptable engineering limits.

#### 4.6.1.2 Power Supplies

The power supply for the toroidal field coil provided a current of 65.3 kA. The coil inductance was 6.8 mH with a resistance 4.2 mΩ at 30 ° C. The required voltage was approximately 270 V at the beginning of the flattop. The rectifier had to work only in two quadrants. The choice of a maximum DC voltage of 535 V allowed the use of ordinary thyristors. To obtain a very small DC current ripple to minimize the voltage spikes in the diagnostics, a 12 pulse rectifier was chosen.

Because of the high current required, two bridges were parallel connected through an interphase reactor. The bridges were supplied with two 3 phase systems, 30° shifted, of 230 V (phase to neutral). The safety factor for ordinary thyristors (1200 V) was 2.1. The rectifier was able to provide a negative voltage that was used at the end of the discharge to bring the coil current to zero, thereby reducing coil heating and saving energy. The power supplies were completely conventional with a large margin during standard thyristor operation.

#### 4.6.1.3 Systems and Robustness Analysis

The systems control parameters were therefore well within range, and the toroidal field coils and their power supplies have proved to be a highly robust system, with maximum fault-tolerance, including protection against short circuits. Both power supply thyristors and toroidal field coils were highly modular, and could have been easily replaced in case of a fault.

Robustness trade-offs in design and operation resulted in very high robustness with no obvious fragilities. The performance was optimized to design values and large margins were available for longer pulses at lower toroidal fields, or repetitive operation at maximum design field. The systems characteristics are:

- systems control – The design of the coils was standard, and the control systems for the power supplies were simple and robust given the straightforward nature of the power supply waveform.
- fault-tolerance – The coils were designed to withstand the torques generated by a coil short-circuit, as were the power supplies.
- modularity – The construction of the coils was highly modular leading to easy replacement. The components of the thyristor power supplies were standard and easily available.
- decoupling – The thyristor firing in the power supplies produced very low level fluctuations in the toroidal field current. These were tiny and massively attenuated by the large inductance of the toroidal field coil.

Robustness trade-offs in design and operation:

- robustness – Robustness was achieved by using standard components and power supplies and a relatively low toroidal field.
- fragility – One fragility might be damage to the systems by a major plasma current disruption. Calculations showed that protection of the toroidal coil and power supply system was provided by the vacuum vessel and shaping coils.
- performance – The performance was excellent as the machine was mostly operated at the design maximum magnetic toroidal field. All toroidal field coils had passed their initial tests and are functioning successfully up to today.
- resource demands - The fully integrated and consistent nature of motor generator, transformers and thyristor rectifier power supplies were provided at a minimal cost.

## *4.6.2 Poloidal Field Coils, Power supplies and Systems Analysis*

### 4.6.2.1 Poloidal Coils

Robustness optimization in design and operation was achieved for the poloidal field coils and their power supplies. The time-averaged coil temperatures were between 50° and 60 ° C. The numbers of turns were chosen so as to obtain convenient voltages for the various DC power supplies with a good copper fraction and good mechanical properties for the coils. The poloidal field coils were loaded by large magnetic forces in the vertical and radial directions. The coils were supported at 8 poloidal planes. The maximum stresses were well within the allowable limits. A problem did occur when some rubber inserts supporting these coils de-polymerised and had to be replaced, some with difficult access.

When a calculation was made of a plasma disruption at full current (1.2 MA), the interaction produced a global torque of 1.56 MNm in the toroidal field coils. This torque was reacted upon the mechanical structure. For reasons of diagnostic access, a design was chosen with tilted struts, similar to that used in other tokamaks. The entire structure was built from Type 304 stainless steel.

A "hybrid" system design option was implemented in which the central ohmic heating solenoid was electrically split at the top and bottom of the coil stack. For a given number of volt-seconds, the top and bottom ohmic coils carry the same ampere-turns as before, in order to guide the flux outside of the plasma region. The current that was programmed to flow in the central OH coil was reduced by half, and an equivalent additional current flowed in the inner shaping windings. This optimized design reduced the average current densities and resistive losses.

#### 4.6.2.2 Power Supplies

Detailed simulations were performed to determine the required currents and voltages required in all of the ohmic and shaping coils. These simulations took account of all the mutual and self inductances and resistances of the electromagnetic coupling between all components of the poloidal field system, including the plasma and vacuum vessel.

The choice of numbers of turns for the shaping coils gave absolute voltages of less than 400 V for the shaping coils and less than 1000 V for the ohmic systems, which was convenient for thyristor systems. The maximum current levels chosen were sufficient to produce the racetrack and some of the non-racetrack equilibria.

The use of twelve-phase rectifiers allowed a low dc-voltage ripple, a clean waveform of the primary current, and a short dead time for control of the rectifiers. The "low-cost" thyristors were widely available. In order to have more rapid control of the power supplies, the motor-generator was wired, at no additional cost, to run at the frequency of 120 Hz rather than 60 Hz at the beginning of the pulse along with the 12 pulse rectifiers. The power supply response time was therefore less than a millisecond, considerably faster than the vacuum vessel skin time.

The transformers provided two different three-phase systems with a relative phase-shift of 30 degrees. The solution using elongated delta-windings at the primary allowed building the transformers in the same way, two by two, with a perfect symmetry.

#### 4.6.2.3 Plasma Disruption Analysis

Disruptions were modelled to determine the induced currents in the shaping coils in the final design. They occurred much faster than the vessel time constant, so the plasma current of 1.2 MA had a very high resistivity imposed. The vacuum vessel was modelled by 20 axisymmetric sections, each of the 16 shaping coils and ohmic coils were present. The time evolution of the current vector I and the voltage vector V was calculated As the induced currents in the vessel decayed and the currents in the shaping coils rose, their currents reached a broad maximum at 10 ms. The current surcharge was easily handled by the thryristor supplies.

#### 4.6.2.4 Fast in-Vessel Coils and Power Supplies for High Frequency Stabilization

Two fast stabilization coils were located inside and near the external corners of the vacuum vessel. They were connected in series opposition so as to produce a horizontal dipole field. The power supply had to provide at least 1.2 MW of power, and the frequency response had to be a few KHz or greater. These requirements were fulfilled without major problems with high power transistors. The same coil design and philosophy was being considered for ITER. The coils in TCV proved to be very effective for the most unstable configurations.

#### 4.6.2.5 Systems and Robustness Analysis

Robustness tradeoffs in design and operation of the ohmic and shaping coils and their power supplies resulted in very high robustness with no obvious fragilities. The performance was optimized to reach design parameters, and large margins were provided over a wide range of plasma shapes and currents.

The systems characteristics were:

- systems control – Multiple simulations were used to determine optimum coil currents and required applied voltages, include PID feedback controllers so the coils could feedback on plasma position and shape deviations. All power supplies had the same maximum ratings, allowing maximum flexibility.
- fault-tolerance – The coils and power supplies were designed to withstand the torques generated by a coil short-circuit. They were designed to be able to withstand a full current plasma disruption.
- modularity – The construction of the coils and power supplies was highly modular leading to easy replacement if needed. During operation, the poloidal field coils effectively functioned independently of the toroidal field coil system. Loading of the motor generator was small compared to its overall capacity.
- decoupling – The thyristor firing in the power supplies produced very low level fluctuations smoothed by coils and vacuum vessel.

Robustness trade-offs in design and operation included:

- robustness – Robustness was achieved by using standard components and power supplies and highly flexible ohmic and shaping coils, each with its own dedicated power supply. The control systems evolved from a hybrid digital-analogue system to a fully digital system as the speed and capabilities of digital systems advanced over the years. .
- fragility – Protection against disruptions of the shaping and ohmic coils and power supply systems was provided by the vacuum vessel and the power supplies.
- resource demands - Given the fully integrated and consistent nature of motor generator, transformers and thyristor rectifier power supplies, high performance could be achieved at a minimum of cost. During operation, the cost of new experiments and upgrades was relatively small.

### 4.6.3 *Flywheel Motor Generator and Systems Analysis*

#### 4.6.3.1 Motor Generator and Transformers

The TCV maximum power of 177 MVA required a motor generator. The individual DC powers of the converters were used to determine the power and energy requirements at the terminals of the rotating machine at 120–80 Hz with the energy extracted of 100 MJ. The motor was wound so as to generate double the normal frequency of 50–60 Hz for power grids so as to double the response frequency of the shaping coils, in conjunction with 12 phase thyristor rectifiers. The motor generation was a turbo type, 4 pole air cooled with a flywheel. It had a nominal power of 220 MVA at 126 Hz, and could drop as low as 96 Hz.

Each ac/dc converter (12 pulse thyristor bridges) had its own transformer whose reactances were chosen to avoid load interactions due to the commutation notches in the primary voltage. The transformers for each coil system had two active parts, elongated delta/star at +15 degrees and − 15 degrees and were connected with interphase reactor and unidirectional for the toroidal field coils, or series connected and bi-directional for the ohmic and shaping coils.

#### 4.6.3.2 System and Robustness Analysis

The motor generator provided a highly flexible power source for all possible upgrades and modifications to TCV. This capability was obtained at a low price through international bidding. The lowest offer was 1/3 of the most expensive bid, yet met the technical specification. Performance was completely reliable, with only one major downtime for standard maintenance made during other machine upgrades. An industry standard solution provided maximum robustness.

### 4.6.4 *Vacuum Vessel and Vacuum Equipment Systems Analysis*

#### 4.6.4.1 Vacuum Vessel

The vessel was constructed from stainless steel 316 L and had a wall thickness of 18 mm and could be heated to 350 ° C. It had a skin time constant of 8 ms and a toroidal resistance of 0.05 mΩ. Stress calculations, using analytic and 3D finite-element methods showed that the vessel could withstand a wide range of load conditions including disruptions.

The vacuum vessel was also designed to allow maximum access by providing a wide range of access ports on the top, bottom and outside of the vessel. The maximum port diameter was 40 cm at the top of the machine to allow personnel access. The many ports made TCV attractive for research. Obliquely angled ports allowed for diagnostic access, and modified port oblique access allowed installation of tangential neutral beam injection. The ports allowed for vacuum pumping and discharge ionization and cleaning, plasma and magnetic diagnostics and auxiliary heating. A complete system of mounting rails were placed inside the vessel that has supported several generations of protection tiles and more recently, in-vessel divertor baffle structures. Limiters, diagnostics and fast response coils were also installed.

#### 4.6.4.2 Systems and Robustness Analysis

- systems control – The vacuum vessel was a key element in plasma stability control by providing instant plasma feedback by image currents free to flow continuously in both poloidal and toroidal directions.
- fault-tolerance – The mechanically strong conducting vessel with its solid construction provided major protection for the rest of the tokamak from plasma disruptions.
- modularity – On short time scales, the vacuum vessel magnetically isolated the plasma from the rest of the tokamak, allowing easier plasma control and operation.
- decoupling – The vacuum vessel smoothes and isolates high frequency oscillations from the rest of the tokamak.

Robustness trade-offs in design and operation:

- robustness – By choosing a solid stainless steel vacuum vessel with only a few structural welds, maximum robustness was achieved against all perturbations. Ring inserts were used at the ports to strengthen the vessel at those locations.
- fragility – The main fragility was that the thick vessel wall and low voltage initiation might not allow a plasma to be formed. As discussed above, plasmas were formed easily as planned and backup solutions were fortunately not required, since backup solutions were not available for some time afterwards.
- performance – By choosing a highly rugged system which allowed baking, instability protection, diagnostics and maintenance access, the vessel achieved maximum performance for all its design criteria.
- resource demands - The simplified construction principles allowed for an inexpensive system and a wide range of possible external procurement.

### *4.6.5 Management of Procurement, Construction and Initial Testing*

Checklist 1.3b recommends that once final designs were complete, detailed management was necessary, including quality control and calls for proposals. This procedure was followed, and it resulted in successful construction which was on budget and only slightly over planned time. The TCV team was expanded to include responsible officers for each major machine system such as power supplies and the vacuum vessel. Calls for proposals were issued world-wide, in accordance with WTO procurement procedures to obtain the best bids possible, consistent with high quality. The engineering choices of concentrating on simple and proven technology ensured there was a wide range of bidders. The important plasma and machine control system was developed in-house (see below).

The details of the final construction and assembly [18] demonstrated how construction proceeded very carefully with a large amount of testing and quality control. There were several minor problems and need for minor redesigns and repairs, but the testing program insured these were easily corrected.

### 4.6.6 Auxiliary Heating Upgrades for TCV

ECRH was a highly effective and efficient heating system that also provided the possibility of current drive effects (ECCD). A new proposal consisted of 9 gyrotrons delivering 4.5 MW for a 1–2 s duration of the TCV pulse, with 3 MW in X2 (second harmonic X-mode) and 1.5 MW in X3 (third harmonic X-mode), corresponding to 83 and 118 GHz at the nominal 1.43 T of TCV operations.

Over the years, TCV was continuously upgraded in capabilities of auxiliary heating, diagnostics, and plasma shaping control. A major upgrade in neutral beam heating [3] involved a 1 MW neutral beam with 25 keV deuterium injection. A second 1 MW beam at 60 keV of deuterium was installed later, but has yet to be fully integrated, since getting that power level through a small port is not trivial.

### 4.6.7 Machine and Plasma Control and Data Acquisition

All of the initial plasma modelling and control simulations provided a firm basis for successful machine start-up, supplying control philosophies and matrices for converting plasma position information into voltage signals. Unlike most tokamaks, the coils were not pre-connected to generate particular shaping fields, leaving full flexibility to the control systems. The plasma position and control system originally implemented [17] operated a complex multi-variable feedback system based on a hybrid digital-analogue matrix computer [19]. TCV control was an in-house development coherent with the extreme machine flexibility envisaged. The 1 kHz reaction time for nearly 200 signals using 1980's technology comprising the plant control and timing distribution was also well ahead of its time and has performance that the TCV team is still finding it hard to replace.

One method for generating the control matrices [7] for active stabilization and control was to superpose four independent feedback systems of coil combinations that generated a radial field (RF), quadrupole field (QF), octopole field (OF), and vertical field (VF). These systems fed back on sums and differences of the flux measurements on the flux loops. The current vectors for each of the control systems were determined by selection of coils near the control-flux loops and added to pre-programmed currents to obtain a particular shape or shape evolution. The inductance matrix between these coil groups and the control-flux loops was inverted to obtain the control-current vectors.

The control of machine operation data collection was performed by a distributed network of computers which controlled CAMAC crates for the data acquisition and microprocessors which in turn controlled all subsystems in real time.

There were such rapid advances in electronics and computers during TCV operation that control systems evolved rapidly as discussed in the scenarios below.

## 4.7 TCV Scenarios

A few of the most important TCV scenarios, based on summary reference and topical papers, are chosen below to contribute to the tokamak design space in Chap. 2 and to illustrate the very wide range of operation and experiments. Together, they demonstrate how the base design of TCV was sufficiently strong to be able to achieve all of its initial design parameters and much more, especially since TCV became supported at the European level as a EUROfusion medium-sized tokamak with hundred of collaborators.

### *4.7.1 High Elongation Plasmas and Ohmic H-Modes*

High vertical elongation allows high plasma current and ohmic heating, which can give access to high confinement regimes. The first plasma current in TCV was obtained at the end of 1992 and the first results reported in 1995 [20]. During its first year of operation, TCV achieved a wide variety of limiter and divertor plasma shapes and reached 0.81 MA plasma current and elongation over 2.0. Ohmic H-modes were regularly produced with a maximum confinement time of 80 ms and a maximum normalized beta of 1.9.

### *4.7.2 Fully Digital Plasma Control*

The speed of modern digital electronics allowed a comprehensive real-time analysis and control system to be implemented. A fully digital controller system, apart from the fast in-vessel coils, was developed [21] and also applied to large tokamaks [22]. The TCV implementation [23] involved a high order multivariable controller using a fully digital processor. The fast in-vessel control systems is still driven by analogue matrix control. Most recently, a DeepMind single artificial intelligence neural network [24] has been used to control all the plasma shaping coils, producing several different configurations.

### 4.7.3 Fully Non-inductive Steady-State Plasmas with Electron Cyclotron Current Drive

Steady state operation becomes possible with auxiliary heating and current drive. A steady-state, fully non-inductive plasma current [25] of 0.123 MA lasting 2 seconds was sustained for the first time in a tokamak using electron cyclotron current drive only. Careful distribution across the plasma minor radius of the power deposited from three 0.5-MW gyrotrons was essential for reaching steady-state conditions. With central current drive, up to 0.153 MA of current was fully replaced transiently for 100 ms. It was even possible to partially recharge the ohmic transformer.

### 4.7.4 Limits of Operating Space for High Plasma Elongation

Vertical axisymmetic instabilities place a limit on the maximum possible elongation of plasmas. An experimental campaign extended the limits of the operating space [26]. The vertical position control system had been optimized with extensive model calculations to operate at the lowest possible stability margin. The growth rate of the axisymmetric instability was minimized by optimizing plasma triangularity and squareness and by operating close to the $n = 1$ external kink mode. Record vertical elongation $= 2.8$ and normalized current $I_N = 3.6$ were achieved.

### 4.7.5 High Power Electron Cyclotron Heating and Bootstrap Current Drive

Bootstrap currents are induced by high plasma pressure and can give a significant contribution to the total plasma current. High-power (4.5 MW) real-time controllable EC systems combined with flexible plasma shaping [27] allowed steady-state discharges to be achieved in which the current was entirely self-generated through the bootstrap mechanism, a fundamental ingredient for ITER steady-state operation. The discharge remained quiescent over several current redistribution times.

### 4.7.6 Electron Internal Transport Barriers with EC Current Drive

Internal transport barriers can significantly improve plasma energy confinement. Electron internal transport barrier regimes sustained by EC current drive were produced and explored [27]. MHD activity was crucial in scenarios characterized by large and slow oscillations in plasma confinement. The observed dependences of the electron thermal diffusivity on triangularity and collisionality were reproduced by non-linear gyro-kinetic simulations and governed by Trapped Electron Mode (TEM) turbulence.

### 4.7.7 H-mode Regimes with Ohmic Heating in H, D and He

In plasmas with high current densities and heating, transitions to high confinement regimes can occur. The transition into high confinement H-mode regime [28] was studied with hydrogen (h), deuterium (d), and helium (He) as majority ion species in identical plasma configurations with a single-null X-point. L–H transitions were obtained with ohmic heating alone. For d and He, the measured power levels were as expected. In the case of h, transitions into H-mode were observed at power levels of about 80% of the expected threshold power, implying a dependence on the electron temperature at the separatrix.

### 4.7.8 Neutral Beam Current Drive and Heating

Directed neutral beam injection can induce currents in the plasma. At 1 MW of tangential neutral beam injector [29] ion temperatures reached 2.5 keV. Toroidal rotation of 250 km/s and neutral-beam current drive were observed.

### 4.7.9 Multi-machine ITER Scenario to Maximize Edge Pedestal Height

ITER-relevant scenario development was performed on TCV, JET and MAST [30]. The key element was initial L-mode operation in a magnetic configuration with a high power threshold for the L–H transition. Once the core pressure and pedestal height reached saturation, the other X-point was activated and triggered an L–H transition. The ballooning stabilization resulted in a lower ELM frequency, which in turn acted to increase the pedestal pressure further. A higher stored energy was then observed which lasted through several ELM cycles.

### 4.7.10 *Exhaust Control and Detachment in Snowflake and Super-X Divertors*

Experiments on exhaust physics and control [29] studied divertor flow control and detachment in various configurations including the super-X divertor and the snowflake divertor, shown in Fig. 4.4.

A conventional single null (SN) was investigated with varying poloidal or total flux expansion shown in Fig. 4.4(a)–(c). Poloidal flux flaring near the target resulted in a configuration termed the X divertor. When total flux expansion was achieved by moving the target to a larger major radius, a super-X divertor was obtained. In the super-X divertor in Fig. 4.4(d), a decrease in parallel heat flux was confirmed. In the snowflake in Fig. 4.4(e), a meaningful connection-length increase was achieved. Nitrogen seeding was applied and it enhanced impurity radiation in the region between the two X-points.

### 4.7.11 *Vessel Wall Conditioning with ECRH*

Superconducting tokamaks need to develop vessel cleaning techniques other than by pulsed voltage discharges. Effective wall conditioning was achieved with second harmonic ECRH in He [31] in support of the superconducting JT-60SA. Poloidal fields were applied and tuned to maximize the discharge homogeneity and wall coverage. Conditioning was verified by successful deuterium plasma breakdown.

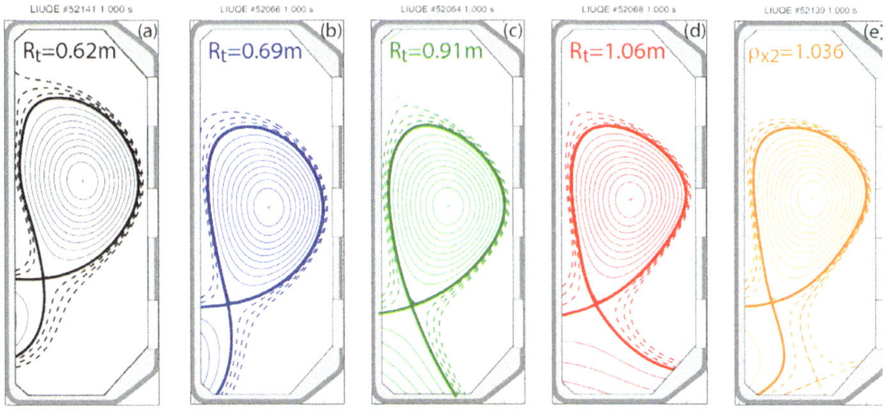


**Fig. 4.4** Flux surfaces for divertor configurations in TCV varying major radius of strike point. (From [29], Fig. 4, permission from EURATOM)

### 4.7.12 Real Time Plasma Control Systems

The distributed digital system was constantly evolving to implement real time plasma control [32]. The basis of several control schemes was the real-time, sub-ms equilibrium reconstruction code RTLIUQE, which was employed as a generalized position and shape controller. The PI controller relied on a singular-value decomposition approach to limit the controlled parameters. A time-invariant version was successfully tested on a variety of plasma shapes.

### 4.7.13 Disruption Avoidance by Real-Time Locked Mode Prevention with ECCD

Disruption avoidance by real-time locked mode prevention or unlocking with electron-cyclotron current drive (ECCD) was thoroughly documented [33] using magnetic and radiation triggers. The focus was on disruptions caused by abnormal impurity inflow, precipitating a disruptive event. The impurity inflow was simulated by a massive, controlled injection of a noble gas. The application of ECCD on the $q = 2$ surface prevented mode locking altogether or could unlock and stabilize the mode, still preventing the disruption.

### 4.7.14 Elimination of Disruption Runaway Electron Current with Current Ramp Down

Runaway electron beams generated by disruptions need to be controlled to avoid machine damage. A controlled ramp-down was used to mitigate a runaway electron (RE) beams [34] generated by noble-gas injection to create a disruption. Upon detection of the current quench, a hybrid fast controller took over to produce a current ramp-down via the ohmic transformer. Assistance was provided by Massive Gas Injection (MGI) to limit the RE beam's energy. The ramp-down rate was kept below a threshold to avoid the appearance of MHD instabilities. A total conversion of RE current into thermal plasma current was observed, thereby mitigating the runaway's effects.

### 4.7.15 High-$\beta N$ Fully Noninductive Scenarios with Combined EC and NBI

High plasma pressure and the ratio $\beta$ are important fusion goals which can be achieved by auxiliary heating and current drive. The 1 MW neutral beam injector was used to study nearly noninductive H-mode discharges sustained by electron cyclotron and neutral beam current drive [35]. Values for $\beta_N$ up to 1.4 in L-mode and 1.7 in H mode were obtained in plasmas with zero loop voltage.

### 4.7.16 Reduced Heat Flux with Grassy ELM's at High Triangularity

Adjusting the properties of ELMS provides a means of regulating power flow into the divertor. Grassy ELMs were obtained at high triangularity and steady fuelling, replacing the type-I ELMs at a triangularity $\delta = 0.54$ [36]. The auxiliary heating was 1 MW of NBI plus 0.75 MW X3 ECRH into a double-null shape. The heat flux in the grassy-ELM regime was reduced tenfold with respect to type-I ELMs.

### 4.7.17 High-Shared-Flux Doublet Configuration with Separate Lobe ECRH

Doublet shaped plasmas could have important confinement advantages, but Doublet III only succeeded in making two circular plasmas connected by a small amount of shared flux, which did not alter the plasma properties. A doublet plasma featuring an internal X-point and a high fraction of shared flux between the two lobes was achieved in TCV [1]. The details of plasma breakdown and initiation in both single and double core plasmas were investigated [37]. A doublet was maintained for 30 ms, with a current of 0.27 MA and peak electron temperature 1.3 keV in both lobes. A transport barrier was probably established in the negative-shear mantle just outside the internal separatrix. Power and deposition scans also showed that the scenario was surprisingly robust against coalescence.

### 4.7.18 Removable Gas Baffles Separating Main and Divertor Chambers

Gas baffles in the divertor region can control the gas pressure both inside and outside the divertor. Removable gas baffles that separated the vessel into main and divertor chambers were installed and operated [38]. The baffles increased the divertor neutral pressure as relevant to ITER operation. Ohmically heated diverted discharges confirmed code predictions with up to a five times increase in divertor neutral pressure. Increasing this pressure in the baffled divertor facilitated access to detachment. Experiments also demonstrated that the onset of detachment was determined by divertor neutral density rather than electron density.

## 4.8 Lessons Learned About Systems Forward Engineering

TCV is the first example in this book of forward engineering using a systems approach, based on Checklists 1.3A and 1.3B, and it illustrates the advantages of a staged approach to machine design at different levels of complexity, including:

- Getting started with preliminary goals and scoping major options
- Review and prioritize goals and evaluate variants
- Produce definitive proposals and engineering designs
- Develop proposals at different stages of complexity
- Proceed to final engineering design and construction after funding approval
- Continuously plan for additional possible funding and upgrades

The history of TCV shows how the original goals and designs of the machine evolved during the scoping process to focus on the most interesting avenues to explore. The fundamental design of the machine was meant to be as flexible as possible, yet at the same time to be optimized to explore new physics regimes. Design choices were also made to minimize the cost of a high performance machine while at the same time maximizing the robustness and taking measures to mitigate several apparent fragilities.

The complexity of TCV is far less than a fusion reactor and considerably less than more recent superconducting coil machines, but its forward engineering demonstrates the general principles that can be used in successful and flexible design of an experimental tokamak.

## References


1. S. Coda and the TCV Team, Physics research on the TCV tokamak facility: from conventional to alternative scenarios and beyond. Nucl. Fusion **59**, 112023 (2019)
2. F. Piras, Extremely Shaped Plasmas to Improve the Tokamak Concept, Doctoral thesis No.5015, Ecole Polytechnique Federale de Lausanne, Switzerland (2011)
3. A. Fasoli et al., An upgraded TCV for tokamak physics in view of ITER and DEMO, 40th EPS Conference on Plasma Physics, P2.104 (2013)
4. A.D. Cheetham et al., The TCA tokamak project report 1979, report LRP 162/80. Ecole Polytechnique Federale de Lausanne, Lausanne: CRPP (1980)
5. F. Hoffman, L. Bighel and J.M. Peiry, Belt pinch experiments, 7th European Conference on Controlled Fusion and Plasma Physics, Vol. II, 137 (1975)
6. F. Hoffman, F.B. Marcus et al., The TCV tokamak proposal phase 1, CRPP-EPFL report LRP 240/84, Lausanne (1985)
7. F.B. Marcus, S.C. Jardin, F. Hofmann, Controlled evolution of highly elongated tokamak plasmas. Phys. Rev. Lett. **55**(21), 2289 (1985)
8. F. Hofmann, F.B. Marcus, A.D. Turnbull, Axisymmetric stability of highly-elongated tokamak plasma – a new method and initial results. Plasma Phys. Control. Fusion **28**(4), 705 (1986)
9. F. Troyon et al., MHD limits to plasma confinement. Nucl. Fusion **26**(1A), 209 (1984)
10. F. Hofmann, A.D. Turnbull, F.B. Marcus, Passive stabilization of axisymmetric modes in an elongated Tokamak. Nucl. Fusion **27**(5), 743 (1987)

11. F. Hofmann, F.B. Marcus, MHD equilibrium calculations of highly-elongated racetracks in a tokamak, report LRP 235/84, CRPP, Ecole Polytechnique Federal de Lausanne (1984)
12. F. Hofmann, F.B. Marcus, Calculation of matrices for shaping and control of elongated plasmas, report INT 147/88, CRPP-Ecole Polytechnique Federale de Lausanne (1988)
13. F.B. Marcus et al., Simulations of control, pressure perturbations, displacements and disruption in highly elongated tokamak plasmas. Nucl. Fusion **30**(8), 1511 (1990)
14. F.B. Marcus et al., Calculation of induced voltages, currents, and forces due to plasma disruptions in the TCV tokamak, using the TSC code, report INT 150/88, CRPP-Ecole Polytechnique de Lausanne (1988)
15. F.B. Marcus, Ch. Hollenstein, Estimation of hard X-ray production from runaway electrons in TCV, report INT 138/88, CRPP-Ecole Polytechnique Federale de Lausanne (1988)
16. D.R. Whaley et al., X- and O-mode electron cyclotron heating breakdown and startup in TCA. Nucl. Fusion **32**(5), 757 (1992)
17. G. Tonetti et al., Design of the TCV tokamak, 16th symposium on fusion technology, London, Great Britain, September 1990, report LRP 413/90, CRPP-Ecole Polytechnique Federale de Lausanne (1990)
18. Fusion technology report, CRPP-Ecole Polytechnique Federale de Lausanne, Report 1991–1992 PSI Fusion technology group (1992). https://inis.iaea.org/collection/NCLCollectionStore/_Public/26/003/26003796.pdf. Last accessed 17 June 2022
19. P.F. Isoz, J.B. Lister, P. Marmillod, A hybrid matrix multiplier for control of the TCV tokamak, 16th symposium on fusion technology, London, Great Britain, September 1990, report LRP 413/90, CRPP-Ecole Polytechnique Federale de Lausanne (1990)
20. J.B. Lister et al., Variable Configuration Plasmas in TCV, Paper IAEA-CN-60/A5–2, pp. 627–632, Plasma Physics and controlled Nuclear Fusion Research 1994, 15th Conference Proceedings, Seville, Spain, 26 September – 1 October 1994, IAEA, Vienna (1995)
21. J.B. Lister et al., The control of tokamak configuration variable (TCV) plasmas. Fusion Technol. **32**, 323 (1997)
22. J.B. Lister, Y. Martin, The control of modern tokamaks, International Conference on Accelerator and Large Experimental Physics Control Systems, 235, Trieste, Italy (1999)
23. M. Ariola et al., A modern plasma controller tested on the TCV Tokamak. Fusion Technol. **36** (1999). https://doi.org/10.13182/FST99-A97
24. M. Sparkes, DeepMind uses AI to control plasma inside tokamak fusion reactor, New Scientist. https://www.newscientist.com/article/2308243-deepmind-uses-ai-to-control-plasma-inside-tokamak-fusion-reactor/?utm_source=nsday&utm_medium=email&utm_campaign=NSDAY_170222. 16 February (2022).
25. O. Sauter et al., Steady-state fully noninductive current driven by electron cyclotron waves in a magnetically confined plasma. Phys. Rev. Lett. **84**, 3322 (2000)
26. F. Hofmann et al., Extension of the TCV operating space towards higher elongation and higher normalized current. Nucl. Fusion **42**(6) (2002)
27. A. Fasoli et al., Overview of physics research on the TCV tokamak. Nucl. Fusion **49**, 104005 (2009)
28. R. Behn et al., Threshold power for the transition into H-mode for H, D and He plasmas in TCV. Plasma Phys. Control. Fusion **57**, 025007 (2015)
29. S. Coda et al., Overview of the TCV tokamak program: scientific progress and facility upgrades. Nucl. Fusion **57**, 102011 (2017)
30. I. Chapman et al., Joint experiments tailoring the plasma evolution to maximize pedestal performance, Nuclear Fusion IAEA Fusion Energy Conference, Kyoto (2016)
31. D. Douai et al., Development of helium electron cyclotron wall conditioning on TCV for the operation of JT-60SA. Nucl. Fusion **58**, 026018 (2018)
32. H. Anand et al., Distributed digital real-time control system for the TCV tokamak and its applications. Nucl. Fusion **57**, 056005 (2017)
33. U.A. Sheikh et al., Disruption avoidance through the prevention of NTM destabilization in TCV. Nucl. Fusion **58**, 106026 (2018)

34. D. Carnevale et al., Runaway electron beam control. Plasma Phys. Control. Fusion **61**, 014036 (2019)
35. C. Piron et al., Extension of the operating space of high-βN fully noninductive scenarios on TCV using neutral beam injection. Nucl. Fusion **59**, 096012 (2019)
36. B. Labit et al., Dependence on plasma shape and plasma fuelling for small edge-localized mode regimes in TCV and ASDEX upgrade. Nucl. Fusion **59**, 086020 (2019)
37. B.P. Duval et al., Singlet breakdown optimization to a doublet plasma configuration on the TCV tokamak, preprint: IAEA fusion energy conference Gandhinagar, India, EX/P1–6 (2018)
38. H. Reimerdes et al., Overview of the TCV tokamak experimental programme, 28th IAEA Fusion Energy Conference, Nice, France (Virtual) 10–15 May 2021 Presentation OV/4–2, Paper IAEA–CN–286-1095, Submitted to Nuclear Fusion (2021).

# Chapter 5
# JET – World's Largest Tokamak and its d-t Fusion Experiments Plus TFTR's

**Abstract** As the world's largest operating tokamak, JET provides a bridge between tokamak experiments and reactors, having demonstrated over 16 MW of fusion power and 59 MJ of energy, power breakeven, plasma current of 7 MA, tritium processing and remote handling. A case study of JET's construction and operation is analysed and evaluated by systems reverse engineering. JET was highly robust and adaptable. A second case study by systems forward engineering is made of the design and operation of its first high power deuterium and tritium experiments. The detailed planning process allowed many successful fusion experiments. A brief summary is also presented of the TFTR tritium experiments.

## 5.1 Systems Analysis Strategy for JET

The Joint European Torus (JET) is the world's largest and highest plasma current tokamak. It holds the record for fusion power of 16 MW generated during the first Deuterium Tritium Experiment (DTE1), and 59 MJ energy during DTE2 in 2022.

This chapter makes a case study which uses Checklist 1.2 from Chap. 1 for systems reverse engineering and evaluation of JET over its entire history of construction and operation. Using Checklist 1.4 for systems forward engineering, a second case study is made of the design and operation of the deuterium-tritium experiments on JET. A summary is also presented of TFTR tritium experiments.

The first items of Checklist 1.2 concern the need to identify form and function and system architecture:



F. B. Marcus, *Systems Approaches to Nuclear Fusion Reactors*, Springer Series in Plasma Science and Technology, https://doi.org/10.1007/978-3-031-17711-8_5

## 5.2 JET Goals, Machine Parameters and Systems Integration

### 5.2.1 *JET Design and Systems Architecting of Highest Level Systems*

Until ITER construction began, the Joint European Torus (JET) was the largest project of the nuclear fusion research programme of Euratom. Paul Rebut was the guiding light behind the JET design following on from TFR (Tokamak Fontenay aux Roses) in France. JET was constructed on time and within budget and met its original objectives [1]. The four main goals were:

- investigating plasma processes and scaling laws at plasma dimensions and parameters approaching those necessary for a fusion reactor;
- examining and controlling plasma wall interactions and impurity influxes in near reactor conditions;
- demonstrating effective heating techniques (particularly, Radio Frequency and Neutral Beam Heating), capable of approaching reactor temperatures in JET, in the presence of the prevailing loss processes;
- studying alpha particle production, confinement and plasma interaction and heating produced as a result of fusion between deuterium and tritium.

The use of tritium and the application of remote handling techniques were particularly reactor-relevant. The JET design parameters and values are shown in Table 5.1. JET is shown schematically in Fig. 5.1, with particular emphasis on the wide range of plasma diagnostics, including a large number of neutron and fast particle diagnostics.

**Table 5.1** JET design parameters

| JET design parameters | Value |
|---|---|
| Plasma minor radius a (horizontally) | 1.25 m |
| Plasma minor radius b (vertically) | 2.10 m |
| Plasma major radius, $R_0$ | 2.96 m |
| Plasma aspect ratio, $R_0/a$ | 2.37 |
| Plasma elongation ratio, $\kappa = b/a$ | 1.68 |
| Flat top pulse length | 25 s |
| Toroidal magnetic field (plasma Centre) | 3.45 T |
| Plasma current, D shaped plasma | 7.0 MA |
| Volt seconds available | 42 Vs |
| Toroidal field peak power | 380 MW |
| Poloidal field peak power | 300 MW |
| Additional heating power (into torus) | 50 MW |
| Weight of vacuum vessel | 108 t |
| Weight of toroidal field coils | 364 t |
| Weight of iron core | 2800 t |

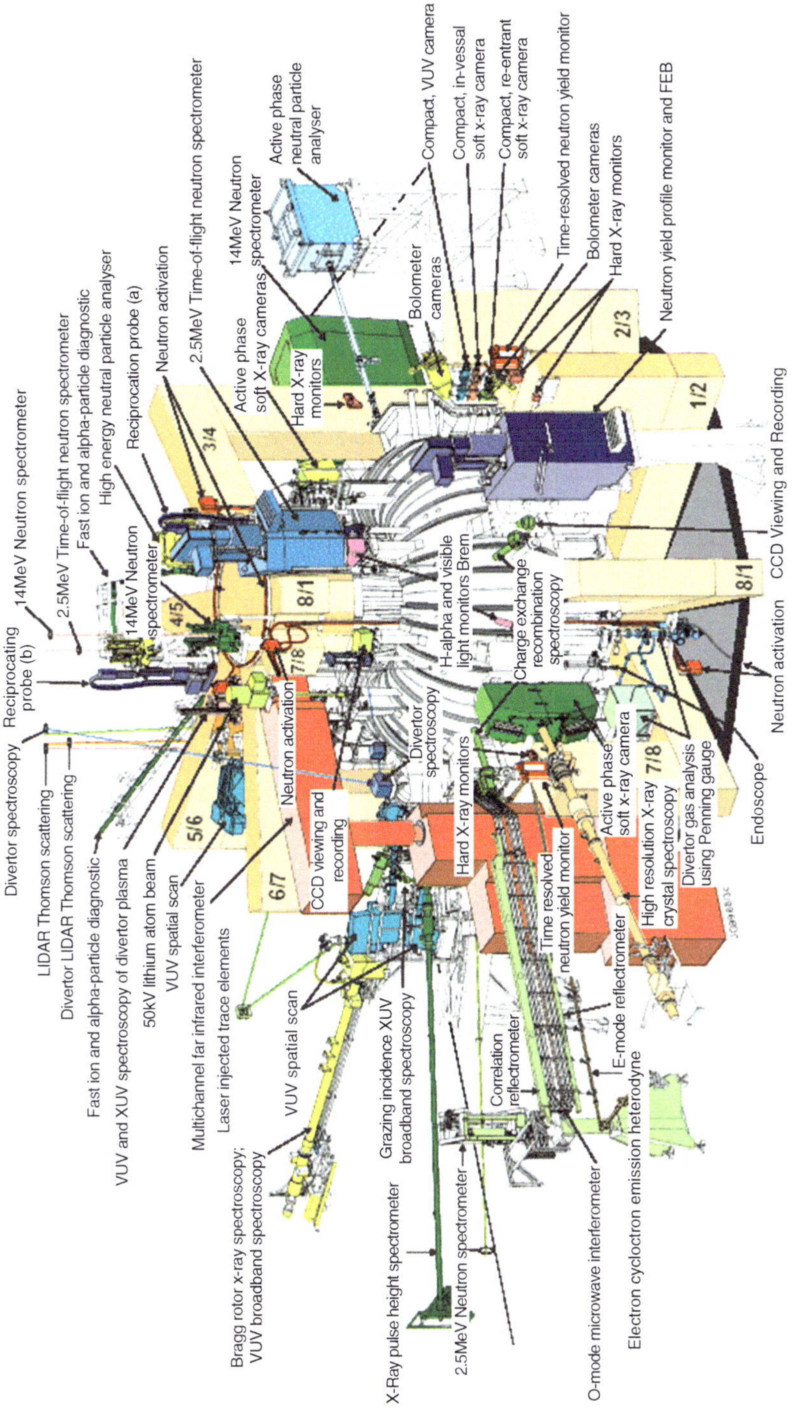


**Fig. 5.1** JET load assembly and diagnostics. (From [2], Fig. 1)

JET has overall dimensions of about 15 m in diameter and 12 m in height. At the heart of the machine was a toroidal vacuum vessel of major radius 2.96 m with a D-shaped cross-section 2.5 m wide by 4.2 m high, and a very low aspect ratio (major to minor plasma radius) of 2.37. The plasma current was produced by transformer action using a massive eight-limbed magnetic circuit. A set of coils around the centre limb of the magnetic circuit formed the primary winding of the transformer with the plasma acting as the single turn secondary. Additional heating of the plasma was provided by radio frequency waves and neutral beams. The toroidal field was provided by 32 D-shaped coils and delivered a full design field at the plasma centre of 3.45 T. The continuously wound poloidal coils were positioned around the outside of the toroidal field coils.

## *5.2.2 Elements of Next Level Systems, Form Relationships and Design Choices*

### 5.2.2.1 Poloidal Coils and Iron Core Transformer Architecture

Since JET was planned to operate near the reactor regime, systems level choices were made consistently with this ambition. In a fusion reactor the coils would need to be made of superconductors and continuously wound. For construction optimization and because of the large size and high field operation, the toroidal coils and poloidal coils in JET were continuously wound and not demountable. The poloidal ohmic and shaping coils had to be located outside of the toroidal field coils.

The poloidal coils system with a divertor installed inside the vacuum vessel is shown in Fig. 5.2, where the plasma boundary of a discharge is shown in blue. The divertor was not part of the original design, but was added later and assembled inside the vessel. Its space reduced the maximum plasma current from 7 to 5 MA.

The ohmic heating coil system included a large iron core and separate return limbs transformer, so the coil system needed to be lowered over this iron core at the centre of the torus. JET's volt-second flux requirement was greater that the 1.5 T saturation limit of iron, so the volt seconds were supplemented by ohmic heating coils creating additional flux that supersaturated the iron core. The expanding flux was returned through outer non-saturated iron transformer limbs, hence reducing stray flux on outside diagnostics and auxiliary heating systems.

The functional interactions at the highest systems level are electromagnetic relationships similar to those shown in Table 2.1 of Chap. 2, with additional interactions with the iron core. The return legs of the iron core exert vertical attraction on the plasma, adding a destabilizing force needing correction by coil currents.

The machine was therefore assembled by creating octants of four toroidal field coils each around four welded-together segments of vacuum vessel. All these octants were moved into place and supported by structural elements, with lower poloidal coils already in place. All the vacuum vessel segments were then welded together.

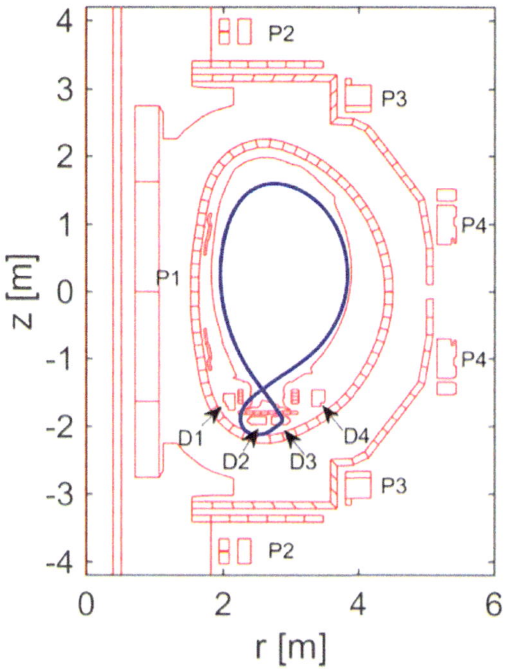


**Fig. 5.2** JET Poloidal coils system and MHD equilibrium with internal divertor coils. (From [3], Fig. 1)

The upper poloidal field coils were then lowered on to the assembly and finally the outer iron core limbs were added. The poloidal coils allow large vacuum vessel ports to permit maximum access to the plasma region.

#### 5.2.2.2 Toroidal Field Coils

The toroidal field coils were designed to achieve the maximum field possible in a large device with low aspect ratio R/a and with water-cooled copper coils. A D-shape was chosen for the low aspect ratio toroidal field coils because it minimized stress in the coils. It was also decided to make the vacuum vessel and plasma the same shape as the coils, to get the best use of the toroidal field and to obtain a higher current in the D-shaped plasma.

#### 5.2.2.3 Vacuum Vessel

The vacuum vessel had a very complicated construction designed for high temperature heating to achieve a high vacuum and high electrical resistivity It allowed a high one-turn plasma breakdown voltage and was structurally strong against plasma disruptions [4]. The vessel was designed as an all metal composite torus of non-circular cross-section. It had a completely welded fabrication of 32 rigid and wedge-shaped sections (equal in number to the toroidal field coils). They were joined together by parallel bellows to form a torus. The bellows linking these rigid sectors determined the electrical resistance of the metallic vacuum vessel the long way around the torus. All forces acting on the vacuum vessel were absorbed by the rigid sectors which also incorporated the ports into the interior of the machine for pumping, diagnostics and plasma heating, etc.

The bellows linking the rigid sectors were welded as inner and outer pairs allowing hot inert gas to raise the temperature of the vacuum vessel to 500 °C for bakeout. The outside of the vacuum vessel was thermally insulated. Inconel 600 was used for the rigid sectors and high resistive Inconel 625 for the bellows.

#### 5.2.2.4 Radiation Shielding from 14 MeV Neutrons

The design goal of megawatts of deuterium-tritium (d-t) fusion power for many seconds had major implications. The entire tokamak structure was surrounded by 2.5 m thick radiation-resistant concrete walls. A separate plant needed to be constructed to contain and reprocess radioactive tritium injected into and recovered from the JET vacuum vessel and the JET neutral beam injection systems. The torus building needed to be made secure against any tritium leakages by maintaining a slight negative pressure. All systems needed to be designed with the possibility of high radiation fluxes occurring. Special diagnostics needed to be developed to measure the 2.45 MeV neutrons from d-d fusion, 14.1 MeV neutrons from d-t fusion, multi-MeV gamma rays and fast alpha particles.

## 5.3 Operational and Experimental Evidence of Success in Machine Operation

### *5.3.1 Initial JET Operation Pre-1989*

The first plasma in JET was obtained in 1984. By 1989 the experimental programme in x-point configuration [5] had been mainly devoted to the study of JET H-modes with high power additional heating. Plasma purity was improved by the use of beryllium gettering. Central values of the ratio of deuterium to electron densities were in the range 0.7–0.9.

#### 5.3.1.1 High Triple Product H-mode Operation at 4.5 MA

Single and double null diverted plasmas were formed without additional coils inside the vessel [5]. In the double-null configuration, it was possible to achieve H-modes with ICRF alone. The global confinement time scaled with power with both ICRF and NBI heating in the range 4–20 MW. Global confinement time did not degrade appreciably by decreasing $q_{95}$. Here $q_{95}$ is the safety factor value q at the flux surface which enclosed 95% of the magnetic flux.

High values of the triple product $n_D \tau_E T_i = 8 \times 10^{20}\ m^{-3}$ keVs and $Q_{d\text{-}d} = 2.5 \times 10^{-3}$ were achieved in deuterium plasmas with deuterium neutral beams in H-mode discharges at 4–4.5 MA. At low density ($3\text{–}4 \times 10^{19}\ m^{-3}$), high values of on-axis $T_i$ of 20–27 keV were produced in the hot-ion mode.

#### 5.3.1.2 High Power Neutral Beam Heating

With neutral beam heating power in excess of 10 MW, the termination of the high performance H-mode was suddenly created by an influx of carbon caused by overheating of the graphite tiles. By sweeping the position of the X-point radially and vertically, it was possible to extend the duration of the H-mode by 50%. The effect of strong gas puffing during the H-mode was to reduce radiation losses and $Z_{eff}$ to achieve an ELM-free phase for longer than 5 s. Very high density regimes were produced by repetitive pellet injection in the H-mode ($n = 3.8 \times 10^{20}\ m^{-3}$). These experiments prepared the way for d-t fusion experiments in JET during the decade 1989–1998.

### *5.3.2 High Performance Plasmas in Preparation for Tritium Experiments*

Important preparations needed to be made to get ready for tritium experiments [6]. Scenarios had to be fully prototyped.

#### 5.3.2.1 Hot-ion H-mode at 5 MA Plasma Current

H-mode operation in JET single-null X-point configuration was extended to plasma currents of 5 MA, toroidal fields of 3.4 T and neutral beam powers of 20 MW with 80 keV and 160 keV deuterium injection. A characteristic of JET H-mode discharges was the absence or very low level of ELM (Edge localized mode) activity. There was a continuous rise in plasma density and a similar increase in bulk plasma radiation, which finally terminated the H-phase after up to 4 s of heating.

The global energy confinement time $\tau_E$ of these H-mode discharges was a factor of 2–3 larger than in comparable limiter discharges. The confinement increased linearly with plasma current and to a lesser extent with toroidal magnetic field but decreased with increasing neutral beam heating power. Part of this degradation could be attributed to poor beam penetration at the higher densities.

Local transport analysis using time-dependent codes indicated that the main contribution to the enhancement came from the plasma edge as manifested by the steep temperature and density gradients (pedestals) observed in this region. A fusion gain factor close to $Q = 1$ was predicted for tritium operation in JET.

#### 5.3.2.2 High Fusion Triple Product $n_D(0)\tau_E T_i(0)$ in Hot-ion H-mode

Plasmas with a fusion triple product $n_D(0)\tau_E T_i(0)$ of $(8\text{–}9) \times 10^{20}\ m^{-3}$ s keV were achieved [7], albeit under transient conditions, which was within a factor of 8 of that required in a fusion reactor. The parameters of a reactor plasma were individually achieved. These results were made possible by a substantial improvements in plasma purity and corresponding reductions in plasma dilution resulting from the use of beryllium as the first wall material facing the hot plasma.

#### 5.3.2.3 Beta Limits with Be Limiters and Gettering in H-modes

High density regimes and beta limits were explored in JET [8]. High-$\beta$ experiments were performed in low toroidal field Double-Null (DN) X-point configurations with Be-gettered carbon target plates. The Troyon limit on $\beta_T$ was reached during double null H-mode discharges. $\beta$-values of 5.5% were obtained at low toroidal fields ($B < 1.2$ T) with additional heating of 10 MW. At high beta, sawteeth were modified and characterised by very rapid heat waves and fishbone-like pre- and post-cursors with strong ballooning characteristics.

#### 5.3.2.4 Possible Confinement Degradation from Fishbone Instabilities

Instabilities [9] appeared during auxiliary heating experiments in JET as periodic bursts of MHD oscillations resembling 'fishbones', which were observed in the signals of several diagnostics. The bursts had repetition times of 10–40 ms and oscillation frequencies ranging from 1 kHz to more than 20 kHz. While the amplitude of these oscillations increased at high poloidal and toroidal beta, they were also observed in plasmas with modest values of beta. They seemed to be due to an instability driven by energetic ions, but were without significant impact on fast ion containment or energy confinement.

#### 5.3.2.5 High Performance with Plasma Current at 7 MA on Be Limiters

Improvements to the performance of JET limiter plasmas [10] were obtained by increasing the plasma current, suppressing sawteeth, peaking the density profile, using beryllium evaporation and employing beryllium limiters. Central electron temperatures $T_e \sim 12$ keV, ion temperatures $T_i \sim 18$ keV and central electron densities $n_{e0} \sim 2.3 \times 10^{20}$ $m^{-3}$ were obtained in limiter plasmas at 3–5 MA. A 7 MA discharge with $q \sim 3.2$ was also demonstrated.

#### 5.3.2.6 High Confinement ELM-Free H-mode and Impurity Generation

Detailed studies were made of the high confinement H-mode [11]:

- The ELM free H-mode was a transient plasma regime.
- The H-mode was marginally limiting on the X point dump plates.
- The global energy confinement time showed an improvement of a factor of 2 over the limiter L mode.
- The radiation collapse which terminated the H-mode determined density limits.
- Improvements were achieved with beryllium gettering of graphite tiles.
- ICRF heating could achieve transient stabilization of sawteeth during H-mode.
- With NBI, a hot ion plasma achieved the highest fusion reactivities.

### *5.3.3 Analysis of Emergent Properties of Systems and Robustness Against Faults*

Given its excellent performance by 1990, JET was already a success. It is appropriate to continue with Checklist 1.2 by evaluating robustness, performance and new knowledge.

#### 5.3.3.1 Major Fragility – Short Circuits in Three Toroidal Field Coils

One important upgrade and one significant repair to JET were carried out during the 1992–1993 shutdown to install a pumped divertor to improve performance.

The fragility [12] was that three of the toroidal field coils developed inter-turn faults due to erosion leaks of the water coolant. These faults, though major, were not catastrophic. It was possible to replace the faulty coils with spare coils available on site in just a few months as part of the scheduled shutdown. The coolant was changed from water to an insulating fluid, and measurements did not detect any new faults on the coils.

#### 5.3.3.2 Robustness Analysis

The overall JET robustness in design and operation after these upgrades is evaluated here.

- systems control – JET was operated by the COntrol and Data Acquisition System (CODAS) [13], a large distributed real-time process control system. The operation was centralized through a fully programmable man-machine interface, based on CAMAC instrumentation and Eurocard signal conditioning. A network of minicomputers was linked to a host mainframe. The plasma shaping system was very flexible, since multiple shaping coil connection configurations were possible by changing cabling before an operating session. The range of shapes included circular, shaped and diverted x-point plasmas. The system was highly reliable.
- fault-tolerance – The vacuum vessel was constructed to withstand very high current disruptions. The machine design in general allowed components to be replaced after faults by including robotic remote handling capabilities.
- modularity – Many of the components were constructed with spares. The installation of internal divertor coils reduced this flexibility.
- decoupling – The continuous vacuum vessel screened the machine.

Robustness trade-offs in design and operation involved:

- robustness – All systems were designed to be highly robust, with careful engineering and pre-installation tests and extremely detailed quality control.
- fragility – At maximum design operation, immense forces were present, especially if a fault was present in a key structural component.
- performance – JET succeeded in reaching all of its goals. Reactor-relevant technologies were demonstrated like remote handling and tritium processing.
- resource demands – Motor generators and the UK grid provided power resources. The project operated within its time and budget constraints.

In general, JET proved to be a highly robust machine, easily upgradable and easy to repair.

#### 5.3.3.3 Overview of JET Performance Metrics

A Summary of JET's Fusion Relevant Performance [14] was provided. A new programme was planned with a pumped divertor system installed inside the JET vacuum chamber. JET plasmas approached those of a reactor plasma:

- Line averaged densities of $1.4 \times 10^{20}\ m^{-3}$ were obtained.
- $\beta$ values up to 5.5% were produced, equal to the Troyon limit at $B_T = 0.9$ T.
- Strongly peaked ion temperatures >28 keV were obtained in low density H-mode discharges. $T_i(0)$ and $T_e(0)$ were simultaneously in excess of 10 keV in higher density discharges.
- Energetic MeV particles, both with isotropic and with transverse velocities, were contained for at least 2 s while they slowed down by classical processes.
- The duration of fusion relevant conditions in JET H-mode discharges was limited to <1 s by influx of carbon impurities from plasma-target plate interaction.
- 100 kW of fusion power in charged particles was produced in $^3$He plasmas.

- $n_D(0)T_i(0)\tau_E$ values of 7 to 9 × $10^{20}$ $m^{-3}$ keVs with $Q_{d\text{-}d} > 2 \times 10^{-3}$ were produced in d-d, equivalent to $Q_{d\text{-}t} \sim 0.7$, $P_{alpha} \sim 2.4$ MW and $P_{fusion} \sim 12$ MW.

## 5.4 Neutron and Fast Particle Diagnostics

### *5.4.1 JET Plasma Diagnostics and the Role of Fusion Product Diagnostics*

As shown in Fig. 5.1, JET had a very wide range of diagnostics, including those for neutron and fast particles. Whole books are dedicated to plasma diagnostics [15–17]. Here the concentration is on fusion product diagnostics, since they were crucial to studying deuterium-tritium experiments and to separating the effects of tritium use from deuterium-only plasmas. The development of these diagnostics is discussed with reference to preparation for tritium experiments.

### *5.4.2 Fusion Products and Diagnostics*

#### 5.4.2.1 Fusion Reactions in JET

The decade from 1989 to 1998 was one of preparing JET to achieve a high production of D-T fusion power. The culmination was the successful production of 16 MW of fusion power, with power gain approaching Q = 1. It was therefore necessary to measure the fusion products, specifically high energy neutrons, charged particles from the d-d, d-t and d-$^3$He fusion reactions and gamma rays.

The main reactions of importance and their fusion products neutron n, proton p, deuteron d, triton (t) and helium $^3$He and $^4$He are:

- d + t –> $^4$He (3.52 MeV) + n (14.06 MeV)
- d + d – > t (1.01 MeV) + p (3.02 MeV) (50%)
- d + d – > $^3$He (0.82 MeV) + n (2.45 MeV) (50%)
- d + $^3$He–> $^4$He (3.6 MeV) + p (14.7 MeV)

The energies quoted are for the reaction products in their centre-of-mass frame of reference. When translated to the 'laboratory' frame, they have a wide spread of energy that reflects the relative velocities of the reacting fusion particles. Measuring this energy spread gives information about the energy distribution of charged particles in the plasma.

The 1 MeV triton from d-d reactions, if it is contained, can slow down in the plasma and undergo d-t fusion and produce 14 MeV neutrons and 3.5 MeV $^4$He alpha particles. If the plasma current is high enough, then the slowing down alpha particles are contained and can heat the plasma. For confinement of most alphas, at least 3 MA is required.

In a fusion reactor, if the alpha heating equals the plasma loss power, the ignition condition $Q = \infty$ is obtained. A reactor could then operate without additional heating power. If deuterium and/or tritium neutral beams are injected at energies of several 10's of keV, they can fuse with plasma ions and with each other and produce more fusion products.

These fusion reactions allowed a whole range of diagnostics to be developed for JET that could measure the total production, spatial and velocity spectrum distribution of these fusion products.

#### 5.4.2.2 Diagnostics Overview

The JET neutron, fast particle and gamma ray detectors [18] for the first tritium experiments are summarized in Table 5.2.

**Table 5.2** Overview of JET neutron fast particle and gamma ray detectors in JET

| Particle nature | Measurement | Diagnostic system |
|---|---|---|
| **Fusion products at birth** | | |
| Total strength | Total neutron yield | Three pairs of absolutely calibrated fission chambers |
| Emission profile | Neutron cameras | Two cameras, d-d and d-t neutrons and gamma rays |
| Energy | d-d neutron spectrum | Spectrometers: $^3$He ion chamber and time of flight |
| Energy | d-t neutron spectrum | Tandem proton recoil, associated particle, mag. Proton recoil |
| **Fast and confined particles** | | |
| All $v_{perp}$ | Energy content | Diamagnetic loop |
| All | ICE emission | ICE antenna |
| All | Fast neutrals | High energy neutral particle analysers |
| Tritons (burnup) | 14 MeV n counters | Silicon diodes: High efficiency system |
| Tritons (burnup) | 14 MeV n profile | Neutron camera profile monitor with NE213 detectors |
| RF driven (alphas) | Gamma ray spectra | NaI, BGO |
| RF driven and alphas | Wave scattering | Gyrotron, Alfven waves |
| **Slow and confined particles** | | |
| Helium Ash | Charge exchange spectroscopy | Profile system + optical link |
| **Fast and lost particles** | | |
| All | Neutrons, gamma and X-rays | Fast data acquisition channels |
| Alpha particles | Electrical current | Five faraday cups |
| Alpha particles | Electrical current | Diamond detectors |

### 5.4.3 *Measurements with Neutron and Fast Particle Diagnostics*

#### 5.4.3.1 Overview of Measurements

Fast particle behaviour in JET [19] was investigated. The earliest observations concerned escaping 15-MeV protons from the d-$^3$He reaction; they were detected in the form of spikes at the time of sawtooth crashes. Subsequent observations with a neutron camera showed that sawteeth redistribute NBI 80-keV deuterons away from the central region of the plasma.

Extensive use was made of the detection of gamma rays created when ion cyclotron resonance frequency (ICRF)-driven fast ions reacted with plasma fuel ions and with the main plasma impurity ions carbon, oxygen, and beryllium. Threshold reactions showed that ICRF-driven ions can exceed energies of 7.5 MeV. Using ratios of gamma-ray intensities, tail temperatures in the mega-electron-volt range were diagnosed. The energy content of these ions exceeded 1 MJ.

The measurement of 14-MeV neutrons emitted during the burnup of tritons generated by the d-d reaction indicated that the single-particle behaviour of 1-MeV tritons was classical within 20%, which implied similar behaviour for 3.5-MeV alpha particles in d-t plasmas.

#### 5.4.3.2 Neutron Profile Monitor Measurements

The spatial profile of neutron emission was measured with two orthogonal neutron cameras with a total of 19 channels. This neutron profile monitor was an upgrade of a previous version [20]. In order to improve its shielding and operational flexibility, the larger shields were double in weight, increasing to 53 tons and 25.5 tons. The dynamic range of the system was extended by a factor of 20 by using pairs of remotely controlled rotatable collimators. Each line of sight was equipped with three in-line detectors:

- a CsI(TI) (thallium activated cesium iodide) photodiode assembly for measuring hard x-rays from fast electrons in the 80–300 keV energy range
- an NE213 liquid scintillator with pulse shape discrimination electronics for measuring 2.5 MeV neutrons, low level 14 MeV neutrons (mainly from triton burnup) and gamma rays ($E > 500$ keV)
- a fast BC 418 scintillator for measuring high levels of D-T 14 MeV neutrons.

The scale of the neutron camera is comparable to that of the JET vacuum vessel and provided an indication of what is involved in a high level of neutron shielding. In a fusion reactor, all heat and radiation sensitive components need to be protected from the plasma by a one-meter thick radiation shield.

#### 5.4.3.3 The 2-D Measurements of 2.5 MeV Neutron Sawtooth Crashes

Measurements from the neutron profile monitor were analysed tomographically to deduce the 2-D spatial distribution of neutron emissivity during NBI heating and before and after sawtooth crashes [21]. In a discharge where the 2.5 MeV global neutron emission decreases slightly to 5/6th of the pre-crash rate of $10^{16}$ n $s^{-1}$, the axial emissivity drops to 1/6th of the pre-crash value. The sources of neutron emissivity were analysed with a Fokker-Planck formulation which calculates the velocity distribution of fast ions slowing down in the plasma. The main change in neutron emissivity was due to fast ion redistribution during a sawtooth crash and a decrease in beam-beam emissivity.

Similar measurements were made of both 2.5 and 14 MeV neutron emissivity before and after sawtooth crashes in high d-d neutron yield, hot ion H-mode plasmas [22]. Deuterium-deuterium (d-d) fusion produces 2.5 MeV neutrons and 1 MeV tritons (t) at nearly equal rates from its two reaction channels. An upper limit of 10% was placed on the net fraction of fusion product tritons which were displaced from the plasma axis by the sawtooth crashes. Which was a much smaller net fraction than that typically observed, 35–55%, for displaced injected neutral beam deuterium ions. The implication of these observations was that 3.5 MeV alpha particles in JET would be relatively resilient to sawtooth crashes when their spatial width was sufficiently large.

#### 5.4.3.4 Neutral Beam Injection of 120 keV Helium

Experiments were carried out in JET using 120 keV helium NBI, with the goal of making neutron emission profile and spectroscopy measurements from thermal plasma fusion without beam-plasma and beam-beam interactions [23]. Injected power levels up to 5 MW with $^3$He and 7 MW with $^4$He, lasting up to 3 s were achieved. The helium NBI produced efficient ion heating and similar global and local energy confinement times to those obtained with d-NBI, in both limiter and X-point plasma geometries and L-mode and H-mode plasma confinement regimes. These measurements extended the range for determining ion temperatures with neutron diagnostics to fusion reactors with dominant alpha ($^4$He) heating.

#### 5.4.3.5 Thermal Plasma Neutron Production and Ion Thermal Diffusivity

Ohmically heated plasmas without additional heating achieved ion temperatures high enough to produce considerable neutron fluxes which were measured [24] by the JET neutron emission profile monitor. The profiles of $T_i$ and ion thermal diffusivity ($\chi_i$) were derived under steady-state conditions. The inferred $\chi_i$ for $0.3 < r/a < 0.5$ was in the range 0.5–2.5 $m^2\ s^{-1}$.

In further studies [25], the spatial profiles of the neutron emission from deuterium plasmas were observed to relax following the termination of strong heating with NBI. The profiles of $\chi_i$ for minor radii in the range $0.25 < r/a < 0.55$ were found to be flat and to take values between 0.3 and 1.1 $m^2 s^{-1}$ for H-mode plasmas.

#### 5.4.3.6 Gamma Ray Emission from High Fusion Power from RF Accelerated Helium-3 ($^3He$) Minority Fuel Ions

Helium-3 minority fuel ions have been observed to be accelerated to MeV energies by ion cyclotron radio-frequency (ICRF) heating. These energetic ions undergo nuclear reactions with impurity $Be^9$ and $C^{12}$ ions in the plasma and characteristic gamma radiation was emitted. This gamma ray emission has been detected with the profile monitor [26]. Two-dimensional profiles of the gamma radiation emission corresponded to a cross-section weighted density of $^3He$ ions having energies of 2 or more MeV. As expected for ICRF heating, the observed spatial profile of the gamma radiation showed that access by the accelerated ions to the plasma volume on the high field side of the RF resonance layer is effectively inhibited. The profile appeared to demand the presence of a group of passing particles that are localized in poloidal extent. These features are reproduced by a model that combined a 2-D bounce averaged Fokker-Planck calculation with an orbit-following code.

#### 5.4.3.7 High Energy Gamma Rays from Runaway Electrons

High energy gamma rays can be produced by runaway electrons. The perpendicular x-ray emission up to a few MeV of runaway electrons was measured in JET low-density ohmic discharges by means of the profile monitor [27]. A diffusion model simulating the temporal evolution of the line-integrated x-ray signals was used to determine the runaway radial transport coefficient in the central region of the plasma ($D_r \approx 0.2\ m^2 s^{-1}$ for $r/a < 0.5$). A comparison was made with the predictions of turbulent transport theories and limits on the level of radial magnetic field fluctuations were determined.

#### 5.4.3.8 Neutron Spectrometry Systems

The original neutron spectrometry systems [28] were designed for the d-d phase of operations, including a complementary set of compact neutron spectrometers housed in a shielding block in the torus hall. The collimation system with adjustable aperture defined a horizontal line of sight significantly tangential to the plasma axis. The spectrometers consist of a $He^3$ gridded ionisation chamber, organic liquid scintillators, and a proton recoil telescope. They covered a neutron emission intensity from $1 \times 10^{14}$ n/s to $5 \times 10^{15}$ n/s. The proton recoil telescope was insensitive to gamma rays and allowed a high counting rate for high power d-d fusion experiments. The spectrometry systems were operated at JET for temperature measurements in suitable discharges and were particularly useful in studies of NBI and in where a substantial fraction of the neutron yield is of non-fusion origin.

### *5.4.4 Robustness Analysis of Neutron and Fast Particle Diagnostics*

Robustness optimization in design and operation of neutron and fast particle diagnostics is crucial, not only for the understanding of fusion plasmas, but also for the operation of the entire JET facility. The radioactive activation of the JET vacuum vessel and other components by d-d produced neutrons and especially by d-t produced neutrons must be measured accurately and continuously, for safety and for vessel access. The neutron diagnostics operate in conjunction with other plasma diagnostics such as those that measure plasma density and temperature. Robustness analysis indicates:

- systems control – In a fusion reactor, the neutron and fast particle diagnostics will be a key part of the fusion power feedback-control system.
- fault-tolerance – There are several global neutron generation detectors and all of their signals were compared with each other for fault detection. The profile monitor provided an independent measurement of neutron flux.
- modularity – The neutron systems were internally redundant and also functioned as a separate but linked system with other plasma diagnostics.
- decoupling – The multiplicity of diagnostics meant that small fluctuations in particle production can be smoothed out. The calibration of the diagnostics took place at a wide range of positions around the JET vacuum vessel.

Robustness trade-offs in design and operation:

- robustness – The diagnostics were modular in both different types and multiple copies, hence they had a very high robustness. The dynamic measurements were backed up by sample irradiation as a continuous check on total activation. The various systems, especially the electronics and detectors, could be replaced relatively easily.
- fragility – The system had fragilities in that systematic errors in calibration could make the overall measurements invalid. Great care was taken to overcome this fragility by multiple calibration methods. Also, multiple detectors and methods were used to protect against a single localized fault.
- performance – The neutron diagnostics on JET covered a wide dynamic range over the full set of JET parameters, both in d-d and d-t fusion operation.
- resource demands – The diagnostics were well supported.

## 5.5 Preliminary Tritium Experiments (PTE)

The design of the Preliminary Tritium Experiments (PTE) carried out in November 1991 was described in a summary [29] and in four papers in a Nuclear Fusion special topic grouping [30–33]. A second case study on JET is presented here on the planning, execution and analysis of these PTE and the highest power DTE1 experiments, with emphasis on systems forward engineering. From Checklist 1.4, the first steps are to establish goals, review lessons learned and examine options.

The fourth original JET goal of studying alpha particle physics was begun with d-d or ICRH accelerated fast ion fusion, but much higher levels of fusion power were required. Therefore, the formal goals of PTE were:

- to produce more than 1 MW of fusion power in a controlled way (Actually, 1.7 MW was produced);
- to validate transport codes and provide a basis for predicting accurately the performance of d-t plasmas from measurements in d plasmas;
- to determine tritium retention in the torus systems and to establish the effectiveness of discharge cleaning techniques for tritium removal;
- to demonstrate the technology related to tritium usage;
- to establish safe procedures for handling tritium in compliance with the regulatory requirements.

There were also a set of unwritten goals which were equally or more important:

- Demonstrate MW levels of d-t fusion power before the TFTR tokamak.
- Make a demonstration of JET's capabilities as the world's leading tokamak.
- Test the neutron and fast particle diagnostics and demonstrate their capability in advance of the full power DTE1 experiments.
- Justify continuing to fund the JET programme, which had to be actively renewed every 3–4 years, especially for the proposed pumped divertor upgrade.

The tritium experiments had no precedent. The neutron and fast particle diagnostics had been tested on deuterium plasmas and on the 14 MeV neutrons produced by tritons from d-d reactions slowing down, causing fusion at low levels.

### *5.5.1 Systems Design of Preliminary Tritium Experiments (PTE)*

#### 5.5.1.1 Preliminary Scoping of PTE

The preliminary scoping of PTE involved the Checklist 1.4 items of setting goals, prioritizing models, producing test plasmas, performing necessary upgrades, and revise goals as planning and events dictate.

#### 5.5.1.2 Operation Limitations and Hardware Upgrades for PTE

Since vessel access would be required for the pumped divertor upgrade during 1992/1993 immediately after the end of PTE, the vacuum vessel activation limits for safety required a neutron budget of less than $1.5 \times 10^{18}$ neutrons. The tritium reprocessing plant was not yet operational, so tritium had to be shipped to JET in uranium beds that could be heated, limiting the amount of tritium available to 0.2 g, corresponding to 2000 Curies.

To ensure the optimum usage of tritium and to maximize beam-beam and beam-plasma fusion, the tritium would be injected from 2 of the 16 Positive Ion Neutral Injection (PINI) sources at the lower energy of 78 keV. The remaining 14 PINIs were operated in deuterium, with 12 at 135 keV and 2 at 75 keV.

An important characteristic of neutral beam injectors was that ion sources produce a combination of monatomic, diatomic and triatomic singly-charged molecules that enter a gas neutralizes, and head towards the plasma as single atoms at full, ½ and 1/3 the injection energy. The neutralization efficiency is higher for lower velocities. Since tritium is heavier, a very high neutral fraction of 79% of the injected tritium was at the full energy, giving optimum plasma penetration. The triton injection total power was 1.5 MW into the plasma.

The tritium particle fuelling rate was 13% of the total deuterium plus tritium fuelling rate. The total tritium consumption for a 2 s heating pulse was 0.012 g. Most of the non-injected tritium was deposited on cryopanels in the injectors, which was collected by a cryogenic gas system especially installed.

#### 5.5.1.3 Diagnostics and Performance Estimates

Over 30 plasma diagnostic systems were in operation for the experiment, including the neutron and fast particle diagnostics and many of the other types of plasma diagnostics shown in Fig. 5.1. Essential plasma diagnostics [34] included magnetic measurements for MHD equilibrium calculations, electron cyclotron emission for $T_e$, an infrared interferometer for $n_e$, LIDAR Thomson scattering for $T_e$ and $n_e$, active charge exchange recombination spectroscopy for $T_i$ and impurity concentrations, and visible Bremsstrahlung for $Z_{eff}$. Beam injection and transport consistency was examined using the TRANSP code [35] which had been continuously improved over the years to model all aspects of the plasma in a 1-D radial model. TRANSP was used to analyse discharges in pure deuterium and extrapolate performance to the tritium experiments, which indicated that over a MW of fusion power would be produced. It was used to analyse the data from the PTE measurements.

#### 5.5.1.4 Target Plasmas for PTE

For technical reasons related to tritium, the current pellet injection and ICRH systems were not used. To obtain the highest fusion rate, hot-ion H-modes were chosen as the target plasma, produced with a single null X-point diverted onto the upper carbon target. The toroidal field was reversed to produce ion drift away from the divertor target to more evenly spread the heat load. The discharges chosen had a current of over 3 MA.

### *5.5.2 Preliminary Deuterium-Tritium Fusion Experiments (PTE)*

From Checklist 1.4, the next steps were to carry out the experiments, and to coordinate publications. The PTE experiments were carried out in four stages:

- Carry out several hot-ion H-mode discharges in pure deuterium, and choose the best performance shot among them in terms of d-d neutron yield.
- Repeat the conditions for the best pure deuterium discharge using 1% tritium in one PINI to test the 14 MeV neutron diagnostics and the tritium systems.
- Use 100% tritium in 2 PINIs for at least two repeated discharges.
- Observe removal of trapped tritium from the vacuum vessel walls.

#### 5.5.2.1 Pure Deuterium Hot-ion H-mode Target Plasma for Tritium Experiments

A hot-ion H-mode [29] was established by first allowing the plasma density to fall during the evolution from a limiter defined plasma to x-point diverted. This resulted in a moderately peaked electron density profile, also maximizing axial neutral beam deposition. At 12 s into the discharge, the beam power was increased to 15 MW, leading to an H-mode transition. During the next second, sawteeth were stabilised and centrally-peaked heating produced peaked temperature profiles, with ion temperature reaching 18.6 keV and electron temperature 10.5 keV. The electron density profile became rather flat at $3.5 \times 10^{19}$ m$^{-3}$ and the global energy confinement time was 1.2 s, all nearly reactor relevant parameters. The high performance phase was terminated at 13.4 s by a rise at the plasma edge in carbon and deuterium light emission (carbon bloom), followed by a sawtooth collapse of the central temperatures. TRANSP simulations of the discharge, including calculations of the contributions of beam-beam, beam-plasma, and thermal d-d fusion gave excellent time-dependent agreement with measured total 2.5 MeV d-d neutron flux, with the thermal component providing about half the total at the peak. The toroidal beta value was 2.2% and the triple product was $9.0 \times 10^{20}$ m$^{-3}$ keV s.

#### 5.5.2.2 Neutral Beam Injection of 1%T in One PINI in Hot-ion H-mode

The previous discharge was repeated with 1% tritium in one PINI. Both 14 MeV neutrons from d-t fusion and 2.5 MeV neutrons from d-d fusion were simultaneously measured time-dependently by the global fission counters, the neutron profile monitor and neutron spectrometers. Again, there was excellent agreement with TRANSP simulations.

#### 5.5.2.3 Fusion Power of 1.7 MW from Neutral Beam Injection of 100% Tritium in 2 PINIs into Hot-ion H-mode

To minimize tritium usage and neutron production, only two discharges were attempted with 100% tritium in 2 PINIs [7]. A peak fusion rate of 14 MeV neutrons was $6.0 \times 10^{17}$ n/s, corresponding to 1.7 MW of total fusion power. Excellent agreement was obtained in the TRANSP simulation of both the time-dependent neutron yield and a profile modelling of the signals from all of the 10 horizontal and 9 vertical channels of the neutron profile monitor.

#### 5.5.2.4 Neutron Emission Profile Measurements During PTE

During the PTE the temporal evolution and the two dimensional (2-D) spatial profiles of the 2.5 and 14 MeV neutron emissivities from d-d and d-t fusion reactions were inferred [30] from measurements with the JET neutron emission profile monitor. The data were analysed using NEUTOMO, a tomographic analysis algorithm [36] modified for the neutron profile monitor.

In discharges with 1% tritium beam injection, neutron emissivity ratios showed that approximately the same overall deposition profiles resulted from tritium as from deuterium beams.

A thermalized tritium diffusion experiment was performed in which the t to d density ratio was found to be spatially constant across the plasma; in conjunction with similar particle source profiles, this indicated that deuterium and tritium have similar particle transport properties.

Figure 5.3 shows the time dependent two-dimension emission profile of 14 MeV neutrons was determined during the maximum fusion power plasma which included a sawtooth crash at 13.30 s which redistributed beam ions and thermal emission. The profiles also correspond exactly to the birth profile of 3.5 MeV alpha particles, which would have additionally heated the plasma as they slowed down and thermalized, being well confined in these high current plasmas.

During the tritium cleanup phase, residual tritium entering the plasma produced a spatially constant ratio of tritium to deuterium, confirming the similarity of their particle transport properties. These experiments were analysed used ratio methods [30] applicable over a wide range of discharge types.

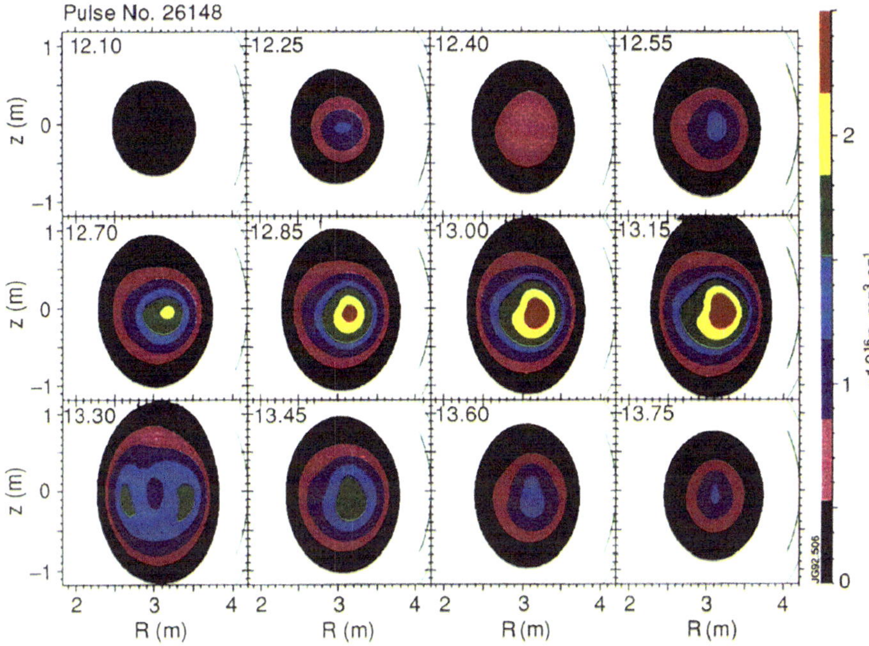


**Fig. 5.3** Neutron camera tomographic reconstruction of 14 MeV neutron emission during PTE, including sawtooth crash. (Reused with permission from [30], Fig. 9)

#### 5.5.2.5 Particle and Energy Transport During PTE

The particle and energy transport properties of the high fusion performance PTE pulses were studied using TRANSP [31]. The particle diffusion coefficient of tritium was determined by using the neutron profile monitor to observe the decay of a small quantity of injected tritium in a deuterium background plasma. A good simulation of the measured 14 MeV neutron emissivity profile was obtained during the decay phase if the mixing of the two species is described by a model where the tritium diffusion coefficient was similar to deuterium.

The energy confinement of these low density, hot ion, H-mode discharges was found to have both improved central and edge confinement over the conventional medium to high density H-mode discharges, regardless of the presence or absence of tritium in the discharge. As the tritium concentration of these d-t discharges was small (varying from <1% to 10%), no isotopic dependence was expected and indeed none was observed. Enhancement factors of at least twice the value predicted by H-mode scaling expressions were observed but only transiently. A local transport analysis was completed to try and establish the reason for the improved confinement and its transient nature. Similarities between these pulses and the DIII-D VH-mode discharges were observed.

#### 5.5.2.6 Ion Cyclotron Emission Measurements During PTE

During PTE, the ion cyclotron emission (ICE) spectrum [32] contained superthermal emission lines which corresponded to cyclotron harmonics of alpha particles at the outer midplane. The rise and fall of the neutron flux during a single discharge was followed the ICE signal delayed by the alpha slowing-down time. Some alphas born in the core made orbital excursions to the outer midplane edge and were unstable to fast Alfven-ion Bernstein waves.

#### 5.5.2.7 Release of Tritium from the Carbon First Wall of JET During PTE

Experiments were performed to deplete the torus of tritium [33] to quantitatively compare the effectiveness of different methods. The overall effectiveness was such that the normal tokamak programme was resumed 1 week after the PTE. The release of tritium from the vessel was found to scale with the deuterium release from the vessel, suggesting that dilution and mixing of the hydrogen isotopes in the vessel walls is important.

High density, disruptive tokamak discharges were found to be the most successful plasma pulses for tritium removal. Purges with deuterium gas were also effective and had the advantage of operational simplicity. Tile release of tritium was simulated using a model that incorporated implantation into a thin surface layer as well as diffusion of tritium into and out of the bulk material.

## 5.6 Full Power Deuterium-Tritium Experiments (DTE1)

### *5.6.1 JET Optimization in Preparation for DTE1*

Having completed the PTE experiments, the validity of the original goals of JET were confirmed. Following Checklist 1.4, it was time to perform the necessary machine upgrades to obtain maximum performance from JET and develop new reactor-relevant scenarios.

### *5.6.2 Machine Upgrades*

#### 5.6.2.1 Pumped Divertor and Vessel Interior Upgrade

Impurity influxes had limited performance during hot-ion H-modes and prevented the realisation of steady state conditions at high heating powers. To address impurity control, plasma fuelling and helium ash exhaust, an axisymmetric pumped divertor installation [37] was planned for JET. The configuration should have allowed operating conditions close to those of a reactor.

It was decided to build the divertor coils and cryopump inside the vacuum vessel system [38]. Four water-cooled copper coils were added at the bottom inside of the vacuum vessel to produce divertor plasmas. In the factory, the coil parts were manufactured and process techniques qualified. Assembly took place inside the JET vacuum vessel by people aided by robotics, including welding into Inconel casings and impregnation with epoxy resin. This installation was predicated on no further replacement of toroidal field coils.

There were a total of three divertor configurations using the same coils inside the vessel, referred to as Mark I, Mark IIA/AP and Mark IIGB (Gas Box), with different pumping and dump configurations.

Further upgrades over the years during shutdowns included new ICRH and lower hybrid antennas and a new interior vessel wall. The components were also designed to avoid the destructive effects of halo currents in JET [39, 40].

#### 5.6.2.2 Neutron Diagnostics Upgrades

In preparation for DTE1, the neutron profile monitor was upgraded [41] to provide improved radiation shielding, install remotely adjustable collimation to extend the dynamic range and provide for three in-line detectors within each collimation channel. The number of channels and their viewing directions remained unchanged. These detectors (see more detail in 5.4.3.2) were:

- CsI(T1) photodiode for fast electron bremsstrahlung detection;
- NE213 scintillator for d-d neutrons and low intensity D-T neutrons;
- BC418 small plastic scintillator for high intensity D-T neutrons and for the detection of low intensity D-T neutrons during strong d-d neutron fluxes.

The neutron profile monitor upgrade also led to a proposed design for a neutron camera for ITER [42]. Other diagnostics were radiation hardened where possible.

### *5.6.3 Experimental Preparation for $Q = 1$ Energy Breakeven Experiments at High Power*

JET entered 1994 [43] nearing the end of the longest and most extensive modification since the initial assembly of the device. Faulty toroidal magnetic field coils had been replaced. The new machine capabilities included:

- a new pumped divertor configuration (Mark I);
- an upgraded neutral beam system with 7.8 MW of power at 140 keV d and 13.6 MW at 80 keV d;
- new ICRF heating antennae designed to couple 20 MW;

- a new LH launcher designed to couple 10 MW;
- a system for plasma current and shape control;
- a fast radial field amplifier for vertical position control;
- eight saddle coils inside the vacuum vessel for disruption control and the study of TAE modes;
- an extensive array of new diagnostics, especially for divertor measurements.

During the time until the DTE1 experiments in 1997, a new range of scenarios was developed for DTE1. These included:

- ELM-free hot-ion H-modes with a pumped divertor [44];
- Lower hybrid current drive with high bootstrap current [45];
- Power step-down experiments [46, 47].

### *5.6.4 Systems Approaches to Planning Full Power Deuterium-Tritium Fusion Experiments*

A systems approach to planning DTE1 is given from Checklist 1.4, including assembling a task force, reviewing goals, proposing experiments and developing a prioritized programme.

A DTE1 group was constituted by including task force (experiment operations groups) and topic group (scientific areas) leaders and special experts, led by the Chairman Alan Gibson and the Scientific Secretary (this author). It produced a report [48] which covered the full planning process and is described here.

#### 5.6.4.1 DTE1 Goals and Constraints

To establish the goals for DTE1, JET scientific management made presentations to the JET Scientific Council and the JET Council which decided on the goals [49]:

- The physics mission of DTE1 would have the crucial objective of studying the isotopic effect on confinement scaling and H-mode threshold power in D-T plasmas. These would be the first experiments of this kind in the geometry appropriate to ITER and including a divertor. These experiments would be essential to determine whether the d-t performance improvements observed in the circular cross-section TFTR tokamak are also realised in the D-shaped cross-section of JET and ITER.
- Furthermore, the H-mode threshold power in d-t plasmas would be determined for the first time in these JET experiments. This would allow more accurate assessments of the ignition margin and the heating requirements for ITER.
- In addition, JET's capability for long pulse operation and impurity control should permit some 10 MW of fusion power for several seconds (typically with 50% tritium). The α-particle heating would then make a significant contribution to the

plasma power balance and this would allow the effects of α-particle heating (confinement and thermalization of alpha particles and stability of toroidal Alfven eigenmodes in the presence of α-particles) to be studied and experience gained for ITER.
- The operating conditions foreseen for ITER, namely long pulse ELMy H-mode detached radiative divertor plasmas, could also be studied in d-t, albeit at reduced levels of fusion power. These results could provide important information for the design of the ITER divertor.

These goals were separated into mission categories for the purpose of identifying the principal contributions of the various experiments, as follows.

- Alpha Particle Heating – Significant Contribution to Power Balance (High Fusion Q)
- 10 MW of Fusion Power
- Isotopic Effect on H-mode threshold in ITER
- Isotopic Effect on Confinement Scaling in ITER
- Long Pulse ELMy H-mode detached radiative divertor plasmas for ITER divertor
- Neutron Calibration and Cleanup – Necessary Preparation Experiments
- Alpha Particle Heating Physics – Confinement and Thermalization, Instability Effects
- RF coupling to reactor relevant divertor plasmas
- Alpha Particle Heating -TAE Modes

DTE1 also had key Technology Missions:

- DTE1 would also have a technology mission with the objective of carrying out and demonstrating key ITER and reactor-relevant technologies, such as tritium handling and processing, remote handling and control, and heating systems operating in d-t.
- Specifically, DTE1 would provide a first test of a large scale technology for processing tritium in conjunction with an operating tokamak.

In line with the above objectives, the proposed tritium requirements for the experiments were double the proposed site inventory, so the tritium plant called the AGHS (Advanced Gas Handling System) would be crucial for operations. The site inventory of tritium was planned to be 37 bar-litres ($10^{24}$ molecules of $T_2$).

For heating systems operating in d-t, a specific physics item includes the demonstration of RF heating in d-t plasmas. The 160 keV neutral beam box with eight PINIs was to be operated in tritium, to take advantage of higher neutralisation efficiency of tritium at a given energy and higher tritium beam with beam and thermal deuterium fusion rates.

The experiments were also strongly constrained by the activation limits on the vacuum vessel of $2 \times 10^{20}$ neutrons and an experimental duration length of 3–4 months. Remote handling and control would be essential for future changes to the divertor target assembly.

#### 5.6.4.2 Preparing, Assembling and Prioritizing Experimental Proposals

The DTE1 experimental proposals were prepared mostly on the basis of extrapolating previous JET discharges to d-t conditions, so that there was a firm experimental background for most of what was proposed. It was necessary to project the requirements in terms of d-t neutron reduction and tritium requirements for the beams and gas introduction systems, and of beam power and particle currents obtained from 160 keV tritium NBI. A series of memos to aid in these projections was prepared and made available to the DTE1 group, as follows:

- Tritium beamline gas consumption and beam particle currents
- Sample DTE1 outline proposal and technical information
- Prediction of d-t performance from d-d discharges using ratio methods [30]
- Determination of the tritium fuel ratio profile with the JET neutron profile monitor
- Summary of gas usage in the 1994–1995 experimental campaign

Each proposal was submitted on the basis of a standard form. An initial summary was tabulated of the requirements to perform all the experiments proposed. The total requirements for all proposed experiments were more than double the neutron budget and over quadruple the tritium budget. It was clearly necessary to choose among these proposals and to avoid overlaps. To make the best use of the group's expertise, a voting procedure was established to determine priorities.

The philosophy behind the voting was as follows. Each "elector" was encouraged to propose a programme that would best fulfil the guidelines set by the physics mission of DTE1. Electors were requested to choose 12 main experiments in priority order with 12 points given to the highest priority experiment, 11 to the next, etc., to roughly correspond to 12 weeks of the experimental programme, plus additional time for two experiments declared to be essential, neutron diagnostic calibration and tritium cleanup.

In order to accomplish the programme with maximum efficiency, it was decided to associate experiments with these main experiments as "piggyback" experiments. These experiments were subordinate to a particular main proposal, hence parasitic, with no separate allocation of neutrons.

Subordinate experiments, typically of relatively small neutron budgets, were related to the main experiment and helped it to fulfil its primary role. The planning of these was fully integrated with the main experiment, but with lower priority.

Parasitic experiments mostly required additional measurements and as such were expected to be attempted with all suitable experiments. Detailed perusal showed that they had hoped for optimised conditions for special diagnostics and for dedicated time. Up to 3 piggyback experiments were allowed per proposal.

The chosen proposals were as follows:

Proposal titles 12 main and 2 necessary experiments

- Maximum fusion power in hot-ion ELM-free H-mode
- Beam power step-down to obtain high fusion Q and alpha particle heating
- ELMY H-mode and $\rho^*$ scaling
- Maximum fusion power in a steady-state ELMy H-mode
- H-mode power threshold measurements in d-t plasmas
- Assessment of ICRF heating methods in reactor-relevant d-t plasmas
- Isotope effect on H-mode power threshold
- $\alpha$-particle heating experiment
- Maximum fusion performance with combined heating at high toroidal field
- Tritium transport
- Isotope mixture control in hot ion H-modes
- Use shear reversal configurations to increase performance of hot ion modes
- Neutron diagnostics calibrations using: "1%" tritium; "10%" tritium
- Tritium clean-up methods using: plasma; gas soaking; glow discharge

Piggyback-subordinate experiments

- Study of divertor closure/retention with trace tritium
- Isotope mixture control in high density discharges
- Tritium beam fuelling efficiency in low-recycling hot-ion H-mode (1% tritium)
- Deuterium to tritium change-over characterisation
- Tritium fuelling efficiency
- High power radiative divertor
- ITER-relevant H-mode scenario

Piggyback – Parasitic

- Saturation dependence on isotope
- Direct measurements of TAE
- Study TAE excitation during the decay of high fusion yield d-t plasmas
- Limitations during high power d-t discharges
- Study transport of a-particles as function of sawteeth
- Optimised experiments for measurements of distribution function of a-particles

Each elector designated the appropriate neutron budget for each experiment. As a result of the voting procedure, a total of 27 experiments (Main, Piggyback and Calibration) were selected from the total 44 proposals. The voting indicated that $183 \times 10^{18}$ neutrons should be allocated to the 12 main experiments, and that $9 \times 10^{18}$ d-t neutrons should be allocated to the 7 subordinate experiments. All the neutron allocations were consistent with requirements of the individual programs.

The proposed programme was also found to satisfy all requirements of the physics mission. The number of discharges and tritium requirements were well within the project time and technical constraints.

The above picture seemed ideal. However, there were several important complications to detailed and successful implementation of the programme. With care, it was hoped that it still would be possible accomplish the programme.

#### 5.6.4.3 Proposed Experiments NOT Selected

A number of proposals were not selected, even though they were excellent:

- Maximum $Q_{d\text{-}t}$ in a steady-state ELMy H mode
- Steady-state elm-free regimes with $T_i \sim T_e \sim 15$ keV
- TF ripple losses of alpha particles
- Assess α-physics properties by comparing NBI + ICRH plasma in d-d and d-t plasmas.
- Hot-ion ELM-free H-mode with max heating power
- Alpha particle pressure enhancement in high Te discharges
- High power radiative divertor
- H mode threshold in RF only H-modes in a 5 MA d-t plasma
- Control the MHD activity through q variation from 2.5 to 4.5 and study of MHD activity on confinement and slowing down of a-particles
- Test of the effectiveness of $T_2$ soaking for T wall loading and control the t/d in the plasma
- ELM free H-mode
- Determination of the poloidal distribution of tritium in the walls after selected t-fuelled discharges (both for NBI and gas)
- Study influence of momentum and plasma rotation on stability of the outer MHD modes
- Study stability of plasmas with high levels of α-particles against disruptions
- Effect of the vessel temperature on the d-t changeover and cleanup
- Investigation of instability associated with d-t plasmas
- Use of $t_2$ (or d/t) glow discharge in stagnant gas for t wall loading and control of the t/d in the plasma

The main reasons many were not selected were that they would have required far too large a neutron budget or a tritium use budget or that there was some overlap with other proposed experiments. In addition to the ripple loss experiment, perhaps the most regretted omission was the maximum $Q_{d\text{-}t}$ in a steady-state ELMy H mode.

It was already possible to run ELMY H-mode discharges in steady-state [50] with d-d for 5 s with a power input of 20 MW. This proposed experiment for DTE1 would have produced over 10 MW of fusion power for 5 s, which was indeed part of the mandate of the DTE1 experiments, but would have consumed well over ¼ of the neutron budget. These experiments were later performed successfully in DTE2.

#### 5.6.4.4 Reliability and Margins in Neutron Budget and Time

- All the experiments first needed to be demonstrated in the preceding d-d phase, with high reliability and performance. Any experiments that were unsuitable would have been eliminated at that time.
- The experiments were in priority order, which was an indication of which experiments to eliminate if problems arose with neutron budgets or time.
- It was desirable to intersperse d-d operating time into the d-t phase to tune discharges to optimum before the main d-t experiments were carried out. The feasibility of such interspersing depended upon how efficiently the change-over could be carried out in terms of time, neutrons and tritium fractions.
- Experiments with 12 discharges of similar initial conditions had a relatively high probability of success because of common discharge features.
- In the 6 discharges in three experiments relying on RF coupling and heating, these were preceded by extensive d-d work on RF coupling and heating, and by experiments to optimise and demonstrate RF coupling.

#### 5.6.4.5 Prerequisites for the High Neutron Yield Programme

- For planning purposes, the 6 experiments with 18 discharges and 80% of the neutron budget were defined as the high neutron yield programme.
- Before the high yield programme, a number of experiments had to have been done first. These included neutron diagnostics calibration, establishment of cleanup methods and establishment of mixture control techniques, and 1% tritium mixture experiments, all with low neutron yields.

#### 5.6.4.6 Order of Experiments

- Once the basic prerequisite experiments were done, there were important choices to be made between the order of high and low neutron yield experiments and the relative order within these two categories.
- One argument was that all low budget experiments should be done first, since they had a high benefit to neutron cost ratio, and since the high budget experiments could be done with full knowledge of their results.
- The counter argument was that it was best to intersperse these experiments, so we would have known early in the programme if there were major problems with the high yield discharges.

#### 5.6.4.7 d-d Simulations and Development During DTE1

- If possible, it was highly desirable to run a series of discharges in d-d just before each high yield experiment, to obtain optimum conditions for d-t.
- The tritium from the previous experiment had to be reduced to the level of 0.33% tritium in the plasma, to reduce the d-t neutron rate down to the d-d rate. One d-d neutron at 2.5 MeV produced about 1/3 of the activation of one d-t neutron at 14 MeV. This cleaning had to be accomplished quickly and with low neutron production.
- The d-d series of experiments had to have been run without using a significant number of d-t neutrons.
- The mixture control experiments had to show how to go immediately from a pure deuterium environment to an approximately 50–50 d-t environment in the plasma on the next pulse. Otherwise, many neutrons would have been wasted at intermediate concentrations of d-t.
- The total time of this cycle should not have compromised the overall programme duration. It was instructive to observe the long time spent during the PTE phase to optimise the two hot ion discharges.

#### 5.6.4.8 Disruptions, Instabilities, Damage

- If a discharge carried risk of in-vessel damage, should it be attempted early, when personnel intervention was still possible, or later, when most of the other experiments have been done?
- Even if damage was not done, there was a certain recovery time from a disruption. Great care had to be taken to not waste d-t neutrons during the recovery from a disruption or other events.

#### 5.6.4.9 Planning of the d-d Phase Before DTE1 Begins

- DTE1 was an ambitious programme, with many elements to be demonstrated in the d-d phase before it began. If there was not time for these demonstrations, there was no point in including some items in the DTE1 programme.
- The preceding d-d phase included plasma restart and characterisation of the Mark II divertor. This already restricted the time available. Any requirement for an in-vessel intervention just before DTE1 begins would involve loss of time and some additional risk.
- In addition to demonstration of DTE1 discharges, there were a number of experiments that had to be done in h or d during the d-d phase, so as to have useful scaling data for the tritium experiments.
- It was probably true that the DTE1 planning programme would determine most of the experiments in the preceding d-d phase, apart from Mark II characterisation.

- The preceding d-d phase would probably allow 3 months at most for DTE1 development and demonstration. This programme would therefore become very similar to the actual DTE1 programme in scheduling.
- It was necessary to determine specific goals in terms of data performance for each experiment during the d-d phase in order to qualify for the d-t phase.

#### 5.6.4.10 The 1996 d-d Programme

It was recognised that this proposed DTE1 programme relied on the intended d-t discharges having been previously developed and validated in the 1996 d-d Programme. This was true for all the proposals, but applied with particular emphasis to the high performance programme. It was clear that large tritium fractions should not be introduced into JET until the DTE1 experiment conditions had been demonstrated and developed as reliable and robust scenarios in d-d.

Some experiments that were chosen for DTE1 were omitted from the actual programme because of technical problems or because the preliminary d-d programme experienced difficulties. Two major disappointments occurred. One was that the power step-down experiments proved too unreliable in maintaining the plasma free of instabilities after power reduction, so it was not possible to sustain discharges at the highest performance level. Quasi-steady discharges were obtained at somewhat lower power levels. The other was that lower hybrid heating and current drive was not used during the high power radio frequency heating experiments, even though part of the original proposals.

#### 5.6.4.11 Conclusions on DTE1 Proposal

The Group produced a proposed programme for DTE1 which addressed the programme as set out in [49]. The proposed group of 27 experiments was selected from the 44 proposals considered. Nevertheless, it was apparent that the programme was still very demanding in terms of the time and neutron budget available, so inevitably, some elements would have to have been curtailed or dropped. Decisions were influenced by the d-d programme results.

### *5.6.5 Deuterium-Tritium High Power Nuclear Fusion Operational Scenarios and Results in JET*

The Checklist 1.4 steps relevant to the full power DTE1 experiments were carried out successfully in 1997 including prototype experiments in deuterium, trace tritium tests, the main program adjusted according to intermediate results and publications of the excellent results.

#### 5.6.5.1 Trace Tritium Experiments in DTE1

Trace amounts of tritium [51] were introduced in short gas puffs into steady state deuterium plasmas in JET. The transport properties of the edge and core plasma and the time evolution of the fraction of tritium in the recycling flux were determined in space and time from the 14 MeV measurements by the neutron profile monitor. For H-mode discharges, the particle transport was best described by gyro-Bohm scaling in the core and Bohm scaling near the edge.

Trace amounts of tritium were also injected into JET ELMy H-modes [52] to study density limits. Analysis with the neutron profile monitor and extraction of the particle transport coefficients allowed distinction between broad classes of mechanisms which were suggested as explanations for the H-mode density limit.

#### 5.6.5.2 D-T Fusion Power of 16.1 MW and Q ≈ 1 in Hot-ion H-modes with Divertor in DTE1

High fusion power of 16.1 MW was produced [53] using d-t mixtures in a 4.2 MA/3.6 T ELM-free H-mode and optimized shear regimes. The transient value of the fusion amplification factor was 0.95 ± 0.17, consistent with the high value of $n_D \tau T_i(0) = 8.7 \times 10^{20}\ m^{-3}$ s keV. These conditions were maintained for half an energy confinement time until excessive edge pressure gradients resulted in discharge termination by MHD instabilities. The d-t fusion power showed the expected ratio of 210 to similar d-d plasmas. Plasma mixture control was achieved by loading the vessel walls with the appropriate d-t mix. Magnetic fluctuation spectra showed no evidence of Alfvenic instabilities driven by alpha particles, in agreement with theoretical model calculations.

#### 5.6.5.3 Alpha Particle Heating of 1.3 MW in DTE1

Plasma heating by alpha particles generated during the DTE1 experiments was clearly demonstrated [54]. The highest electron temperature was clearly correlated with the maximum alpha particle heating power and the optimum d-t mixture; the maximum increase was 1:3 ± 0:23 keV with 1.3 MW of alpha particle heating power, consistent with classical expectations for alpha particle confinement and heating. Alpha particle heating was unambiguously separated from possible isotope effects by varying the tritium concentration. There was no detectable isotope effect on the energy confinement time.

#### 5.6.5.4 Internal Transport Barriers and High Triple Product in DTE1

In the optimized shear regime [55], clear internal transport barriers were established for the first time in d-t, with a power similar to that required in d-d. The ion thermal conductivity in the plasma core approached neoclassical levels. Real time power control maintained the plasma core close to limits set by pressure gradient driven MHD instabilities, allowing 8.2 MW of d-t fusion power with a triple product of $10^{21}$ $m^{-3}$ s keV. Quasi-steady-state discharges with simultaneous internal and edge transport barriers were produced with high confinement and a fusion power of up to 7 MW. These double barrier discharges showed great potential for steady state operation.

#### 5.6.5.5 Bulk Ion Heating of Deuterium and Q ≈ 0.22 with ICRH in DTE1

Reactor relevant ICRH scenarios [56] were demonstrated during d-t experiments using H-mode divertor discharges with ITER-like shapes and safety factors. Deuterium minority heating [57] in tritium plasmas was demonstrated for the first time. For 9% deuterium, an ICRH power of 6 MW gave 1.66 MW of fusion power from reactions between suprathermal deuterons and thermal tritons. The Q value of the steady state discharge reached 0.22 for the length of the RF flattop (2.7 s), corresponding to three plasma energy replacement times. The Doppler broadened neutron spectrometer measurements showed a deuteron energy of 125 keV, which was optimum for fusion. Strong bulk ion heating was obtained at the same time as high fusion efficiency. Deuterium fractions around 20% produced the strongest ion heating together with a strong reduction of the suprathermal deuteron tail. The ELMs had low amplitude and high frequency and each ELM transported less plasma energy content than the 1% required by ITER. The energy confinement time, on the ITERH97-P scale, was 0.90, which was sufficient for ignition in ITER.

#### 5.6.5.6 Bulk Ion Heating with $^3$He Minority ICRH in DTE1

The $^3$He minority heating [57] in approximately 50–50% d-t plasmas with up to 10% $^3$He demonstrated strong bulk ion heating. Central ion and electron temperatures of 13 and 12 keV were obtained. The normalized H-mode confinement time factor was 0.95.

#### 5.6.5.7 Second Harmonic Tritium ICRH in DTE1

Second harmonic tritium heating [57] produced energetic tritons above the critical energy which heated the electrons in JET, whereas in ITER the lower power density

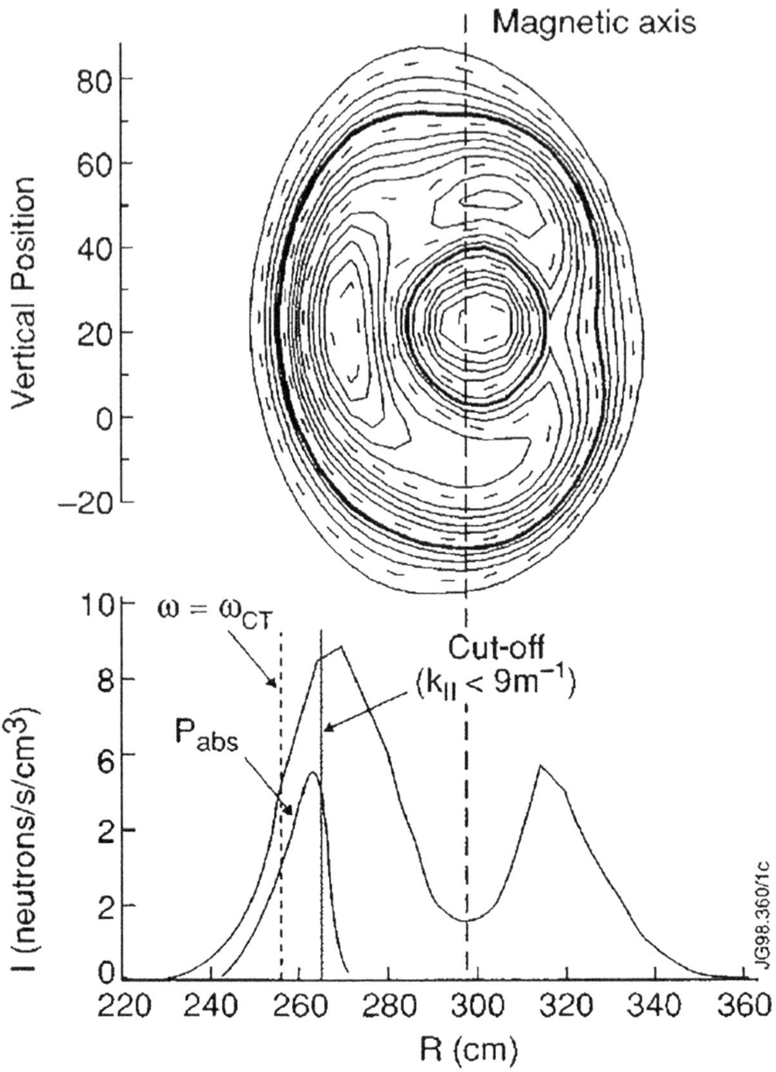


**Fig. 5.4** Neutron emission profiles for off-axis tritium minority heating. The bold solid curve corresponds to half-height. (Reused with permission from [57], Fig. 22)

would allow mainly ion heating. The neutron spectrometer measurements gave a triton perpendicular temperature of 35 keV and a parallel temperature of 7 keV. Tomographic reconstruction of the neutron emission profile shown in Fig. 5.4 displayed an annular profile with the maximum emissivity occurring close to the peak power absorption as given by the ISMENE code. The asymmetry of the profile was probably due to the longer time spent by the tritons on the high field side of the magnetic axis, verified by orbit calculations.

#### 5.6.5.8 Combined NBI and ICRH in ELM-free H-modes During DTE1 Giving Record Stored Energy and Fusion Power

High power combined NBI plus ICRF heating experiments [58] were carried out in the JET Mark IIa divertor configuration in the hot ion ELM-free H-mode regime in d-t 4.2 MA/3.6 T plasmas. ICRF heating power of 9.5 MW was tuned to the fundamental hydrogen minority and combined with NBI of 22 MW. ICRF heating contributed to a fusion power of 16.1 MW and stored energy of 17 MJ.

#### 5.6.5.9 Tritium Recycling, Retention and Clean-up in DTE1

DTE1 allowed a detailed study of tritium recycling and retention [59] in a pumped divertor configuration relevant to ITER. There were two inventories of tritium:

- a dynamic inventory which controlled the fuelling behaviour of a discharge
- a continually growing inventory which played a small role in the particle balance of a single discharge but dominated for a long experimental campaign

The JET retention scaled-up to ITER proportions suggests that ITER could reach its tritium inventory limit in less than 100 pulses with wall and divertor conditions similar to JET in DTE1.

During 1997 JET operation with d-t plasmas, 35g of tritium were introduced into the torus, mainly by gas puffing. The torus tritium inventory accumulated at a rate of about 40% of the input. After tritium operation ceased, the experimental program continued with deuterium and hydrogen fuelled experiments, during which time the tritium inventory decreased to about 17% of the total input. Techniques aimed at clean-up and tritium removal were employed [60]. The plasma tritium fraction was reduced to below the 1% in a few days, but the tritium inventory became steady at 6 grams.

## 5.7 Post-DTE1 Operation in JET – Preparation for DTE2

For the post DTE1 period from 1998 to 2021 and DTE2, JET operated with a wide range of conditions and upgrades in support of ITER, DEMO and the upcoming DTE2. Following Checklist 1.4, the next steps for upgrading JET to a reactor-level behaviour and carrying out DTE2 are preparing prototype experiments, upgrading the machine and optimizing plans.

Following are some of the highlights and especially preparations for DTE2, including the 2009–2010 installation of a new beryllium/tungsten plasma facing wall so as to simulate ITER conditions. DTE2 especially planned for high fusion performance during longer periods.

- Divertor and SOL studies in the Mark I, II, IIA/AP, IIGasBox divertors [61]
- Lower hybrid current drive (LHCD) and negative shear with internal transport barrier [62]
- Variable toroidal field ripple experiments [63]
- Safe and flexible operations and plasma control [64]
- Plasma breakdown with a carbon and ITER-like wall [65, 66]
- Improved confinement at high beta with an ITER-like wall [67]
- Plasma ramp-up, ramp-down and operation scenarios for ITER [68]
- Wall conditioning for ITER [69]
- Operational scenarios in support to ITER [70]
- Diagnostic enhancements for d-t scientific campaigns [71]
- Assessment of the baseline scenario at $q_{95} \sim 3$ for ITER [72]

## 5.7.1 Planning for DTE2 Experiments in JET

Extensive preparations were made for the d-t operation DTE2 in JET with the ITER like-wall (ILW) [73]. Experiments for DTE2 would be the first d-t plasmas with the ITER mix of plasma-facing component materials. Preparation involved review of the physics basis for the d-t operational scenarios, including:

- fusion power predictions through first principles and integrated modelling;
- the impact of isotopes in the operation;
- physics of d-t plasmas such as thermal and particle transport, high confinement H-mode access, Be and W erosion, fuel recovery.

This effort also required improving several aspects of plasma operation for DTE2, including:

- real time control schemes;
- heat load control;
- disruption avoidance and a mitigation system including the installation of a new shattered pellet injector;
- novel ion cyclotron resonance heating schemes such as three ions;
- new diagnostics (neutron camera and spectrometer, active Alfvèn eigenmode antennas, neutral gauges, radiation hard imaging systems;
- calibration of the JET neutron diagnostics at 14 MeV for accurate fusion power measurement.

The central scenario planning was based on two high power scenarios, extrapolated to DTE2 operation with tritium.

### 5.7.2 *Extrapolated Baseline Scenario*

The ITER baseline scenario proposed for ITER was a plasma current of 15 MA to achieving its goals of $Q = 10$, fusion power of 500 MW at normalised pressure, $\beta_N = 1.8$ and normalised confinement $H_{98}(y.2) = 1$. It was based on existing plasmas as in JET [72] relying on a stationary ELMy H-mode at $q_{95} \sim 3$. The data showed that operating at high plasma density, with line-averaged density at 85% of Greenwald density was achievable. Encouraging results were obtained on JET [73] for the baseline scenario at 3 MA, 2.8 T and a line-averaged density of nearly $7 \times 10^{19}\ m^{-3}$. With an injected power of 28 MW of NBI and 5 MW of ICRH, a neutron rate of $3 \times 10^{16}\ s^{-1}$ could be achieved for more than five energy confinement times (~1.5 s). The equivalent fusion power of ~7 MW was computed with TRANSP, composed of 40% beam–target and 60% thermal–thermal fusion power, giving a $Q_{d\text{-}t}$ of 0.21.

### 5.7.3 *Extrapolated Hybrid Scenario at High Normalized Beta*

Based on the hybrid scenario [67], experiments were conducted in JET [73] at 2.2 MA, 2.8 T, $q_{95} = 4$, $\beta_N = 2.8$ normalised confinement $H_{98}(y.2) = 1.3$ and a line-averaged density of $5 \times 10^{19}\ m^{-3}$. The injected power was 28 MW of neutral beam injection and 5 MW of ICRH. A d-d neutron rate of $3 \times 10^{16}\ s^{-1}$ was achieved for more than five energy confinement times (~1.5 s). The equivalent d-t fusion power of ~7 MW was computed with TRANSP, composed of 65–70% beam–target and 30–35% thermal–thermal fusion power, giving a $Q_{d\text{-}t}$ of 0.21.

### 5.7.4 *Completed Preparations for d-t Experiments*

As of May 2021, preparations had been completed for the DTE2 experiments in JET [74] with full d-t experiments planned to begin in July 2021, as follows:

- Preparation for safe operation with tritium was completed.
- ITER-Like-Wall (ILW) operation led to new physics:
  - W screening in hot pedestal
  - small ELMs scenarios
  - Ne compared favourably to $N_2$ as seed gas
- JET Shattered Pellet Injection (SPI) experiments provided ITER with important information for preparing for disruption mitigation.

### 5.7.5 Successful DTE2 Operation

In 2022, DTE2 was performed, and it was reported [75, 76] that JET had produced 59 megajoules of fusion energy from d-t, lasting for 5 seconds and corresponding to a sustained average fusion power of 11 MW, even better than the 7 MW projected from the d-d simulation phase. The importance of this result is that the long pulse will give confidence to the technological choices used for the divertor and first wall materials in ITER for power handling and impurity control.

In summary, JET continued to be one of the principle experiments helping to prepare for ITER and is operating as part of the EUROfusion programme.

## 5.8 Fusion Power Experiments in TFTR

### 5.8.1 Tokamak Fusion Test Reactor (TFTR) Design

The main goal of TFTR was to provide a demonstration of energy breakeven in d-t fusion. TFTR was initially the Two Component Torus (TCT) planning to inject co- and counter- circulating neutral beams in a low density plasma to optimize beam-beam fusion [77]. Considerable improvements in performance were obtained by higher density resulting in higher thermal and total fusion power.

The performance of the Tokamak Fusion Test Reactor (TFTR) and also JET were summarized [78]. TFTR was used to study transport, magnetohydrodynamic stability, and energetic particle confinement. Significant advances were made in the reduction of turbulent transport by means of internal transport barriers, identification of the bootstrap current and study of energetic fusion alpha-particles.

Parameters for TFTR were $B_T < 6$ T, $I_p < 3$ MA, a = 0.9 m, and R = 2.6 m. The nearly circular poloidal boundary of the TFTR plasma was set by a limiter on the inboard (high field) side composed of graphite and carbon fibre composite (CFC) tiles mounted on a water-cooled Inconel backing plate. The limiter could withstand heat outflow of ~30 MJ. The influx of carbon in high performance discharges was the dominant impurity and resulted in $Z_{eff}$ values of about 2.

The TFTR neutral beam system was composed of four beam lines, each with three positive-ion sources. The ion sources operated either in deuterium or in tritium. The maximum operating voltage was 120 kV and a maximum injected power into a d-t discharge was 40 MW.

The TFTR ICRF system used four antennas to launch magnetosonic waves. Hydrogen and $^3$He minority, second harmonic tritium, and second harmonic deuterium heating were studied. In addition, fast wave conversion into ion Bernstein waves was used to heat and drive currents in multiple ion species plasmas.

TFTR operated routinely and reliably from the start of d-t operations in November 1993 to the completion of experiments in April 1997 including 1090 d-t shots. Tritium was used in both the neutral beam and gas injection valves to fuel the plasma. A low inventory tritium purification system was installed and operated in the final phases of the d-t campaign to decrease shipments of tritium.

## 5.8.2 *TFTR Experimental Results*

### 5.8.2.1 Supershots with Maximum Fusion Power

Experiments with d-t in TFTR [79] produced a maximum fusion power of 10.7 MW, using 39.5 MW of neutral beam heating in a supershot discharge. The energy confinement time was observed to increase by 20% in d-t relative to d plasmas. The $n_i(0)T_i(0)\tau_E$ product increased by 55%. Extensive lithium pellet injection increased the confinement time to 0.27 s and enabled higher currents.

### 5.8.2.2 Alpha Prticle Physics

Alpha particles in TFTR [80] were generally well confined in MHD-quiescent discharges, and alpha heating of electrons was observed. The theoretically predicted toroidicity-induced Alfvén eigenmode was seen in discharges with weak magnetic shear. The loss of energetic alpha particles to a detector at the bottom of the vessel was well described by the first-orbit loss mechanism. No loss due to alpha particle driven instabilities was observed.

### 5.8.2.3 Tritium Retention

The short term retention of tritium in the plasma facing components and co-deposited layers ranged from 51% in high performance neutral beam heated supershot discharges to >90% in L-mode discharges. After extensive removal efforts, the long term retention was ~16% [81]. For ITER after one hundred shots, the in-vessel tritium inventory limit would be reached using carbon.

## 5.8.3 *Robustness and Fragility Analysis*

Although TFTR exceeded its original budget and proposed construction time, it had a successful deuterium phase and fulfilled its main goal of demonstrating significant fusion power during 3 years of safe and successful tritium operation. Machine availability was comparable to that during operation in deuterium. Nearly a thousand kCi (99g) of tritium were processed. The tritium purification system operated in a closed cycle during the final tritium run. TFTR therefore demonstrated successful maintenance and operation of an activated and tritium loaded machine. The main fragility was limitation of plasma pulse duration time by impurity blooms from limiters.

## References


1. JET Project, The JET Project, JET Report EUR-JET-R7, Culham (1775)
2. A. Wojenski et al., Data quality monitoring considerations for implementation in high performance raw signal processing real-time systems with use in tokamak facilities. J Fusion Energy **39**, 221 (2020)
3. B. Viola et al., The impact of poloidal flux expansion on JET divertor radiation performance. IAEA Fusion Energy Conference, Ahmedabad, India, IAEA–CN–258–144 (2018)
4. G.H. Rappé, The J.E.T. (joint European torus) vacuum vessel. Revue de Physique Appliquee **12**(10), 1735 (1977)
5. JET Progress Report 1989. JET Report EUR 12808 EN, EUR-JET-PR7 (1989)
6. M. Keilhacker, JET Team. The JET H-mode at high current and power levels, Proceedings of the 12th International Conference on Plasma Physics and Controlled Nuclear Fusion Research, Nice 1988, vol. 1, IAEA, Vienna; 1989. pp. 159–181 (1988)
7. P.H. Rebut, JET Team. Recent JET results and future prospects. Proceedings of the 13th International Conference on Plasma Physics and Controlled Nuclear Fusion Research, Washington DC, 1990, vol. 1, IAEA, Vienna (1991). pp. 27–52
8. P. Smeulders, JET Team. High density regimes and beta limits in JET. Proceedings of the 13th International Conference on Plasma Physics and Controlled Nuclear Fusion Research, Washington DC, 1990, vol. 1, IAEA, Vienna (1991). pp. 219–227
9. M.F.F. Nave et al., Fishbone activity in JET. Nucl. Fusion **31**(4), 697 (1991)
10. P. Lomas, JET Team. Optimisation of performance in JET limiter Plasmas. Proceedings of the 13th International Conference on Plasma Physics and Controlled Nuclear Fusion Research, Washington, DC, 1990, vol. 1, IAEA, Vienna (1991). pp. 425–435
11. A Tanga, JET Team. High performance H-modes in JET, Proceedings of the 13th International Conference on Plasma Physics and Controlled Nuclear Fusion Research, Washington DC, 1990, vol. 1, IAEA, Vienna (1991). pp. 261–276
12. J.R. Last et al., JET TF coil faults-detection, diagnosis and prevention. IEEE Trans. Magn. **30**(4) (1994)
13. H. Van der Beken, CODAS: the JET control and data acquisition system. J. Fusion Technol. **11**(1) (1987)
14. A. Gibson, JET team, Fusion relevant performance in JET. Plasma Phys. Control. Fusion **32**(11), 1083 (1990)
15. P.E. Stott, *Diagnostics for experimental thermonuclear fusion reactors* (Springer, 1998)
16. I.H. Hutchinson, *Principles of Plasma Diagnostics* (Cambridge University Press, 2005)
17. A.J.H. Donne, Introduction to plasma diagnostics. Trans Fusion Sci. Technol. **49**, 349 (2006)
18. G.J. Sadler, JET neutron group, Fast particle diagnostics at JET: status and plans. Nucl. Fusion. **35**(12), 1609 (1995)
19. G.J. Sadler et al., Investigations of fast-particle behaviour in joint European torus plasmas using nuclear techniques. Fusion Technol. **18**, 556 (1990)
20. J.M. Adams et al., The JET neutron emission profile monitor. Nucl. Inst. Meth. Phys. Res. A **329**(1–2), 277 (1993)
21. F.B. Marcus et al., JET neutron emission profiles and fast ion redistribution during sawtooth crashes. Plasma Phys. Control. Fusion **33**(4), 277 (1991)
22. F.B. Marcus, JET neutron group, Effects of sawtooth crashes on beam ions and fusion product tritons in JET. Nucl. Fusion **34**, 687 (1994)

23. F.B. Marcus et al., JET experiments with 120 keV 3He and 4He neutral beam injection and neutron diagnostics applications. Plasma Phys. Control. Fusion **34**(8), 1371 (1992)
24. B. Esposito, F.B. Marcus, et al., Ohmic ion temperatures and thermal diffusivity profiles from the JET neutron emission profile monitor. Plasma Phys. Control. Fusion **35**, 1433 (1993)
25. M. Sasao, JET neutron group, Determination of the ion thermal diffusivity from neutron emission profiles in decay. Plasma Phys. Control. Fusion **36**, 1 (1994)
26. O.N. Jarvis et al., Gamma ray emission profile measurements from JET ICRF-heated discharges. Nucl. Fusion **36**(11), 1513 (1996)
27. B. Esposito et al., Runaway electron measurements in the JET tokamak. Plasma Phys. Control. Fusion **38**, 2035 (1996)
28. N.P. Hawkes et al., A 2.5 MeV neutron spectrometry system with a tangential line of sight for the D-D phase at the JET tokamak. Nucl. Inst. Meth. Phys. Res. A **335**(3), 533 (1993)
29. P.H. Rebut et al., Fusion energy production from a deuterium-tritium plasma in the JET tokamak. Nucl. Fusion **32**, 187 (1992)
30. F.B. Marcus et al., Neutron emission profile measurements during the first tritium experiments at JET. Nucl. Fusion **33**(9), 1325 (1993)
31. B. Balet et al., Particle and energy transport during the first tritium experiments on JET. Nucl. Fusion **33**(9), 1345 (1993)
32. G.A. Cottrell et al., Ion cyclotron emission measurements during JET deuterium-tritium experiments. Nucl. Fusion **33**(9), 1365 (1993)
33. P. Andrew et al., Experiments on the release of tritium from the first wall of JET. Nucl. Fusion **33**(9), 1335 (1993)
34. Keilhacker M et al. Review of diagnostic systems 1986, JET Report JET-IR(86)16 (1986)
35. R.J. Goldston et al., New techniques for calculating heat and particle source rates due to neutral beam injection in axisymmetric tokamaks. J Comput. Phys. **43**, 61 (1981)
36. R.S. Granetz, P. Smeulders, X-ray tomography on JET. Nucl. Fusion **28**, 457 (1988)
37. Keen BE, Watkins ML. The pumped divertor – the new phase of JET, Europhysics News, January (1992)
38. Tesini A et al. Construction and testing of the JET divertor coils inside the vacuum vessel. 15th IEEE/NPSS Symposium on Fusion Engineering – Supplement (1993)
39. F.B. Marcus et al., Simulations of control, perturbation, displacement and disruption in highly elongated tokamak plasmas. Nucl. Fusion **30**, 1511 (1990)
40. M.A. Pick, P. Noll, P. Barabaschi, F.B. Marcus, L. Rossi. Evidence of halo currents in JET. IEEE Plasma Physics Conference (1992)
41. O.N. Jarvis, J.M. Adams, F.B. Marcus, G.J. Sadler, Neutron profile measurements in the joint European torus. Fusion Eng. Des. **34–35**, 59 (1997)
42. F.B. Marcus et al., A neutron camera for ITER. Rev Sci Instrum. **68**, 514 (1997)
43. JET Progress Report 1994. JET Report EUR 16474 EN (EUR-JET-PR12); 1995
44. P. Lomas and the JET Team, High fusion performance ELM-free H-modes and the approach to steady operation. 16th IAEA Fusion Energy Conference, Montreal, Canada (1996)
45. F.X. Soldner et al., Lower hybrid current drive in JET and reactor applications, plasma physics and controlled nuclear fusion research. IAEA 15th Conference Proceedings, Seville, Spain, Paper IAEA-CN-60/A3-2 (1994). p. 423
46. F.B. Marcus et al., Evolution and optimization of neutron emissivity profiles in JET. Europhysics Conference 19C, Part II, 53 (1995)
47. F.B. Marcus et al., A power step-down approach to extended high performance of ELM-free H-modes in JET. Nucl. Fusion **37**(8), 1067 (1997)
48. A. Gibson (Chairman) and F. Marcus (Scientific Secretary). Report of the group to consider experimental proposals for DTE1 in JET, vol. 1: Report, JET Report JET-TN(96)01, vol. 2: Detailed proposals, JET Report JET-TN(96)02 (1996)
49. JET Council paper EUR FU 95/JC 60/4.1 (rev.) (1995)
50. R. Sartori et al., Confinement of high-current steady-state ELMy H-modes with the JET mark II divertor. Plasma Phys. Control. Fusion **40**, 757 (1998)

51. K.-D. Zastrow et al., Trace tritium transport in JET. Nucl. Fusion **39**(11Y), 1891 (1999)
52. G.F. Matthews et al., Trace tritium and the H-mode density limit. J Nucl. Mater. **266–269**, 1134 (1999)
53. M. Keilhacker et al., High fusion performance from deuterium-tritium plasmas in JET. Nucl. Fusion **39**(2), 209 (1999)
54. P.R. Thomas et al., Observation of alpha heating in JET DT plasmas. Phys. Rev. Lett. **80**(25), 5548 (1998)
55. M.L. Watkins et al., Physics of high performance JET plasmas in DT. Nucl. Fusion **39**(9Y), 1227 (1999)
56. D.F.H. Start et al., D-T fusion with ion cyclotron resonance heating in the JET tokamak. Phys. Rev. Lett. **80**(21), 4681 (1998)
57. D.F.H. Start, Bulk ion heating with ICRH in JET dt plasmas. Nucl. Fusion **39**, 321 (1999)
58. F.G. Rimini et al., Combined heating experiments in ELM-free H-modes in JET. Nucl. Fusion **39**, 1591 (1999)
59. P. Andrew et al., Tritium recycling and retention in JET. J. Nucl. Mater. **266–269**, 153 (1999)
60. P. Andrew et al., Tritium retention and clean-up in JET. JET Report JET-P(99)01; (1999)
61. R.D. Monk et al., Recent results from divertor and scrape-off layer studies at JET. Nucl. Fusion **39**(11Y), 1751 (2002)
62. C. Gormezano et al., Overview of JET results in support of the ITER physics basis. Nucl. Fusion **41**(10), 1327 (2001)
63. Saibene G et al. Results of the variable toroidal field ripple experiments in JET (INIS-XA–08N0893). International Atomic Energy Agency (IAEA) (2008)
64. Sartori F et al. JET operations and plasma control: a plasma control system that is safe and flexible in a manageable way. Proceedings – symposium on fusion engineering. 2009. https://doi.org/10.1109/FUSION.2009.5226522.
65. P.C. De Vries, Characterisation of plasma breakdown at JET with a carbon and ITER-like wall. Nucl. Fusion **53**(5), 3003 (2013)
66. D. Mueller, The physics of tokamak start-up. Phys. Plasmas **20**, 058101 (2013)
67. C.D. Challis et al., Improved confinement in JET high β plasmas with an ITER-like wall. Nucl. Fusion **55**, 053031 (2015)
68. A.C.C. Sips et al., Progress in preparing scenarios for operation of the international thermonuclear experimental reactor. Phys. Plasmas **22**, 021804 (2015)
69. D. Douai et al., Wall conditioning for ITER: current experimental and modelling activities. J. Nucl. Mater. **463**, 150 (2015)
70. X. Litaudon et al., Overview of the JET results in support to ITER. Nucl. Fusion **57**, 102001 (2017)
71. J. Figueiredo et al., JET diagnostic enhancements testing and commissioning in preparation for DT scientific campaigns. Rev Sci Instrum. **89**, 10K119 (2018)
72. Sips ACC, et al. Assessment of the baseline scenario at q95 ~ 3 for ITER. Nucl. Fusion 2018;58 (12):126010.
73. E. Joffrin et al., Overview of the JET preparation for deuterium–tritium operation with the ITER like-wall. Nucl. Fusion **59**, 112021 (2019)
74. J. Mailloux et al. Overview of JET results for optimizing ITER operations. 28th IAEA Fusion Energy Conference, Nice, France (Virtual) 10–15 May 2021, Presentation OV/1-2, Paper IAEA–CN–286-1080, Submitted to Nuclear Fusion; 2021.
75. EUROfusion NEWS. 2022. https://www.euro-fusion.org/news/2022/european-researchers-achieve-fusion-energy-record/
76. T. Donne Highlights from the JET D-T campaign. IN: IAEA webinar on building stars: breakthroughs in fusion R&D, March 24 2022. https://youtu.be/5KkVGASnPJ0. Accessed 20 July 2022
77. R.M. Kulsrud, D.L. Jassby, Neutralised colliding beam torus. Nature **259**, 541 (1976)

78. Hawryluk RJ et al. Results from D-T experiments on TFTR and implications for achieving an ignited plasma. Philos Trans., vol. 357, no. 1752, The Approach to Ignited Plasma; 1999. pp. 443–69
79. M.G. Bell et al., Overview of DT results from TFTR. Nucl. Fusion **35**, 1429 (1995)
80. S.J. Zweben et al., Alpha-particle physics in the tokamak fusion test reactor DT experiment. Plasma Phys. Control. Fusion **39**(5A), 275 (1997)
81. C.H. Skinner et al., Tritium experience in large tokamaks: application to ITER. Nucl. Fusion **39**(2), 271 (1999)

# Chapter 6
# Superconducting and Long-Pulse Tokamaks for Prototyping Reactor Technology

**Abstract** As a means of prototyping the technology and long pulse aspects of a fusion reactor, several tokamaks were constructed using superconducting windings. Systems reverse engineering is used to analyse them and their experiments by understanding their goals, reviewing their construction and deriving lessons learned. Systems approaches are applied to identify trade-offs and operational space options for superconducting and steady-state tokamaks, which contribute to the reactor design space. Some of these tokamaks have demonstrated pulse lengths of several hundred seconds in high power H-mode discharges. Many of the technologies needed for ITER are being tested, such as actively-cooled plasma-facing components. The robustness and fragilities of these complex and highly inter-connected superconducting machines are evaluated.

## 6.1 Systems Analysis Strategy for Prototyping Reactor Subsystems

In the initial chapters, much of the emphasis was on machines that achieved plasma physics goals of density, temperature, confinement time, heating and power exhaust. As a result, the understanding of the plasma physics has grown greatly and increased confidence in constructing tokamak fusion reactors. In this chapter and the next on ITER, the focus is shifting towards new physics and new technologies that will be required in an operating fusion reactor, including steady-state operation and compatible machine components.

A paper on the design strategy for EU DEMO reactor components [1] commented that:

- The plant design should drive R&D.
- Technical solutions should be based on proven design features to minimize technological risks.
- A traceable design process with a rigorous systems engineering approach is needed.



F. B. Marcus, *Systems Approaches to Nuclear Fusion Reactors*, Springer Series in Plasma Science and Technology, https://doi.org/10.1007/978-3-031-17711-8_6

These principles are maintained in several ways by the tokamaks described in this chapter. Tokamaks such as KSTAR and JT-60SA are modelled on the design for ITER except for the use of tritium and its breeding blanket and neutron shielding of the coils. The fourth recommendation of [1], that fusion is a nuclear technology and will be assessed as such, is not taken into account.

This chapter emphasizes the systems engineering aspects of developing and analysing these machines. The analysis uses Checklist 1.5 for using systems reverse engineering to analyse these reactor prototype machines by establishing their goals and determining their systems architecture and results in order to examine robustness and fragilities and to learn lessons. The choices made of variants of systems architecture will be examined in depth.

### 6.1.1 *Goals and Systems Architecture Choices for Superconducting and Steady-State Tokamaks*

Superconducting coil tokamaks represent major technological advances that are vital for prototyping fusion reactors and for gaining operating experience. The largest currently existing machine is the joint Japanese-European tokamak JT-60SA shown in Fig. 6.1 [2]. The machine is about half the size of ITER and is described in detail later. The most notable features of JT-60SA are the superconducting coils for the

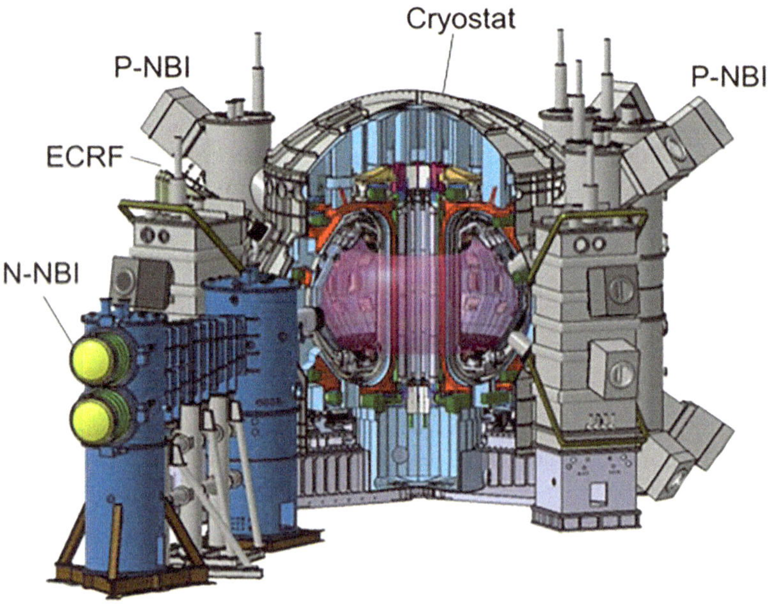


**Fig. 6.1** Cross section of the JT-60SA tokamak. (Reused with permission from [2], Fig. 3)

toroidal and poloidal field systems, the divertor, bell-jar cryostat enclosing the entire machine and auxiliary heating systems.

#### 6.1.1.1 Parameters of Superconducting Toroidal Field Coil Tokamaks, Without and With Normal Conductor Poloidal Field Coils

Parameters of tokamaks with superconducting toroidal field coils and normal conducting poloidal field coils are shown in Table 6.1. Those with superconducting poloidal field coils as well are shown in Table 6.2. The non-superconducting large

**Table 6.1** Parameters of tokamaks with superconducting toroidal field coils and normal conducting poloidal field coils

| Parameter/tokamak | T-7, HT-7 | TRIAM-1 M | T-15 | TORE SUPRA | WEST |
|---|---|---|---|---|---|
| Start year | 1979, 1995 | 1986 | 1988 | 1988 | 2016 |
| Max plasma MA | 0.30 | 0.42 | 2.30 | 1.70 | 1.00 |
| R(m) | 1.22 | 0.80 | 2.43 | 2.25 | 2.50 |
| a(m) | 0.27 | 0.12 | 0.70 | 0.70 | 0.50 |
| Btor (T) | 2.5 | 8.00 | 5.00 | 4.50 | 3.70 |
| Max pulse (s) | 400 | 18,000 | 5 | 1000 | 1000 |
| Circular or div? | Circular | Elong+div | Circular | Circular | Divertor |
| #TF coils, conductor | 24, NbTi | 16, $Nb_3Sn$ | 24, $Nb_3Sn$ | 18, NbTi | 18, NbTi |
| PF coil conductor | Copper | Copper | Alu, $LN_2$ | Copper | Copper |
| Auxiliary heating | LH, IBW | LH | NBI, EC | LH, EC, IC | LH, EC, IC |

**Table 6.2** Parameters of tokamaks with superconducting toroidal and some or all superconducting poloidal field coils, plus normal JT-60U

| Parameter/ tokamak | SST-1 | EAST | KSTAR | JT-60 U | JT60-SA | ITER |
|---|---|---|---|---|---|---|
| Start year | 2013 | 2006 | 2008 | 1985 | 2021 | 2025 |
| Max plasma MA | 0.22 | 1.00 | 2.00 | 4.00 | 5.50 | 15.00 |
| R(m) | 1.10 | 1.85 | 1.80 | 3.40 | 2.96 | 6.20 |
| a(m) | 0.20 | 0.45 | 0.50 | 1.0 | 1.18 | 2.00 |
| Btor (T) | 3.00 | 3.50 | 3.50 | 4.00 | 2.25 | 5.30 |
| Max pulse (s) | 1 | 100 | 300 | 60 | 100 | 600 → 3000 |
| Circular or divertor | Divertor | Divertor | Divertor | Divertor | Divertor | Divertor |
| #TFCoils, conductor | 16, NbTi | 16, NbTi | 16,$Nb_3Sn$ | 18, Copper | 18,NbTi | 18,$Nb_3Sn$ |
| PF coil conductor | NbTi, Cu | NbTi | $Nb_3Sn$, NbTi | Copper | $Nb_3Sn$, NbTi | $Nb_3Sn$,NbTi |
| Auxiliary heating | LH, EC, IC, NB | All | All | All | EC, NB | All |

tokamak JT-60U was added to the table since it contributed steady-state operation scenarios and provided the basis for the superconducting JT-60SA.

Note that many of the tokamaks with superconducting toroidal and poloidal coils are large, with major radius greater than 1.8 m and high plasma currents of greater than 1.0 MA. They also include a divertor and plan long pulses. None use tritium except ITER, added for comparison. Although the machines are highly versatile, the scenarios described here generally concentrate on their long-pulse experiments.

### 6.1.2 Key Elements and Constraints of Steady-State Operation

The main reason for the use of superconducting coils in a reactor is to remove the need for hundreds to thousands of MW of electrical power just to sustain the toroidal and poloidal coil systems, making net energy gain in a reactor very difficult. A steady-state tokamak reactor needs a source of current drive, although tokamak design can optimize the inherent bootstrap current drive of the plasma. These current drive sources are expensive and they need electricity, so they require a high degree of optimization and must run steady-state. Neutral beam injectors with positive ions are only a few tens of percent efficient due to low neutralization efficiency. Negative ions are much more efficient. RF heating can be very efficient [3].

To achieve steady-state operation or at least high duty cycle, all components need to reach constant values and all cooling must be active. Tokamaks are currently by their nature pulsed, and it is still an open question if a tokamak, once started, can operate continuously for months or years. If not, can a tokamak discharge be started up, shut down and restarted with a high duty factor? Perhaps the greatest challenges of steady-state operation are in the domain of plasma-wall and plasma-divertor physics interactions.

## 6.2 Design Space for Superconducting and Long-Pulse Tokamaks

### 6.2.1 Goals for Using the Design Space

The superconducting tokamaks that have operated or are under construction so far, with the exception of ITER, did not include high power fusion of d-t as an objective. Only the most recent and largest superconducting tokamak JT-60SA has a plasma current that could fully confine fusion alpha particle (possibly also KSTAR) but neither of them uses tritium. Their primary goals included:

- Demonstrate the technology of using superconducting toroidal field coils.
- Demonstrate the technology of using superconducting poloidal field coils.

- Save electricity costs and minimize the requirements for motor generators.
- Fully prototype the machine integration of superconductors.
- Achieve very long pulse or steady-state in
  - toroidal and poloidal field operation;
  - plasma duration and plasma and impurity parameters;
  - plasma diagnostic operation such as magnetic probes;
  - feedback control of plasma and other systems;
  - auxiliary heating and current drive equipment;
  - plasma-facing wall, limiter and divertor components;
  - plasma-facing and vessel wall hydrogen isotope inventory;
  - pumping and gas handling systems.
- Establish highest plasma properties possible in quasi-steady regimes.
- Achieve "advanced tokamak" operation which includes high beta, high bootstrap current and steady state current drive.
- Study phenomena that finally terminate discharges.
- Examine restarting after discharge termination.

Achieving all these goals simultaneously is a major challenge.

### 6.2.2 *Technical Aspects of Superconducting Magnets*

Designing a tokamak with some or all superconducting coils involves several important design choices and constraints:

- Conductor type: NbTi, Nb3Sn, high temperature superconductor yttrium-barium-copper-oxide and their properties, such as malleability.
- Critical temperature for superconducting transition, and critical field versus temperature and maximum safe operating temperature
- Stabilizer against quench: Copper or Al matrix and stabilizer ratio to conductor
- Coolant (gas or liquid or super-cooled, supercritical He)
- Superconducting and normal connectors to room-temperature electrical cables.
- Maximum allowed rate of current and magnetic field change
- Quench properties and risk to magnets, especially during plasma disruption
- Coil heating and curing, casing, electromagnetic shielding, thermal capacity before quench, quench recovery
- Susceptibility to neutron damage when used in a fusion reactor

In addition to superconductor characteristics, certain system architectures in conventional tokamaks are not easily available with superconductors, in particular demountable toroidal field coils, although high temperature superconducting joints are under development (see Chap. 8). All superconducting coils used so far were delivered to the construction site as fully wound coils. Therefore, superconducting poloidal field coils, also fully wound, needed to be located outside toroidal field coils.

A fundamental choice in systems architecture of superconductors is whether or not to enclose the entire magnetic system and vacuum vessel of a tokamak inside a "bell-jar" cryostat which provides a vacuum and thermal protection, or whether to have cryostats for individual or groups of coils. Important considerations include how the mechanical stresses and thermal conduction to mechanical structures will be dealt with, as well as thermal shielding to protect components at liquid helium temperatures. Most modern superconducting tokamaks use a giant cryostat that covers the tokamak. Auxiliary heating systems such as neutral beam injectors would be located outside the cryostat.

## 6.2.3 Systems Architecture of Choices and Trade-Offs

There are many architectural choices available among form and function, that depend both on the nature of the superconducting coils themselves and also on the other subsystems with which they interact. Some of the main questions determining form are:

- Are all coils superconducting or not?
- Are all coils inside a single cryostat?
- Are there poloidal coils inside the toroidal field coils?
- Are there poloidal coils inside the vacuum vessel?
- Is the plasma circular or elongated with a divertor?
- Is a divertor present, what is its form, and does it have active pumping?
- What methods of auxiliary heating and current drive are chosen?
- What is the overall size, aspect ratio, plasma current and toroidal field of the tokamak?

The main requirements of function are to:

- demonstrate technological solutions;
- produce steady-state operation;
- determine equilibrium values of all quantities during steady-state operation;
- identify and quantify the lessons for building a fusion reactor.

The design space of a superconducting tokamak involves all the choices mentioned in the Chap. 2 design space, but particular choices involve special trade-offs. Following are some of the most important choices and trade offs.

### 6.2.3.1 Magnetic Field Coils

- Toroidal field coils: All but one (JT-60 U) of the machines discussed in this chapter have superconducting toroidal field coils. A completely water-cooled

copper coil machine such as JT-60U which was designed for high fields, was run at lower currents to achieve very long pulse operation.
- Poloidal field coils: If poloidal field coils are superconducting, then there is a possibility of the tokamak operating at steady-state at the full design field. However, if water cooled copper or nitrogen cooled aluminium conductors are used, then operation is inertially or cooling limited. Such conductors could be wound inside the toroidal field coils or vacuum vessel for a divertor or fast stabilization coils. Such coils could withstand higher heat and radiation fluxes than superconducting coils and require less space.

#### 6.2.3.2 Superconductor Material, Maximum Field and Coil Temperature

- NbTi: Niobium titanium is the most commonly used superconductor, with critical temperature of about 10 K and critical field of 15 T. In the Large Hadron Collider, the NbTi coils are cooled to 1.9 K to allow safe operation at 8.3 T.
- $Nb_3Sn$: Niobium tin has a critical temperature of 18.3 K and a critical field of 30 T. Its typical maximum field in use is 12 T.
- High field, high temperature superconductor: It is claimed [4] that the use of Rare-Earth Barium Copper Oxide (REBCO) would allow a maximum conductor field of 25 T. However, this is relatively new technology and has not been used in any operating tokamak so far. Current leads for ITER coils were fabricated from high temperature superconductor bismuth strontium calcium copper oxide (BSCCO) 2223 tape with critical temperature 108 K.
- Summary: NbTi offers lowest cost and easiest manufacture and cooling requirements, and can best resist pulsed fields. $Nb_3Sn$ offers higher fields, but at higher cost and difficulty of manufacture. REBCO, once proven, could have great advantages (See Chap. 8).

#### 6.2.3.3 Plasma Heating and Current Drive Pulse Length

Auxiliary heating and current drive systems, such as NBI, ECRH, ICRH and LH were initially short pulse and high power. All are now being designed for longer pulse at high efficiency, which will require continuous power from the electricity grid at steady state. In a fusion reactor, some of the heat from fusion products would need to be converted to provide this electricity, at a thermodynamic efficiency depending on the coolant temperatures.

#### 6.2.3.4 Low Voltage Start-Up

Many of the earlier superconducting tokamaks were designed with an iron core and/or a central solenoid with water cooled copper. It could deliver a high starting voltage and a large number of volt-seconds to ionize filling gas, increase the plasma current and supply a low but constant voltage to maintain the plasma current as long as flux was available. Since then, tokamaks such as TCV have shown that a low starting voltage is possible, and that plasma current can be ramped up and maintained by external auxiliary heating and current drive. Tokamaks with superconducting ohmic solenoids and poloidal field coils must use low voltage start-up to avoid self-heating from eddy currents that could quench the coils.

#### 6.2.3.5 Ohmic Heating Transformer Coil and Aspect Ratio

With the advent of low or zero voltage plasma start-up and current drive, there is a temptation to design a tokamak without a central solenoid or iron transformer core. It would greatly ease design of the machine, and allow lower aspect ratio as in spherical tokamaks. However, no tokamak has yet taken this option. It would be advantageous to gain high confidence that a central transformer is not needed.

#### 6.2.3.6 Plasma Shape and Divertor

It is a general consensus that a tokamak reactor will need some plasma shaping and have an x-point separatrix leading to a divertor to handle plasma exhaust. The shaping coils to accomplish these tasks have much higher current when these coils are outside rather than inside the toroidal field coils. Thus this architecture choice is fundamental for machine design.

#### 6.2.3.7 Reactor Conditions, Plasma Facing Components and Divertor

The key questions are what conditions should be modelled to see if the experiment is relevant to a reactor. A florescent light is a steady-state plasma, but its low temperature and density means it is a poor model. Some of the key parameters for similarity to a reactor are:

- Heat flow per unit area;
- Wall and divertor material;
- Wall conditioning;
- Divertor and gas pumping geometry;
- Component trapped gas inventory;
- Lost fast ions and neutrals from heating and fusion products;
- Fully steady-state conditions;
- In the plasma and scrape-off layer, the densities and temperatures of ions, electrons and impurities;
- Resistance to plasma disruptions, runaway electrons and halo currents;
- Plasma start-up and ramp-down capabilities.

#### 6.2.3.8 High Performance Versus Long Pulse Operation

The ideal goal of a machine is high performance at steady state in fusion reactor conditions with tritium operation and tritium breeding in a commercial fusion reactor. Even without tritium, high performance means a large and expensive machine with components all operating near their limits. Machines such as JT-60SA and ITER aim at high performance and long pulse.

Some machines that produce very long pulses have shown that there are questions of wall recycling and wall inventory that can only be answered by steady-state operation, which needs to be extended to reactor conditions.

In what follows, the options actually chosen in various tokamaks are analysed. Their operational scenarios are described and added to the Chap. 2 design space.

## 6.3 HT-7 (Formerly T-7)

### *6.3.1 Design*

The HT-7 superconducting tokamak [5] was rebuilt in China from the Kurchatov Institute T-7 to improve access to the plasma, allow more diagnostics and to add wave launchers for high power LHCD and ICRH. Modifications were made for long pulse operation with higher plasma parameters and finished in 1994. HT-7 has a layered structure, starting from the plasma and going outwards:

- inner bakeable liner
- vacuum vessel
- inner copper shell screen with liquid nitrogen cooling
- superconducting magnet with forced flow helium cooling
- outer copper shell screen with liquid nitrogen cooling
- cryostat
- poloidal field and coils and ohmic heating transformer
- iron core transformer with return legs

Four units of rigid segments with bellows connections make up the vacuum vessel, with 34 ports for diagnostics, wave launchers and other experimental equipment. The toroidal resistance of the discharge chamber is 2.3 mΩ. A changeable stainless steel liner is assembled inside the vessel as the plasma facing first wall for improving the capabilities for heating, cooling and wall conditioning. It also protects the bellows from the plasma bombardment. The vessel can be baked by poloidal field induced current and cooled by water flow.

The toroidal field coils, which in T-7 consisted of 48 NbTi-Cu superconducting windings with nine coolant channels, were rearranged into 24 coils for the improvement of accessibility with still acceptable ripple. Two layers of copper screen with 15 mm thickness and the cryostat were remade to increase the number and size of the ports, and the current lead was reconstructed to minimize the cryogenic dissipation. The design parameters of HT-7 are given in Table 6.1.

### *6.3.2 Scenarios*

#### 6.3.2.1 Long Pulse Operation of 240 s with LHCD

Long pulse operation of up to 240 s [6] used lower hybrid waves to sustain a plasma with 1 keV electron temperature. The discharge was terminated due to recycling which caused an uncontrollable density rise.

#### 6.3.2.2 Long Pulse Operation of 400 s with Cooled Belt Limiter

New techniques were used to raise the plasma discharge length to 400 s [7] achieving 100% current drive at 50 kA plasma current, without flux from the ohmic transformer. The temperature on the belt limiter of the circular plasma was controlled during this time. The limitation for an even longer pulse came from the uncontrollable density rise due to outgassing from the first wall, which was not actively cooled.

### *6.3.3 Robustness, Fragility and Relevance to Steady-State Operation*

As a first demonstration of the use of superconducting toroidal field coils, T-7 and its successor HT-7 were successes. The long pulse discharges demonstrated the physics of long pulse current drive, which was limited by a non-cooled vacuum vessel wall.

## 6.4 TRIAM-1M

### *6.4.1 Design*

The Japan TRIAM-1M tokamak [8, 9] has as objectives:

- Development of the high-field superconducting magnet for fusion reactor and establishment of SC magnet technology
- Development of steady-state tokamak operation using non-inductive current drive
- Investigation of impurity dynamics and hydrogen recycling

- Feedback control of non-circular plasma for steady-state operation
- Additional heating for high performance plasma

At the time that these objectives were originally formulated in the mid 1970s, they were already highly ambitious and highly relevant to fusion reactor development and were all achieved in subsequent experiments.

TRIAM-1M combined very high field superconducting toroidal field coils with an elongated and divertor plasma. There were 16 toroidal field coils made of $Nb_3Sn$ to continuously contain a plasma with parameters of $R = 0.8$ m, $a \times b = 0.12$ m $\times$ 0.18 m. The maximum field reached 11 T at the windings and 8 T at the plasma centre. Operation was begun in 1986. Continuous operation of the superconducting magnets system longer than 100 days was demonstrated.

The plasma chamber was made of stainless steel SUS304L and the poloidal D-shape limiters and toroidal divertor plate were made of Mo. The surface of the plasma chamber was cleaned effectively using electron cyclotron resonance discharge cleaning (ECR-DC) without baking the chamber. In ohmic heating (OH) experiments, the maximum plasma current of 420kA was achieved and the elongation of the D-shaped plasma cross section reached 1.4/1. In order to realize continuous steady-state operation, the non-inductive current drive experiments used 2.45GHz, 50 kW LHCD from a klystron and a 4 × 1 grill launcher.

### *6.4.2 Scenario: 5 h Fully Non-inductive Steady-State Operation with LHCD*

A 1-h [9], a 2-h and then over a 3-h discharge [10] was produced. Current ramp-up without using the centre solenoid coils was investigated by using ECRH for breakdown and LHCD for current ramp-up. The very long pulse regime was realized in a low electron density and low power (< 10 kW) discharge. The global particle balance in long duration discharges was investigated and 30 min after the plasma initiation, the gas supply was turned off, and a constant plasma density was maintained. After that the density was sustained by the recycling flux alone until the end of the discharges.

A steady-state, fully non-inductive plasma driven completely by LHCD that lasted 5 h [11] was operated in a low power regime that allowed accurate measurements of heat balances. It was confirmed that the total heat to the plasma facing components equalled the input power.

### 6.4.3 Robustness, Fragility and Relevance to Steady-State Operation

This highly robust and adaptable machine was able to explore extremely long pulses to investigate recycling and power balances. The relevance to steady-state operation is important in the regime of particle balance, but low power scenarios do not give the full picture of power handling and particle control together.

## 6.5 T-15

### 6.5.1 Design

The T-15 [12] was a large, circular plasma tokamak with R = 2.43 m, a = 0.70 m and a toroidal field 5 T, allowing plasma currents up to 2.3 MA. Its main feature was the toroidal field super-conducting winding using forced flow He-cooled $Nb_3Sn$ conductor. The poloidal field coils were made of liquid nitrogen cooled aluminium. The entire magnet system was placed inside a common cryostat.

### 6.5.2 Scenario: Ohmic Heating in T-15 and Test of Superconducting Windings

Stable plasma discharges using ohmic heating were established for 1 s at a current up to 700 kA. It was shown that the tokamak could operate under normal plasma production conditions.

### 6.5.3 Robustness, Fragility and Relevance to Steady-State Operation

The technical demonstration of large superconducting coils made of $Nb_3Sn$ was a success. Operation was limited due to funding problems and political upheavals.

## 6.6 Tore-Supra, Later WEST, a Case Study of Prototyping a Superconducting Tokamak Reactor

Following Checklist 1.6 from Chap. 1, the form and function options for prototyping superconducting coils for a fusion reactor have been examined. In the case study made of the French Tore-Supra, a large scale tokamak with superconducting toroidal field coils, its evolution was followed to show the continued relevance to tokamak fusion reactors. The relevance to ITER was to demonstrate that superconducting coils could indeed be used in a large scale tokamak. The later conversion to WEST allowed the exploration of long pulse operation using ITER divertor technology and shape along with tungsten as a plasma facing material. The emergent properties of the system showed how long pulse operation could be obtained under these new conditions.

### *6.6.1 Design*

#### 6.6.1.1 Toroidal Field Coils and Vacuum Vessel

Tore-Supra [13] had a toroidal field magnet (9 T maximum field) that was built with superconducting coils using NbTi conductor and a copper matrix with CuNi barriers, cooled by a superfluid helium bath at 1.8 K. No convection or mass flow was necessary for cooling by taking advantage of the large heat transfer and absorption properties of superfluid helium. The winding was made with a small monolithic conductor and enclosed in a strong welded steel casing which supported most of the electromagnetic forces and acted as a magnetic shield (20 ms time constant) against fast field changes. This casing dissipated most of the induced energy losses and was cooled by a supercritical flow of helium at 4.5 K.

All the coil casings were interlocked in order to have the full torus behave like a solid block, restraining the centripetal forces by compression on the coils and the axial torque coming from the vertical field by insulated shear keys. This design led to a mechanically strong magnet with low heat losses and no difficult support, but required a very precise machining and highly stressed insulators.

The final design used 18 circular coils which was the best solution for convenient access to the plasma, reduced cost and acceptable toroidal field ripple at the plasma. After winding and before being enclosed in their welded heavy case, coils were tested pair by pair at 1.8 K in a solenoidal field of the same average value as in the torus. In case of emergency, the total magnetic energy inside the torus (600 MJ) could be quickly (15 s) dumped into large resistors.

Two vessels provided separate vacuums for the plasma and the toroidal magnet. They were linked together in 18 poloidal planes, either by large access ports (12 sections) or by large plates which housed piping connexions to the inner vessel. The two completely welded vessels were built with a high modularity from rigid sectors joined through bellows. Each sector had a double wall, made from thin shell sheets linked by a sheet lattice, providing good mechanical strength, low weight and very low leak probability. The outer vessel, in spite of high resistivity toroidal bellows, formed a rigid toroid which was restraining in compression.

Two radiation shields were made from two steel sheets and cooled at 80 K by a gaseous helium flow. They protected each side of the toroidal magnet against thermal radiation from the hot vessels. These shields formed a rigid toroid, with 18 bolted insulating gaps. The complete torus was built from six modules, each one enclosing three coils, assembled on site. After building the toroidal magnet and making cryogenic pipe connexions, the nested torii, radiative shields and vessel were successively welded from inside and outside along poloidal cuts and on the six coaxial supports of the toroidal magnet.

#### 6.6.1.2 Poloidal Field Systems

One single system of axisymmetric coils plus an iron core was used to induce the plasma current with a starting voltage of 100 V/turn. It produced a plasma MHD equilibrium through the control of the vertical field amplitude and its curvature index. The iron core magnetic circuit was chosen for its lower cost.

Long pulse operation at maximum plasma current of 1.7 MA for 30 s required a system of power supplies directly connected to the grid. The pulse length limitation to 30 s at maximum performances resulted from:

- the flux swing available for the plasma current flat top
- the coil temperature rise of the water-cooled copper poloidal field coils
- the permitted duration of a strong power pulse supplied from the grid.

The total flux swing available inside the iron core was 14 Volt-seconds, of which 6 were required for plasma current rise up to 1.7 MA, using 3.7 for the flux inside the plasma, 1.4 for the flux between the plasma boundary and the iron core, 1.0 for the ohmic consumption inside the plasma during this phase and 8 for the plasma current flat top (due to residual plasma resistivity at high electron temperature, estimated at 0.3 V, leading to a flat top duration a little below 30 s).

A feature of this system was that at reduced current in the plasma and poloidal field coils, external current drive could maintain the plasma current indefinitely.

Essentially steady-state operation was achieved on Tore-Supra [14]. Long pulse operation had previously been obtained in a large tokamak on JT-60U [15] and on TRIAM [10]. On Tore Supra, high-power steady-state operation required:

- superconducting magnets to limit the energy consumption;
- long pulse capability of heating and current drive systems;
- cooled plasma facing components able to handle the injected power;
- long pulse capability means for fuelling and pumping the discharge;
- diagnostics and real time plasma control.

To meet these goals, a toroidal pumped limiter defining the circular plasma and the entire vacuum vessel wall were actively cooled with high pressure water. All of the plasma heating systems were designed to operate at several megawatts for several hundreds of seconds. Special supersonic molecular beam injector modules were developed to have much higher fuelling efficiency than standard gas puffing. A continuous pellet injector of solid hydrogen was also installed. Full real-time control was developed for plasma current profile and MHD equilibrium.

The result was that steady discharges were developed with zero applied loop voltage for several hundred seconds.

## 6.6.2 *Scenarios*

### 6.6.2.1 High Power, 400 s Long Pulse in Circular High Current Plasma with LHCD and 1 GJ Injected

A high power, 400 s long pulse in circular high current plasma with LHCD and 1 GJ injected [14] was achieved. The plasma parameters were $B_T = 3.4$ T, $I_p = 0.7$ MA, $n_e(0) \leq 3 \times 10^{19}$ m$^{-3}$, q(0) between 1.5–2, $T_e(0) \sim 5$ keV, $T_i(0) \sim 1.5$ keV, zero loop voltage and constant current. This discharge achieved the Lower Hybrid Enhanced Performance (LHEP) regime with a negative magnetic shear within a narrow core region ($r/a < 0.2$). The improvement factor with respect to the ITER L-mode scaling was 1.35.

The in-vessel deuterium particle balance was evaluated, and only half of the injected particles were recuperated in the pumps, the rest was implanted inside in-vessel components. There was no saturation observed in wall inventory, even for multiple long pulses, probably associated with trapping in the carbon components.

### 6.6.2.2 Long Pulse Operation with LHCD and ECCD Synergies

An improvement in the efficiency of ECCD in the presence of LHCD waves was predicted and tested in Tore Supra [16], where 700 kW of LHCD was replaced by 500 kW of ECCD. Zero applied loop voltage was maintained the entire time, even though ECCD was expected to have much lower efficiency.

### 6.6.2.3 Long Pulse 120 s High Confinement Limiter Operation with Combined ICRH and LHCD

Long pulse experiments lasting 120 s using LHCD were carried out in Tore Supra, supplemented by 8 MW of ICRH power, using h-minority heating to obtain efficient bulk ion heating [14]. Acceleration of the plasma in the toroidal co-direction was observed, thereby stabilizing the ion temperature gradient modes and the trapped electron modes. The result was enhanced confinement of about 1.7 times expected L-mode times.

### 6.6.3 *Robustness, Fragility and Relevance to Steady-State Operation*

Tore Supra demonstrated highly reliable operation of superconducting toroidal field magnets during decades of operation. It also showed that long pulse operation could be achieved in this environment along with enhanced confinement regimes in a plasma with a limiter at high power loading. The fragility of limiter plasmas was dealt with by an upgrade to WEST with a divertor.

## 6.7 WEST (Tungsten {Symbol "W"} Environment in Steady-State Tokamak)

### 6.7.1 *Design*

The divertor installation conversion of Tore Supra to WEST was planned to provide an integrated platform for testing ITER divertor components under combined heat and particle loading in a tokamak environment [17]. A cross section of the new WEST divertor configuration, constructed with water-cooled copper coils inside the vacuum vessel, is shown in Fig. 6.2 [17]. The plasma volume and maximum plasma current were therefore reduced compared to Tore Supra.

The upper divertor was made of actively-cooled, tungsten-coated copper tiles, and the lower divertor had actively cooled ITER-like units, complemented with inertially cooled tungsten coated units [18]. Side limiters were all tungsten coated. Boronization was used for wall treatment.

### 6.7.2 *Scenario: Plasma and Impurity Confinement in a Long Pulse, High Power, Tungsten Divertor Environment*

A 5 MW pulse was examined with respect to the existing confinement scaling laws with aspect ratio at A = 5.6 [18] and it was found that it did not scale as expected. A 33 s electron-heated, low-torque plasma was examined and throughout the pulse, the W concentration remained stable around $3 \times 10^{-4}$. The modelled W peaking was twice lower than that of the electron density.

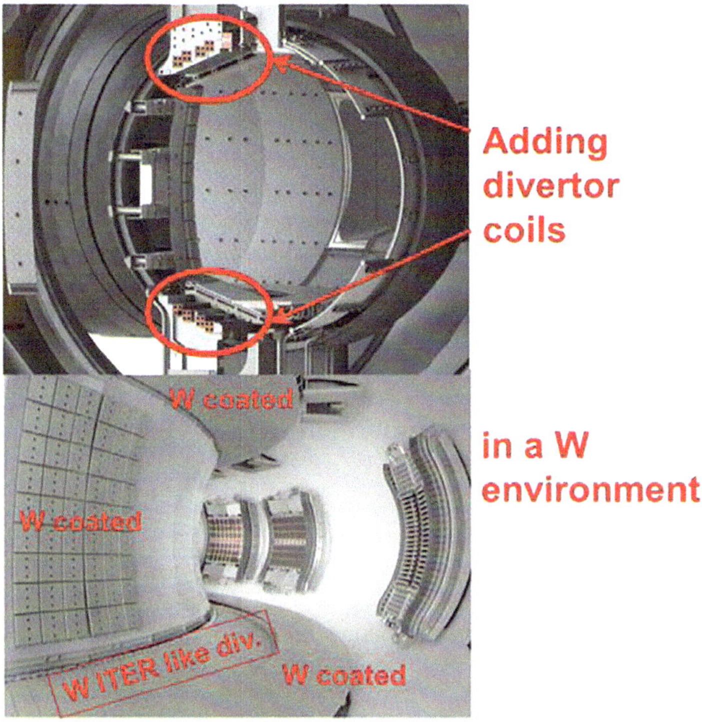


**Fig. 6.2** CAD views of WEST. (From [17], Fig. 1)

The Phase I experiments on WEST with a full tungsten environment and inertial cooling were continued [19]. Long pulse discharges with duration of 30 s to 1 min were routinely performed, obtaining:

- Up to 9 MW of combined ICRF/LHCD power was coupled to the plasma, without W accumulation.
- No failures occurred on ITER-grade plasma-facing unit prototypes, but evidence of damage was observed after several hours of plasma at moderate heat fluxes.
- Hot spots were observed.
- Cracks and local melting occurred at moderate heat flux on misaligned components.

Phase II with the full ITER-grade divertor system aiming at 1000 s duration plasmas started in summer 2021.

### 6.7.3 Robustness, Fragility and Relevance to Steady-State Operation

The highly robust Tore Supra machine was successfully converted to a divertor machine by installing normally conducting copper coils and divertor plates inside the vacuum vessel, and long pulse operation was obtained.

A major fragility was that since the divertor coils now link the toroidal field coils, any toroidal field coil fault would be very difficult to repair. Such a danger is mitigated by the previous demonstration of the reliability of these TF coils. The high power and long pulse operation are highly relevant to reactor conditions.

## 6.8 SST-1 (Steady-State Superconducting Tokamak)

### 6.8.1 Design

The India SST-1 tokamak [20] is shown in Fig. 6.3. Superconducting NbTi conductor was used for all toroidal field coils and some poloidal fields. A high vacuum cryostat enclosed all the superconducting coils and the vacuum vessel. Liquid

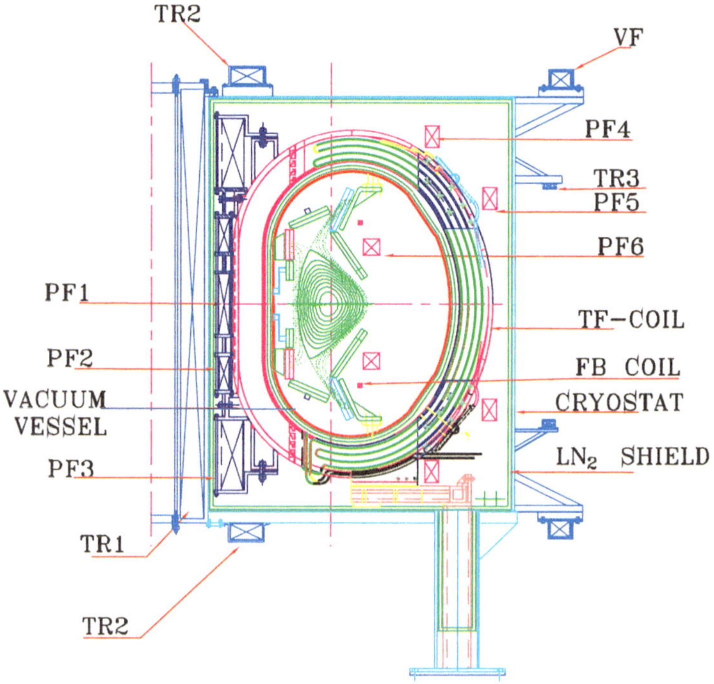


**Fig. 6.3** SST-1 Cross section. (Reprinted with permission from [20], Fig. 1)

nitrogen cooled thermal shields were placed between the vacuum vessel and the coils as well as between the cryostat and the coils in order to reduce the radiation heat load. A normal conductor ohmic transformer system was provided to initiate the plasma and sustain the current for the initial period. A pair of external vertical field coils was provided for circular plasma equilibrium at the start-up stage of the plasma. A set of saddle coils were placed inside the vacuum vessel to provide fast vertical control of the plasma, while poloidal field coils are to be used for shape control.

Other subsystems included RF systems for pre-ionization, auxiliary current drive and heating, NBI, cryogenic systems at liquid helium (LHe) and $LN_2$ temperatures, and a chilled water system for heat removal from the various subsystems. The divertor system occupied a large amount of space inside the vacuum vessel, leaving room for a small plasma compared to the toroidal coil area.

### *6.8.2 Scenario: LHCD*

Experiments using LHCD in SST-1 [21] have had discharges lasting up to 1 s. The cryogenic systems were being upgraded to allow improved cooling [22].

### *6.8.3 Robustness, Fragility and Relevance to Steady-State Operation*

The tokamak SST-1 has demonstrated the technology of superconducting toroidal and poloidal field coils. The divertor system used a significant fraction of space inside the toroidal field coils.

## 6.9 EAST (Experimental Advanced Superconducting Tokamak)

EAST [23–25] was built in China and is shown in Fig. 6.4. The main goal was to explore advanced steady-state plasma operation. The EAST design was similar to ITER in shape and MHD plasma equilibrium, yet aimed at being more flexible.

### *6.9.1 Design*

The features of EAST are as follows:

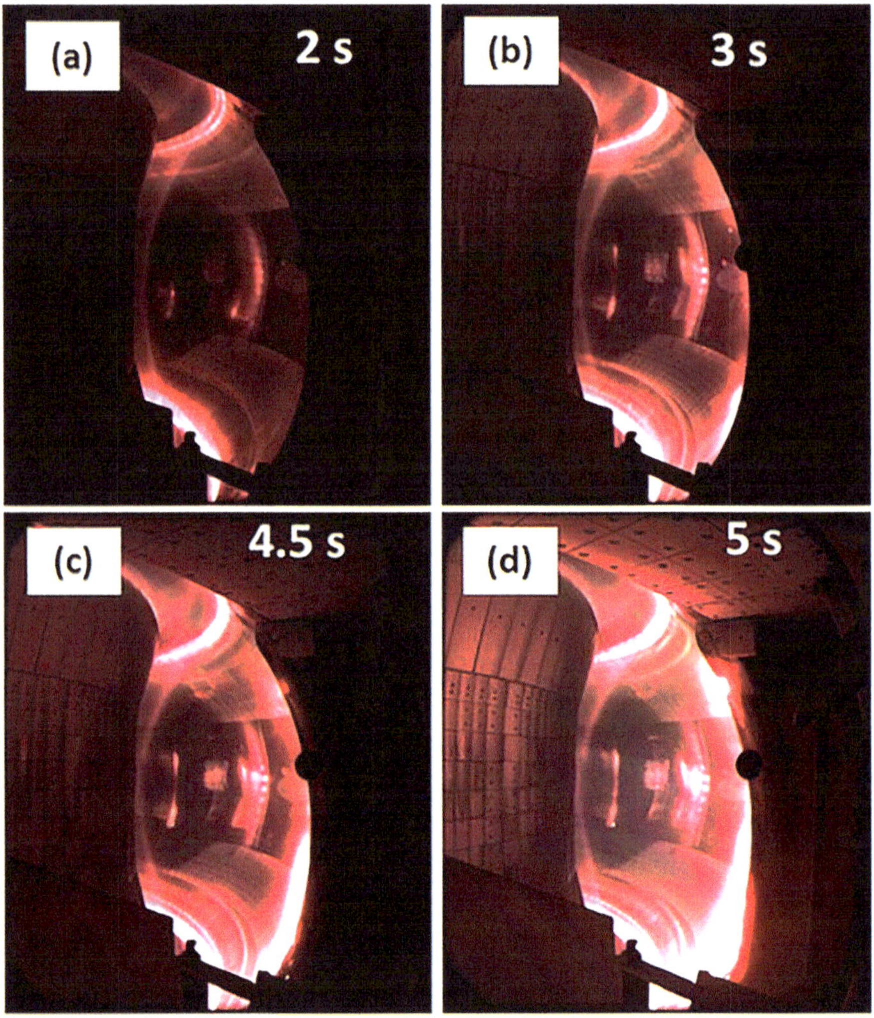


**Fig. 6.4** EAST Plasmas and vacuum vessel (From [25])

- Both toroidal and poloidal field coils used NbTi.
- Non-inductive plasma current drive and heating systems were designed for continuous operation.
- The 12 poloidal field coils each had individual power supplies for maximum control and shaping flexibility, allowing single-null and double null configurations.
- Real time data collection and feedback systems provided steady-state profile control.

- Active cooling, changeable plasma facing components and divertors were used to make power and particle handling studies very efficient.
- Advanced plasma-facing materials (PFM) and wall conditioning techniques were investigated.
- Advanced diagnostic measurements for steady-state operation were developed and installed.

Much attention was devoted to NbTi engineering development of cable-in-conduit conductor (CICC) design, fabrication and testing. Extensive calculations and tests were made of the optimum copper to superconductor ratio and three designs were defined for different coil systems.

## *6.9.2 Scenarios*

### 6.9.2.1 Long Pulse of 100 s with ECRH and LHCD in H-mode

Long-pulse steady-state discharges lasting 100 s were achieved [26] with good H-mode confinement (H98(y2) ~ 1.1). Good control of impurity and heat exhaust was provided by the tungsten divertor. ECRH and LHCD achieved fully non-inductive current drive. The X-point and plasma shape and local gas puffing were optimized to maintain RF power coupling and particle exhaust. On-axis ECRH was superimposed on the LHCD during the H-mode to avoid high-Z impurities accumulating in the core.

### 6.9.2.2 Plasma–Wall Interaction Control for H-mode Operation over 100 s with an ITER-Like Tungsten Divertor

A total power injection up to 0.3 GJ was achieved in EAST long pulse H-mode operation of 101 s with an ITER-like water-cooled tungsten (W) mono-block divertor [27]. The peak temperature of the tungsten target saturated with a heat flux ~3.3 MW $m^{-2}$. The plasma footprint on the divertor was adjusted so that the heat load was broadly dispersed. Active feedback control with deuterium fuelling or divertor impurity seeding was successful while maintaining good energy confinement in the main plasma. Active feedback control of radiative power with neon allowed heat flux reduction to the divertor.

### 6.9.2.3 Improved Long Pulse Scenario Development

Further advances were made in long pulse scenarios [28]:

- Long-pulse high-performance – A minute long-pulse steady-state high beta H-mode plasma on EAST modelled the CFETR reactor (see Chap. 8) by

obtaining $\beta_P \sim 2.0$, $\beta_N \sim 1.6$, bootstrap current fraction $f_{bs} \sim 50\%$ and $H_{98\,(y2)} > 1.3$ at $q_{95} = 6.5 \sim 7.5$. It was sustained by RF heating only with the ITER-like tungsten divertor.
- Active control of divertor radiation – Active control of divertor radiation was integrated into the high beta H-mode ($\beta_P \sim 2.5$, $\beta_N \sim 2.0$, $f_{bs} \sim 50\%$) plasma scenario without a degradation of confinement.
- Grassy-ELM regime – The grassy-ELM regime was extended to the normalized parameter space designed for CFETR.
- ELM suppression – ELM suppression was achieved.
- Liquid lithium limiter – A flowing liquid lithium limiter plate was assembled and tested.
- Helium operation – The H-mode power threshold in a helium plasma was found to be 1.2–2.2 times higher than in deuterium plasma with only RF-heating.
- Tungsten divertor – A lower tungsten divertor with high closure was installed.

### *6.9.3 Robustness, Fragility and Relevance to Steady-State Operation*

EAST has proved to be highly robust and very versatile in achieving a wide range of objectives, without evident fragilities. The operation is highly relevant to future long pulse reactors and is aimed directly at supporting the CFETR reactor.

## 6.10 KSTAR (Korea Superconducting Tokamak Advanced Research)

### *6.10.1 Design*

The Korea KSTAR [29] is shown in Fig. 6.5. The design decision was made to make both toroidal and poloidal coils superconducting and to enclose the entire tokamak system in a cryostat.

The KSTAR cryostat was an 8.8 m diameter, single walled cylindrical vacuum vessel with a dome shaped lid that provided a vacuum boundary to protect the superconducting magnets. Electrical and mechanical penetrations with bellows were designed to restrict displacements of ports due to electromagnetic loads and thermal loads. The vacuum vessel thermal shield and the cryostat thermal shield acted as thermal radiation protection for the coils. The 16 toroidal field coils were wound with $Nb_3Sn$ and the poloidal field coils with either $Nb_sSn$ or NbTi.

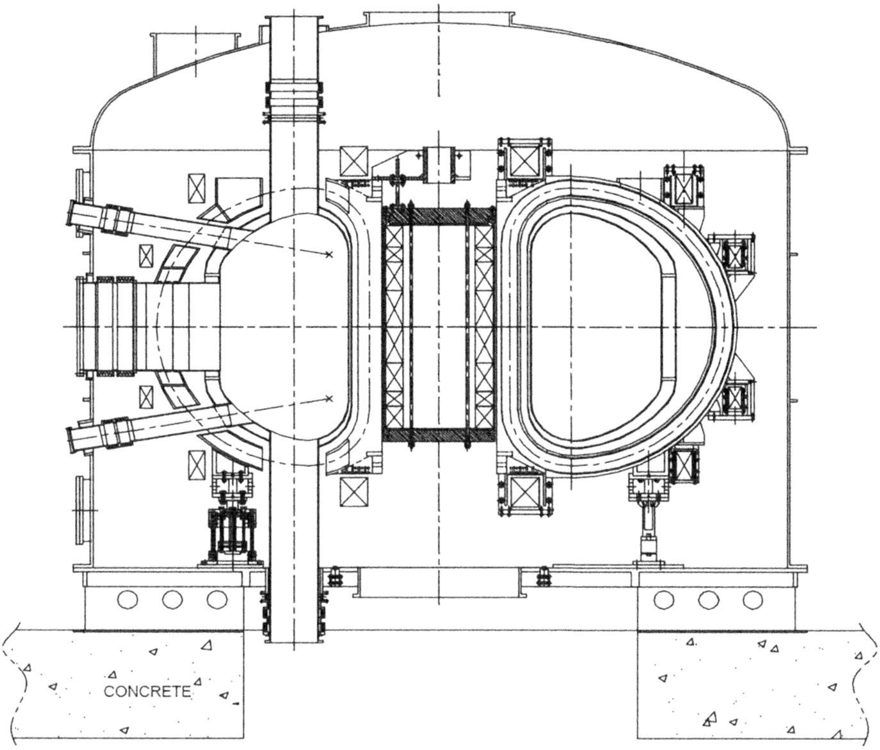


**Fig. 6.5** KSTAR configuration. (Reused with permission from [29], Fig. 2)

## *6.10.2 Scenarios*

### 6.10.2.1 Long Pulse 30 s H-mode Without ELM-Crash Using Control of Resonant and Non-resonant Modes

The in-vessel coil system with n = 1 and 2 resonant toroidal mode numbers to control perturbations was used to suppress ELM-crashes [30]. A long pulse of 34 s H-mode without ELM-crashes was produced using the n = 1 coils with $P_{NBI} \sim 2.8$ MW, $I_p = 0.56$ MA, and $B_T = 1.8$ T. The pulse length of the H-mode discharge was extended to over a minute at a moderate plasma beta.

### 6.10.2.2 Advanced Scenarios Toward Fusion Reactors

KSTAR developed scenarios relevant to fusion reactors [31]:

- Enhanced performance – Improved performance in various operating regimes was obtained and machine parameters were expanded, including early transition to diverted plasmas, sustainment of high ion temperature, hybrid scenarios, stationary high beta discharges and long-pulse H-modes.

- RMP ELM suppression – Key issues for RMP ELM suppression were further resolved concerning the optimal poloidal spectrum, collisionality and real-time control capability.
- Modelling and diagnostics – Cross-validation between advanced diagnostics and modelling provided new insights into fundamental transport process.
- SPI – Symmetric multiple Shattered Pellet Injections (SPIs) were tested for ITER disruption mitigation.

### 6.10.3 Robustness, Fragility and Relevance to Steady-State Operation

The KSTAR tokamak has proven to be a technological and scientific success, without evident fragilities. Experiments were performed on long pulse operation and designing plasma start-up scenarios for superconducting tokamaks [32].

## 6.11 JT-60-U (Japan Tokamak 60 m$^3$ Plasma: Upgraded JT-60)

### 6.11.1 JT-60-U Design and Relevance to JT-60SA

JT-60U was one of the world's large tokamaks, along with JET, TFTR and T-15, but not designed to introduce tritium. It used water-cooled copper coils and was a conversion from JT-60 to improve its divertor configuration. It was able to access H-modes at high toroidal field [33] with 4 T on axis at R = 3.4 m, a = 1.0 m and 2.7 MA plasma current, using up to 25 MW of neutral beam heating power. It concentrated strongly on advanced tokamak scenarios and set a record triple fusion product for steady-state. Its impressive results led to construction of the superconducting JT-60SA.

### 6.11.2 Scenarios

#### 6.11.2.1 Steady-State Advanced Tokamak Operation for 60 s

A 60 s current flat top and a 30 s long H-mode were obtained [34]. A high normalized beta of 2.3 was maintained for 22.3 s and 2.5 for 16.5 s in a high-$\beta_p$ H-mode plasma. An ELMy H-mode plasma was also extended and changes in the wall recycling were observed. Higher $\beta_N$ (~3) was maintained for 6.2 s in a high-$\beta_p$

H-mode plasma. A high bootstrap current fraction ($f_{BS}$) of ~75% was sustained for 7.4 s in a reversed shear plasma. Advanced tokamak operation [35] was further explored by using improved flexibility of NBI and ECCD to study H-mode, pedestal physics, impurity transport, Scrape Off Layer (SOL) and divertor plasmas, plasma wall interactions and active neoclassical tearing mode stabilization.

#### 6.11.2.2 Resolution of Physics and Auxiliary Heating Issues for ITER and JT-60SA

Mechanisms were studied for determining the rotation profile and the effect of rotation on transport and stability, ELM physics and active ELM control and plasma–wall interactions [36].

New records were set for both the beam energy of negative-ion-based neutral beams (507 keV) and the output power of 110 GHz electron cyclotron waves (1.5 MW for 4 s from one gyrotron). These confirmed and extended the heating and current drive capabilities of ITER and JT-60SA.

#### 6.11.2.3 Plasma Formation Without an Ohmic Heating Solenoid

A steady-state tokamak fusion reactor would be greatly simplified and less expensive if it could be built without an ohmic heating solenoid. Without use of the central solenoid [37], plasma start-up was accomplished by electron cyclotron and lower hybrid waves and neutral beam injection. The loop voltage was supplied by the vertical field and shaping coils. The plasma thus obtained had both internal and edge transport barriers, with an energy confinement time of 1.6 times H-mode scaling, a poloidal beta of 3.6, a normalized beta of 1.6, and a large bootstrap current fraction (>90%).

#### 6.11.2.4 Plasma Heating and Current Drive by High Energy Negative Ion Neutral Beam Injection

The current drive capability of negative-ion-based NBI (N-NBI) was extended to the reactor relevant regime [38]. Negative ions have a high neutralization efficiency at the MeV beam energy needed for penetration to the centre of a fusion reactor plasma. With an beam injected power of 3.75 MW at an energy of 360 keV, the beam driven plasma current reached 1 MA in a plasma with central electron temperature Te (0) ~ 10 keV, with a current drive efficiency of $1.55 \times 10^{19} \mathrm{A\ m^{-2}\ W^{-1}}$, approaching ITER requirements.

#### 6.11.2.5 Ferritic Insertion for Reduction of Toroidal Magnetic Field Ripple

Ferritic steel tiles (FSTs) were installed to reduce the toroidal magnetic field ripple and thereby improve the energetic ion confinement [39]. The heat load measurement on the plasma facing components indicated an improved confinement of energetic ions. These results were important for the practical application of reducing ripple in fusion reactors. Experiments were also conducted to examine the effects of ripple on H-modes [40].

#### 6.11.2.6 Highest Fusion Triple Product in a Steady-State ELMy H-mode

JT-60U was able to achieve the highest ever fusion triple product [41] in a steady-state ELMy H-mode discharges of $3.1 \times 10^{20}$ $m^{-3}$ s keV under full non-inductive current drive at $I_p = 1.8$ MA. A high-beta plasma with $\beta_N = 2.7$ was sustained for 7.4 s ($\approx 60\tau_E$), where the duration is determined only by the facility. Destabilization of neoclassical tearing modes (NTMs) was avoided by tailoring the current and pressure profiles. A real-time NTM stabilization system was also developed.

### *6.11.3 Robustness, Fragility and Relevance to Steady-State Operation*

JT-60U convincingly demonstrated the operation of a very large tokamak, and the adaptability of the original JT-60 machine to be upgraded to mitigate its original divertor design fragilities. It succeeded in its goal of delivering a record high fusion triple product under steady-state conditions, indicating that steady state operation does constrain performance below maximum transient records.

## 6.12 JT-60SA (JT-60 Super Advanced)

The JT-60SA tokamak [2] was undertaken as part of the Broader Approach agreement jointly implemented by Europe and Japan. The machine commissioning was carried out together with the ITER team. The mission of the JT-60SA project was to contribute to the early realization of fusion energy by supporting the exploitation of ITER and research towards DEMO by addressing key physics issues. JT-60SA was aimed at producing high-temperature deuterium plasmas at a plasma current of 5.5 MA, with a major radius of ~3 m and a toroidal field of 2.25 T, lasting for a duration of 100 s, which will make it by far the largest superconducting tokamak until ITER begins operation. JT-60SA would aim for full non-inductive steady-state operation with high plasma beta plasmas to establish ITER-relevant regimes.

### *6.12.1 Design*

The cross-section of the JT-60SA tokamak is shown in Fig. 6.6. The superconducting magnet system consisted of 18 Toroidal Field (TF) coils, a Central Solenoid (CS) with four modules, 6 Equilibrium Field (EF) coils. Each TF coil used a rectangular steel-jacketed NbTi cable-in conduit conductor. The PF magnet was composed of the CS and EF coils using $Nb_3Sn$ and NbTi cable-in conduit conductors, respectively. The vacuum vessel (VV) used low cobalt stainless-steel to reduce activation levels. The divertor consisted of inner and outer vertical targets with a V-shaped corner. Cryopanels were installed below the divertor cassette for particle control. Three sets of copper coils consisting of a pair of fast plasma position control coils, 12 error field correction coils and 18 Resistive Wall Mode (RWM) control coils were also installed.

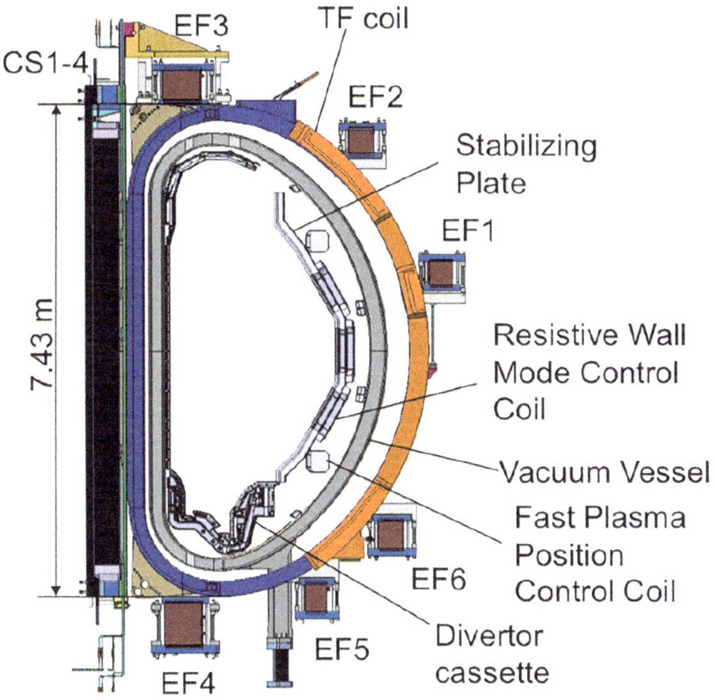


**Fig. 6.6** JT-60SA cross section. (Reused with permission from [2], Fig. 4)

The design includes:

- the capability of many plasma equilibria with divertor configurations and a low aspect ratio of A ~ 2.5;
- long pulse current drive and additional heating up to 41 MW during 100 s;
- a divertor target that can stand up to 15 $MW/m^2$;
- a negative-ion based neutral beam injection up to 500 keV;
- internal RWM stabilizing coils and a stabilizing shell;
- in-vessel components compatible with remote maintenance;
- all components for the device were consistent with the required plasma performance and design requirements;
- cost objectives that were met.

### 6.12.2 Commissioning Until Superconducting Coil Connector Fault

After 13 years, construction was completed on JT-60SA in March 2020. At the end of construction, all main components and their assembly satisfied requirements. The integrated commissioning in conjunction with ITER involved the vacuum systems, superconducting coil cool-down and powering. Successful commissioning was carried out on all systems until a connector to a superconducting poloidal field coil showed melt spots during March 2021 [42]. Commissioning was suspended due to the coil connector incident, and analyses were in progress for recovery and to avoid any repetition.

### 6.12.3 Robustness, Fragility and Relevance to Steady-State Operation

The robustness can only be evaluated after operation. The design is highly ambitious, and provides the possibility of obtaining the same type of plasma performance as the comparably sized JET tokamak, but with truly long pulse operation. The first major fault involving a superconducting coil connector did not cause any damage to the main coil systems, but such a fault could have been very dangerous during full current operation. This fault detection illustrates the contribution of a careful commissioning program towards mitigating fragilities.

## 6.13 Summary Evaluation of Robustness and Fragility in Prototyping Fusion Reactor Subsystems

There has been great progress in the prototyping of long pulse and superconducting engineering. Nevertheless, there are still fragilities which must be mitigated for the success of fusion reactors. The problems can be solved but require demonstrations.

### 6.13.1 Plasma Start-Up and Current Drive

Progress has been made in plasma breakdown and current rise. Plasmas can be initiated with very low starting voltage and even without use of a central solenoid. All tokamaks present and future still include an ohmic solenoid or equivalent poloidal coils in their design. Not including an ohmic solenoid would save critical space in the central region of the tokamak. Leaving out a central solenoid is still a risk for designers to take, but it would greatly ease some design problems, especially for components at small major radius.

### 6.13.2 Resistance to Disruptions

Superconducting tokamaks can operate after plasma disruptions. However, the largest machines are still concerned about the danger and research is in progress to either stop disruptions or reduce their effect if they occur.

Perhaps one of the great obstacles to steady state operation is the recovery from any event that ends a pulse, including a disruption. More work is needed on this fragility, especially on the need to quickly restart operation.

### 6.13.3 True Steady-State Operation

Although there has been excellent progress on long pulses, no tokamak has yet demonstrated truly continuous operation at reactor level plasmas and heat fluxes, nor rapid plasma restart from events that terminate plasmas. This is an ongoing fragility.

### 6.13.4 Long Term Gas Inventory

Although gas retention in the walls has been reduced by hot plasma facing components, it is still not a resolved issue and will be much more complicated when tritium breeding blankets are induced. Gas inventory is crucial not only for reaching true steady state operation, but also for tritium inventory in a reactor.

### 6.13.5 ITER

Many of these questions may be addressed by ITER, discussed in the next chapter.

## References


1. G. Federici et al., An overview of the EU breeding blanket design strategy as an integral part of the DEMO design effort. Fusion Eng. Des. **141**, 30 (2019)
2. S. Ishida et al., Overview of the JT-60SA project. Nucl. Fusion **51**(9), 094018 (2011)
3. R.G. Carter, Radio-frequency power generation. CERN Yellow Report CERN-2013-001, pp. 45–70. 10.5170/CERN-2013-001.45, arXiv:1303.1355 (2013)
4. M. Greenwald et al., The high-field path to practical fusion energy. MIT Report PSFC/RR-18-2 (2018)
5. J.K. Xie, et al., HT-7 superconducting tokamak and its operation. Proceedings of the 16th IAEA Fusion Energy Conference, Montreal, Paper IAEA-CN-64/API-J3 (1996).
6. B. Wan et al., Overview of the latest HT-7 experiments. Nucl. Fusion **45**(10), S132 (2005)
7. B. Wan et al., Recent experiments in the EAST and HT-7 superconducting tokamaks. Nucl. Fusion **49**(10), 104011 (2009)
8. K. Hanada et al., Steady-state operation of the superconducting tokamak TRIAM-1M. Proceedings Technical Committee Meeting, Hefei, China, Report IAEA-TECDOC-1160, p. 19 (1998)
9. S. Itoh et al., Recent progresses on high performance steady-state plasmas in the superconducting tokamak TRIAM-1M. 17th Fusion Energy Conference, Yokohama, Japan (IAEA-CN-69), paper OV2/3 (1998)
10. H. Zushi et al., Overview of steady-state tokamak plasma experiments in TRIAM-1M. Nucl. Fusion **43**(12), 1600 (2003)
11. K. Hanada et al., Power balance investigation in steady-state LHCD discharges on TRIAM-1M. Fusion Eng. Des. **81**, 2257 (2006)
12. B.B. Kadomstev, B.A. Stavisskij, V.S. Strelkov, Tokamak T-15: status and plans. 13th Conference Proceeds Plasma Physics and Controlled Nuclear Fusion Research, Washington, DC, Paper IAEA-CN-53/E-II-3 (1990)
13. R. Aymar et al., Tore Supra: basic design tokamak system. Report EUR-CEA-FC-1068, Fontenay-aux-Roses, France (1980)
14. J. Jacquinot et al., Steady-state operation of tokamaks: key physics and technology developments on Tore Supra. 20th IAEA Fusion Energy Conference, Villamoura (Portugal), Paper IAEA-CSP-25/CD; OV/2-2 (2004)
15. T. Fujita et al., Plasma equilibrium and confinement in a tokamak with nearly zero central current density in JT-60U. Phys. Rev. Lett. **87**, 245001 (2001)
16. G. Giruzzi et al., Synergy of electron-cyclotron and lower-hybrid current drive in steady-state plasmas. Phys. Rev. Lett. **93**(25), 255002 (2004)
17. C. Bourdelle et al., WEST physics basis. Nucl. Fusion **55**, 063017 (2015)
18. C. Bourdelle et al., Ramping up RF power and increasing pulse length in the full tungsten environment of WEST. 46th EPS conference on plasma physics, paper O5.101, Milan (2019)

19. J. Bucalossi et al., Operating a full tungsten actively cooled tokamak: overview of WEST first phase of operation. 28th IAEA Fusion Energy Conference, Nice, France (Virtual) 10–15 May 2021, Presentation OV/2-1, Paper IAEA–CN–286-1038 (2021)
20. Y.C. Saxena, SST-1 team present status of the SST-1 project. Nucl. Fusion **40**(6), 1069 (2000)
21. M. Kumar, Imaging of SST-1 plasma with LHCD power. 27th IAEA fusion energy Conference, Ahmedabad, India, paper IAEA-CN-258-515 (2018)
22. V.L. Tanna et al., Cryogenics system performance enhancement and attempt towards shaped plasma operations in SST-1. 28th IAEA Fusion Energy Conference, Nice, France (virtual) 10–15 May 2021, Presentation TECH/P3-18, Paper IAEA–CN–286-1253, Submitted to Nuclear Fusion (2021)
23. Y.X. Wan et al., Progress of the EAST project in China, Fusion Energy 2004. 20th IAEA proceedings of an international conference held in Vilamoura, Portugal, 1–6 November 2004, C&S Papers Series (CD-ROM), International Atomic Energy Agency (IAEA), Vienna (2005)
24. Chinese Academy of Sciences. EAST (2012) http://english.ipp.cas.cn/rh/east/. Last accessed 17 June 2022
25. Experimental Advanced Superconducting Tokamak, Wikipedia, Creative Commons 3.0 License. https://en.wikipedia.org/wiki/Experimental_Advanced_Superconducting_Tokamak. Last accessed 17 June 2022
26. B. Wan et al., Recent advances in EAST physics experiments in support of steady-state operation for ITER and CFETR. Nucl. Fusion **59**(11), 112003 (2019)
27. L. Wang et al., Advances in plasma–wall interaction control for H-mode operation over 100 s with ITER-like tungsten divertor on EAST. Nucl. Fusion **59**, 086036 (2019)
28. B. Wan et al., Advances in the long-pulse steady-state high-β H-mode scenario with active controls of divertor heat and particle fluxes on EAST. 28th IAEA Fusion Energy Conference, Nice, France (Virtual) 10–15 May 2021, Presentation OV/1-4, paper IAEA–CN–286-1020 (2021)
29. G.S. Lee et al., Design and construction of the KSTAR tokamak. Nucl. Fusion **41**(10), 1515 (2001)
30. H.K. Park et al., Overview of the KSTAR research progress and future plan toward ITER and K-DEMO. IAEA Fusion Energy conference, Ahmedabad, India, Paper OV/2-3 (2018)
31. S.-W. Yoon et al., Overview of KSTAR. 28th IAEA Fusion Energy Conference, Nice, France (Virtual) 10–15 May 2021, Presentation OV/2-3, Paper IAEA–CN–286-1230, Submitted to Nuclear Fusion (2021)
32. J.A. Leuer et al., Plasma start-up design of fully superconducting tokamaks EAST and KSTAR with implications for ITER. IEEE Trans. Plasma Sci. **38**(3), 333–340 (2010)
33. M. Sato et al., H mode of high toroidal field plasmas in JT-60U, Plasma Physics and Controlled Nuclear Fusion Research 1994. 15th Conference Proceedings, Seville, Spain, vol. 1, IAEA-CN-68/A2-9 (1994)
34. S. Ide et al., Overview of JT-60U progress towards steady-state advanced tokamak. Nucl. Fusion **45**, S48 (2005)
35. N. Oyama et al., Overview of JT-60 results toward establishment of advanced tokamak operation. Nucl. Fusion **49**, 104007 (2009)
36. A. Isayama et al., Overview of JT-60U results towards the resolution of key physics and engineering issues in ITER and JT-60SA. Nucl. Fusion **51**, 094010 (2011)
37. S. Shiralwa et al., Formation of advanced tokamak plasmas without the use of an ohmic-heating solenoid. Phys. Rev. Lett. **92**, 035001 (2004)
38. T. Oikawa et al., Reactor relevant current drive and heating by N-NBI on JT-60U. Nucl. Fusion **41**(11), 1575 (2001)

39. K. Shinohara et al., Ferritic insertion for reduction of toroidal magnetic field ripple on JT-60U. Nucl. Fusion **47**(8), 997 (2007)
40. N. Oyama et al., Effect of toroidal field ripple and toroidal rotation on H-mode performance and ELM characteristics in JET/JT-60U similarity experiments. J. Phys. Conf. Ser. **123**, 012015 (2008)
41. A. Isayama et al., Achievement of high fusion triple product, steady-state sustainment and realtime NTM stabilization in high-$\beta p$ ELMy H-mode discharges in JT-60U. Nucl. Fusion **43**(10), 1272 (2003)
42. Y. Kamada et al., Completion of JT-60SA construction and contribution to ITER. 28th IAEA Fusion Energy Conference, Nice, France. (Virtual) 10–15 May 2021, Presentation OV/2-4, Paper IAEA–CN–286-0731, Submitted to Nuclear Fusion (2021)

# Part III
# Prototype Tokamak Fusion Reactors

# Chapter 7
# ITER: A Fusion Proto-Reactor and its Large Scale Systems Integration

**Abstract** The world's largest fusion device ITER is currently under construction, with goals of producing 500 MW of fusion power, obtaining an energy gain Q = 10 and demonstrating reactor technologies such as tritium breeding. A wide range of theory, analysis, systems codes and models are currently being used for ITER. This chapter uses systems reverse engineering to examine the goals leading to the systems architecture and analyse elements and relations to determine emergent properties of robustness and fragilities. Some of the fragilities come from the intrinsic nature of a proto-reactor with large volume and low volumetric heating primarily of electrons. Special attention is paid to the steps taken in ITER to mitigate these fragilities to achieve successful construction and operation.

## 7.1 Systems Analysis Strategy for the ITER Tokamak

### *7.1.1 Overall Systems Analysis Strategy*

ITER (International Thermonuclear Experimental Reactor) is the world's largest and most ambition nuclear fusion machine which incorporates many technologies needed in a fusion reactor. It is a collaborative effort involving over 35 countries. This chapter analyses ITER's design goals, functions and innovative equipment already built. Following Checklist 1.6, this analysis summarizes the design, assigns system elements and relations, uses reverse engineering to determine the systems architecture and examines robustness and fragilities.

### *7.1.2 ITER Goals*

ITER is described as a "critical step" [1] in the development of fusion energy: its role is to confirm the feasibility of exploiting magnetic confinement fusion for the production of energy for peaceful purposes by providing an integrated demonstration of the physics and technology required for a fusion power plant. Rapid progress



F. B. Marcus, *Systems Approaches to Nuclear Fusion Reactors*, Springer Series in Plasma Science and Technology, https://doi.org/10.1007/978-3-031-17711-8_7

is being made in project construction. ITER [2] has ten times the plasma volume of JET and will be a unique experimental tool capable of longer pulse plasmas and better confinement. The machine design goals [3] are to:

- produce 500 MW of fusion power for pulses of 400 s, achieving energy gain $Q \geq 10$ from 50 MW of injected heating power for up to 600 seconds. ITER will not capture the power it produces as electricity;
- demonstrate the integrated operation of technologies for a fusion power plant: ITER will bridge the gap between today's smaller-scale experimental fusion devices and the demonstration fusion power plants of the future;
- achieve a deuterium-tritium plasma in which the reaction is sustained through internal alpha fusion product heating;
- test tritium breeding: One of the missions for later stages of ITER operation is to demonstrate the feasibility of producing tritium within the vacuum vessel;
- demonstrate the safety characteristics of a fusion device: In 2012, the ITER Organization obtained licensing as a nuclear operator in France.

### *7.1.3 ITER Top Level Parameters and Scoping Comparisons*

A systems approach to analyzing ITER begins with comparing its goals and parameters in Table 7.1 to JET and to a possible successor EU-DEMO [1, 4–6]. The major components of ITER are given in Fig. 7.1.

Since ITER is a machine with mostly superconducting toroidal and poloidal coils and actively cooled plasma-facing components, steady state operation is achievable.

**Table 7.1** Extrapolated parameters for JET, ITER and DEMO

| Parameter\Tokamak | JET baseline for ITER | ITER | EU DEMO |
|---|---|---|---|
| Start year | 2021 for d-t experiments | 2025 | 2051? |
| Max plasma current MA | 3 | 15 | 18 |
| R(m) | 2.95 | 6.20 | 9.00 |
| a(m) | 0.9 | 2.00 | 2.90 |
| $A = r/a$ | 3.3 | 3.1 | 3.1 |
| Plasma volume ($m^3$) | 90 | 830 | 2530 |
| $B_{tor}$ (T) | 2.8 | 5.30 | 5.9 |
| Max pulse (s) | 5 | $300–500 \rightarrow \infty$ | 7200 |
| $P_{fusion}$ d-t (calc) MW | 11 | 500 | 2014 |
| Fusion gain Q long pulse | 0.2–0.3 | $10 \rightarrow \infty$ | 40 |
| $Q_\alpha \equiv \alpha$ power/aux heating | 0.05 | $2 \rightarrow \infty$ | 8 |
| Tritium budget/cons. kg | 0.7 (mostly recycled) | 4.0 | 10 kg/month |
| Fusion power duty cycle | $<< 1\%$ | $\approx$15% at full power | $\approx$30% |
| Tritium breeding | No, recycling only | Breeding test only | Self-sufficient |
| Auxiliary Heating (MW) | 28 NBI, 5 ICRH | 33 NB, 20 IC,20 EC | 100 MW? |
| TotAuxHt/Vol ($MW/m^3$) | 0.33 | 0.09 | 0.04 |

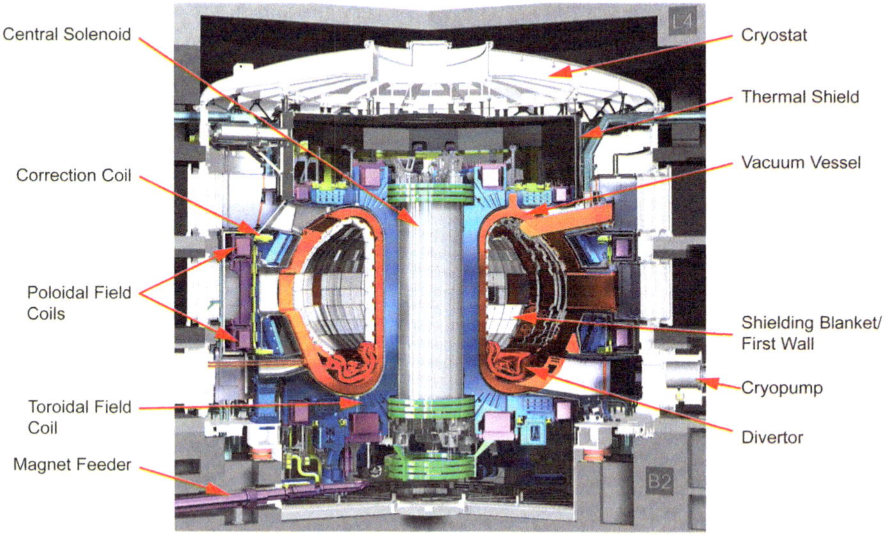


**Fig. 7.1** Cutaway view of ITER with major components. (From [1], Fig. 2)

Therefore, non-inductive current drive allowing at least many 1000's of seconds is possible [1]. With favourable plasma energy confinement, exploring the "controlled ignition regime" becomes possible with Q ≈ 30 or higher. This reactor-relevant regime could be reached, allowing operation at the same fusion power, but with much less auxiliary heating or current drive power.

The approximate and extrapolated parameters shown in this table give a comparison of the scope and goals of each machine with respect to a commercial fusion reactor, which would need:

- fusion gain much greater than 1 (ITER & DEMO);
- plasma heating mostly due to self-heating by alphas (DEMO and ITER);
- true steady-state operation (possibly ITER and DEMO);
- overall plant availability of over 75% (NONE);
- acceptable wall, blanket and shield inventory of tritium (ITER and DEMO);
- tritium breeding with a breeding ratio over 1 (DEMO and tests in ITER);
- high temperature coolant for blanket, shield and divertor (DEMO and ITER?);
- a commercially viable cost (?).

Some reactor concepts with low aspect ratio or high magnetic field discussed in Chap. 8 might operate sooner and at a lower cost. ITER is actually well on the way to taking advantage of these concepts.

The ITER aspect ratio A is 3.1, which is small taking into account the need to shield the inner part of the toroidal field coil from 14 MeV neutrons. The ITER toroidal field coil aspect ratio is 1.8 approaching that of a spherical tokamak without shielding.

The axial toroidal field in ITER is already very high at 5.3 T, with a maximum field at the toroidal field coils of 11.8 T, near the maximum for $Nb_3Sn$ under practical working conditions. Higher fields would involve not only a different conductor, but the forces and stresses would also be much higher. It would therefore be very difficult to design a support structure with materials now available.

ITER has many aspects of a prototype high field and/or a spherical tokamak reactor when the need for 14 MeV neutron shielding is fully taken into account.

Among the characteristics of a working fusion reactor, ITER should achieve Q $>>1$, but does not produce electricity or significant amounts of tritium. ITER hopes to prove that a long pulse tokamak fusion reactor is a viable concept. It will demonstrate vital technologies such as breeding blanket and shield testing for tokamaks and other fusion reactor concepts.

### 7.1.4 Systems Engineering Approaches Used in ITER Design and Construction

Systems engineering and configuration management approaches [7, 8] were used in ITER design and construction and demonstrated their value and necessity in a mega-project. Their scope of systems engineering was to:

- establish the requirements and physical architecture of the project;
- manage its development from conceptual to detailed definition;
- assess and control its performance.

The systems engineering methods chosen involved:

- programmatic objectives;
- parameters selection;
- requirements analysis and functional design;
- analysis of alternative solutions;
- verification and validation.

At the same time, systems engineering was continuously interacting with configuration management:

- requirement management;
- configuration identification;
- design-changes management;
- document and project data control;
- interfaces management;
- risk management.

The scope of configuration management was to ensure that accurate and consistent information on all aspects of the project were always available and continuously updated. The overall project time flow involved:

- conceptual design;
- functional design;
- overall architecture;
- component design;
- systems integration;
- component procurement and installation;
- operation.

The ITER project involved many challenges in both physics and technology. The physics choices and parameters were strongly and interactively linked with the technology, since ITER was first of a kind in many respects.

ITER needed a highly integrated design given that it has a very large number of parts and extremely small clearances for assembly in a very large machine. Few components were made with proven technology.

Project management was an especially complicated challenge, since the project involved international collaboration, a multi-cultural environment, distributed design activities, communications problems, concurrent engineering at several sites. There was a highly complicated procurement scheme among seven international partners, one of which was the multi-national European Union, which provides nearly half of the financial and equipment value contributions.

The complexity and interactivity of all the parts of ITER meant that any proposed change in the design had to be assessed at the systems level. Changes were first examined at the consistency level, involving requirements and design. Size compatibility with either the CAD (computer aided design) model or the actual machine needed checking. Furthermore, the machine level involved integration and impact on performance.

The project made use of the ITER interfaces matrix, which showed the many dozens of systems and the levels of their interaction with each other, all of which needed to be assessed in the original design. All needed to be reassessed in case of a design change.

The project submitted all design components to a risks and opportunities analysis, and assessed the time needed to start action or mitigation of the risks.

The results of the systems approach on ITER were that they succeeded in:

- reviewing working practices;
- clarifying and envisaging future needs;
- assessing different tools available and in use on the market;
- testing and deploying new software;
- introducing procedures for management of design change and documents.

## 7.2 Review of Top Level Systems Architecture and Innovations

### *7.2.1 Overall Systems Architecture*

Figure 7.1 illustrates the systems architecture choices for ITER. Starting from the axis of the fusion plasma itself and working outwards in minor radius, the architecture is listed.

- Plasma - The magnetically confined plasma is elongated with an x-point below, which fills the area inside the vacuum vessel efficiently.
- Instability coils - Nearest to the plasma are a pair of normal conducting in-vessel vertical instability correction coils that are connected in a saddle arrangement. A second set of ELM control coils consisting of 27 coils in a 9x3 arrangement are located inside the vacuum vessel to allow resistive wall mode control and are steel jacketed and water cooled.
- In-vessel diagnostics – Diagnostics need to be near to the plasma surface, often connected to external systems and including reflective surfaces.
- Plasma facing components and neutron shield – The plasma facing surface is a beryllium layer that coats thick stainless steel shields to protect external components, especially superconducting coils.
- Test tritium breeding modules – Modules that test tritium breeding can be easily inserted and removed in outboard ports.
- Vacuum vessel ports – Openings allow access, pumping, diagnostics, heating, breeding modules.
- Divertor – The actively cooled single null divertor is at the bottom of the inside of the vacuum vessel.
- Vacuum vessel - A double-wall stainless-steel water-cooled vacuum vessel is thick to support stresses and to help shield the superconducting coils. It is actively cooled and bakeable to achieve high vacuum.
- Inner thermal shield - A thermal shield protects the superconducting toroidal field coils.
- Toroidal field coils – The $Nb_sSn$ superconducting, 18 D-shaped toroidal field coils and their supporting structure provide a maximum field at the coils of 11.8 T.
- Poloidal field coils - Six superconducting NbTi poloidal field coils and 18 error field correction coils and their support structure are present to shape the plasma.
- Central solenoid - Down the axis of the toroid at small major radius is the central solenoid made of $Nb_3Sn$ designed to reach a field of 13 T. It can provide plasma start-up, current ramp-up and 600 seconds of current flattop by producing voltage to sustain plasma resistive ohmic heating.
- Outer heat shields - Several heat shields are cooled by gaseous helium entering at 80° K.

- Plasma and neutron diagnostics – Various plasma and machine diagnostics are installed.
- Plasma fuelling systems – Gas injection and high speed pellet injectors can fuel the plasma.
- Disruption mitigation systems – A shattered pellet injector is ready if needed.
- Cryostat - A bell-jar type cryostat which is 29 m both in height and diameter, is the largest in the world, with penetrations for external equipment.
- Heating and current drive - Auxiliary systems totalling 73 MW into the plasma will be installed, comprised of 33 MW of NBI using 1 MeV deuterium atoms from negative ions, 20 MW of ECRH at 170 GHz and 20 MW of ICRH at 40–55 MHz, the total upgradable to 130 MW.
- Vacuum pumps - Various vacuum pumping and tritium recycling systems are installed.
- Hot cells and robotics for remote maintenance - Equipment can enter the vacuum vessel for maintenance and upgrades. Components can be removed to a hot cell.
- Biological shield wall – Thick concrete walls surround the experiment.
- Computer systems – They are used for operation, data processing, plant and plasma control and plant operators.
- Balance of plant - It includes power systems, cooling, refrigeration and tritium processing.

## 7.2.2 *Innovation*

Extensive ground-breaking R&D [1] was implemented across a wide range of technology and science areas.

- Significant developments were made in the production of high performance $Nb_3Sn$ superconducting strand and in magnet technologies supporting the construction of the largest superconducting magnets produced to date.
- High heat flux plasma-facing components were fabricated which were capable of sustaining quasi-stationary heat loads of up to 10 MW $m^{-2}$ and transient loads of up to 20 MW $m^{-2}$.
- Fusion nuclear technologies such as remote maintenance and tritium breeding received specific emphasis.
- Diagnostic systems faced particular challenges in the ITER environment, and wide-ranging R&D activities were implemented to develop novel solutions.
- Routine and reliable operation in ITER requires a highly effective capability for the detection, avoidance and mitigation of disruptions and significant science and technology R&D was underway to establish this capability.
- The overall integration of the control requirements for the ITER plasma and facility presents new challenges for fusion control systems, including the need for robust safety and hardware protection.

## 7.3 Plasma with Heating and Current Drive

In this section, the design choices for the plasma with heating and current drive are examined and reviewed for robustness and fragilities.

### *7.3.1 Design Choices*

#### 7.3.1.1 Plasma Properties and Confinement

The plasma is the centre of the tokamak, and many other systems are related to its creation and maintenance. Hence these other systems will be discussed in relation to how they interact with the plasma.

The original technical goals of ITER [9] led to the design constraints on the plasma inductive phase (driven by induced voltage from the central solenoid):

- H-mode scaling law;
- normalized beta, $\beta_N = \beta(\%)$ a B / I(MA) < 2.5
- normalized line-averaged density, $n/n_{GW} = n\ \pi\ a^2\ /\ I(MA) < 1.0$, $n_{GW}$ = Greenwald density $10^{20}\ m^{-3}$)
- safety factor at the 95% flux surface $q_{95} \sim 3$;
- impurity factor $Z_{eff} \leq 2.0$;
- a well controlled, divertor plasma configuration.

At the design parameters of a = 2 m, B = 5.3 T, I = 15 MA at $q_{95} \sim 3$, then the density limit becomes $1.8\text{x}10^{20}\ m^{-3,}$ and beta becomes about 3.5%. The currently planned inductive scenario for JET has a density of $1.13\text{x}10^{20}m^{-3}$ and a $\beta_N$ of 2.0. For H-mode, the confinement time is 3.4 s and the fusion power is 500 MW.

#### 7.3.1.2 MHD Equilibrium and Vertical Stability

Figure 7.2 shows the ITER poloidal coils and the plasma last closed flux surface. The one-turn coil currents for a 15 MA plasma current are given in [10]. The outer poloidal coils have currents opposite to the plasma which provides an equilibrium vertical field. The divertor coils in the inner upper and lower corners have positive currents. The one-turn equivalent current in the lower corner PF6 at 21.84 MA is even larger than the total plasma current in order to create the x-point for the divertor and to elongate the plasma.

The problems of vertical instability are obvious from the coil currents. If the positive plasma current should be displaced vertically downwards from its midplane equilibrium position, then it would be strongly attracted by the positive coil currents in the inner lower corner. Hence rapid response normal conducting saddle coils are

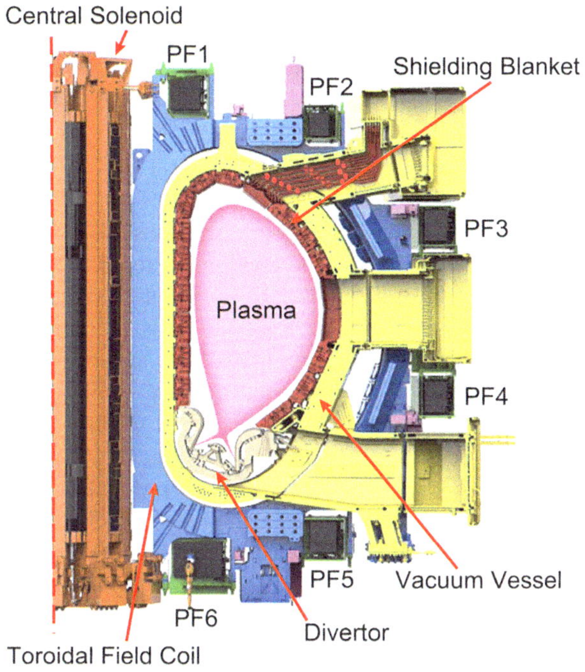


**Fig. 7.2** Poloidal cross-section of the ITER tokamak highlighting the major tokamak components (PF1–PF6 are the poloidal field coils) and illustrating the geometry of the ITER reference plasma. (From [1], Fig. 1)

installed to maintain the plasma in place before vertical displacements become too large to overcome.

#### 7.3.1.3 Alpha and Auxiliary Heating, and Current Drive

ITER initially plans 33 MW of neutral beam heating [11] and 20 MW of electron cyclotron heating and current drive [12]. There would also be 20 MW of ion cyclotron power [13] with the possibility of lower hybrid current drive [14].

At full baseline scenario operation of 500 MW fusion power, the implied alpha heating to the plasma is 100 MW and at $Q = 10$, the implied total auxiliary heating power during constant parameter operation is 50 MW. Hence the total heating power, which exactly balances the loss power in steady-state, is 150 MW. It will be necessary to follow a scenario trajectory where auxiliary heating can bring a plasma close enough to the production of large amounts of alpha heating.

#### 7.3.1.4 MHD Instability

There are a wide range of instabilities that need to be controlled inside the plasma which can cause confinement deterioration or even a disruption. At the edge of the plasma, excessively large ELMs could damage plasma facing components. The tools to control instabilities include plasma current ramp trajectories and external current drive and heating methods that create appropriate current and plasma pressure profiles. External instability coils are also used to control the ELMs.

#### 7.3.1.5 Plasma Start-Up

The physics of plasma start-up as proposed in ITER have been studied in existing tokamaks [15]. Fully superconducting tokamaks produce low loop voltage at the plasma. In ITER, the chosen E ~ 0.3 V/m is at the low end of successful breakdown for existing tokamaks. ECRH heating has demonstrated the possibility of lowering the electric field required for breakdown and can provide power to the plasma during burn-through when the plasma current is low and other heating methods are inefficient. Modelling results indicate that burn-through should be possible in ITER for low filling densities of $1.5x10^{17}/m^3$ and a post avalanche density of $<1.5x10^{18}/m^3$, However, if significant impurities are present, breakdown may not be successful. It may require 5 MW of ECRH for start-up. ITER plans for several MW of ECRH power and to use of some of that during start-up.

#### 7.3.1.6 Plasma Performance Simulations

Detailed simulation of the full plasma cycle from low current after start-up to full plasma current and then ramp down has been experimentally simulated in several smaller tokamaks [16]. Computer simulations have been made for the ITER baseline scenario [17], advanced tokamak scenarios with long pulse and current drive [18], and both [19].

The first two simulations interactively use TRANSP [20], TSC [21] and the integrated modelling code PTRANSP [17] to compute the evolution of the plasma and its MHD equilibrium states. The codes use transport models such as Coppi-Tang [22], GLF23 [23] and MMM95 [24]. The pedestal is modelled with EPED1 [25]. ICRF is modelled by TORIC [26]; the NBI modelled by NUBEAM [27], LHCD by LSC [28] and EC by TORAY [29]. The MHD flat top equilibria are refined with JSOLVER [30], stability against kinks is calculated with PEST-1 [31] and ballooning stability is calculated with BALMSC [32].

All simulations of the baseline scenario used R = 6.2 m, a = 2.0 m, B = 5.3 T, elongation=1.8, triangularity 0.4, with the current ramped up to 15 MA during the first 100 s. Simulations included impurities plus argon injected to radiate power away from the divertor plates. A wide range of plasma and heating parameters were explored, indicating a large number of paths to successful fusion production. Ion and electron temperatures up to 40–60 keV and densities of $10^{20}$ $m^{-3}$ might be realized with energy confinement times greater than 3 s.

Advanced tokamak scenarios begin with a 0.5 MA plasma which is grown to full size and shape by 14 sec, with a diverted x-point plasma and a current of 3 MA. Non-inductive advanced scenarios in ITER will need to operate with ITBs at normalized pressures close to the ideal no-wall limits in order to achieve $H_{98} = 1.6$ and $Q > 5$ at a plasma current of 9MA. It was found that 73 MW of heating power (20 MW each of IC and EC combined with 33 MW NB) can sustain up to 7.4MA of non-inductive current with $Q = 3.6$.

## 7.3.2 *Robustness, Fragilities and Opportunities*

### 7.3.2.1 Low Power Density and Low Direct Ion Heating

In principle, ITER is highly robust for achieving its plasma objectives, with flexibility in configuration and heating scenarios, carefully prototyped divertor and plasma-facing components.

However, fundamental fragilities exist because of the large plasma volume and the reliance on alpha heating power being greater than auxiliary heating power:

- The auxiliary heating power per unit volume is five times lower than in JET.
- Lower heating per unit volume competes against volumetric losses including impurity radiation.
- To achieve its full operating parameters, ITER relies on twice as much alpha heating power as auxiliary heating power.
- Therefore, ITER full parameter operation cannot be simulated in d-d alone.
- In previous experiments, d or t ions which produced fusion had been heated directly by NBI, but in ITER, all the heating methods predominantly give their energy to electrons. Good confinement is therefore needed so that the hot electrons can heat thermal ions to fusion temperatures.
  - The extremely high-energy negative ion NBI at 1 MeV primarily heats electrons.
  - EC heats electrons directly.
  - IC minority ions accelerated to MeV energies heat electrons as would LH. There are some IC scenarios which heat ions preferentially.
  - The 3.5 MeV alpha particles from d-t fusion preferentially heat electrons until the electron temperature exceeds 50 keV [33].
  - Ohmic heating heats electrons and falls as electron temperature rises.
  - Ion heating is therefore dominantly from electron transfer to ions, which is less efficient at high electron temperature.

The fundamental plasma fragility is therefore not only low power per unit volume, but also the heating goes preferentially to electrons which transfer power to ions to produce alpha heating to reach full parameters. ITER provides the opportunity to prove that this fragility can be overcome in fusion reactors. ITER participants have conceived and fully simulated many scenarios supported by detailed calculations to mitigate this fragility, and ITER can likely be expected to be a success.

#### 7.3.2.2 Low Voltage Start-Up

Low voltage start-up is an intrinsic fragility of superconducting tokamaks. The ITER starting voltage per meter is lower than in other machines. A large amount of EC heating is available to mitigate the problem at first harmonic at full toroidal field and 2nd harmonic at half field tests.

#### 7.3.2.3 Large Distance from Plasma to Poloidal Field Coils and Disruptions

The large divertor coil currents as discussed for Fig. 7.2 illustrate the trade-offs and design choices implied in choosing a single null divertor configuration for the plasma with the poloidal field coils outside the toroidal field coils. In case of a vertical instability or disruption towards the divertor coil, a very rapid response would be needed to move the plasma back to equilibrium before it rapidly accelerated. This fragility leads to the mitigation of putting normal conducting saddle coils inside the vacuum vessel for a rapid response that is not attenuated.

#### 7.3.2.4 Long Pulse Operation

ITER has not included LHCD, which produces high-efficiency current drive, in the initial configuration. This omission may prove to be a fragility in advanced tokamak scenarios which could be mitigated by adding this capability later.

## 7.4 Detailed Subsystem Design and Emergent Properties

In this section, the design choices of more major subsystems are examined and reviewed for robustness and fragilities. Analysis concentrates on particular design space choices and their implications.

### *7.4.1 Instability Coils*

#### 7.4.1.1 Design Choices

Experiments on vertical instability control by Alcator C-Mod, DIII-D, NSTX, TCV, and JET have quantified controllability and robustness requirements in the presence of realistic noise and disturbance [34]. Vertical instability saddle coils were designed for ITER [35] and significantly increase the range of plasma controllability. These coils and ELM control coils are shown in Fig. 7.3.

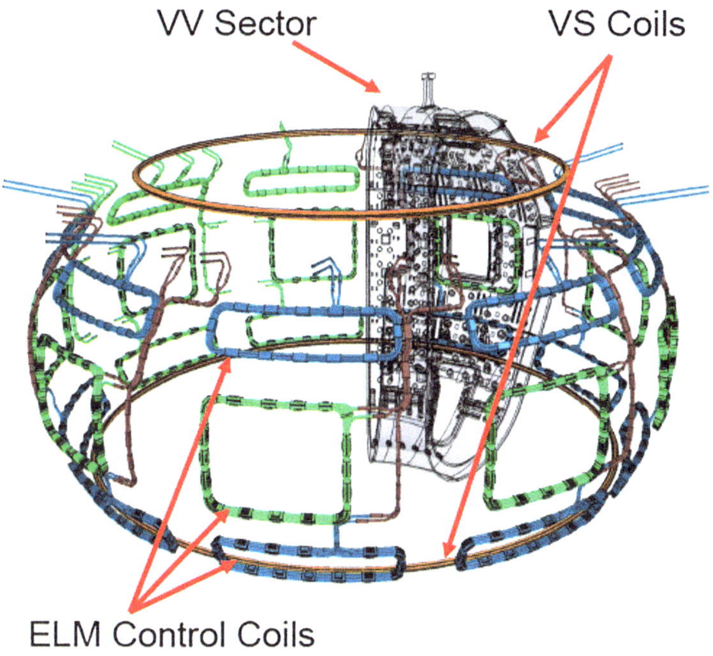


**Fig. 7.3** ITER vertical and ELM stability coils inside vacuum vessel. (From [1], Fig. 3)

The ELM control coils [1, 36] were designed as an array of 27 coils fixed to the wall of the vacuum vessel. These coils provide resonant magnetic perturbations in order to control the plasma so that ELMs are avoided or modified. Each in-vessel coil is wound from about 50 metres of conductor consisting of a 59-mm outer diameter stainless steel jacket, an insulating layer of magnesium oxide to withstand neutron radiation and an inner copper conductor with water cooling.

#### 7.4.1.2 Robustness, Fragilities and Opportunities

The robustness of ITER is increased by these coils which mitigate the dangers from a range of instabilities and open the operating window for better plasma stability and edge control. A fragility of these systems occurs during a disruption and possible induced halo currents that touch the coils. Wall supports and coil protection should suffice. However, repairs would be complicated.

## 7.4.2 *In-Vessel Diagnostics*

### 7.4.2.1 Design Choices

ITER will have about 50 large scale diagnostics systems, many of which are in-vessel or involve in-vessel components. An example of the evolution of the most fundamental diagnostic in tokamak physics is the magnetic field probe. Because of the need for long pulses, the standard measurement techniques of magnetic fields with coil loops are no longer reliable. A whole new technology using the Hall effect [37] had to be designed, developed and tested in a highly hostile environment. Each sensor contains a pair of bismuth Hall probes, themselves an outcome of extensive R&D. This included neutron irradiation (to $10^{23}$ n/m$^2$), temperature cycling tests (73–473 K) and tests at high magnetic field (to 12 T).

### 7.4.2.2 Robustness, Fragilities and Opportunities

The diagnostic was designed to be robust under all circumstances, but there are still fragilities due to thermal expansion or misalignments which have been mitigated by design features and testing.

## 7.4.3 *Plasma Facing Components and Neutron Shield*

### 7.4.3.1 Design Choices

The plasma-facing components as part of the ITER neutron-shielding blanket system posed major challenges [38]. The blanket design was constrained by design and interface requirements from the plasma, vacuum vessel, in-vessel coils and blanket manifolds. The blanket system design consists of 440 blanket modules covering 600 m$^2$, the remaining surface being open ports or the divertor covering 140 m$^2$. A module consisted of a plasma-facing front wall panel made of 8–10 mm thick beryllium on a copper substrate and a 40 cm thick shield block made of austenitic stainless steel (316 L(N)-ITER Grade). Each module was about 1 m × 1.4 m × 0.5 m thick and attached to the vessel by flexible supports, keys and electrical straps. Cooling water (4 MPa at 70 °C) is supplied by manifolds designed to remove 736 MW.

### 7.4.3.2 Robustness, Fragilities and Opportunities

An important systems architecture effect on the thickness of the blanket was that the size of the ITER machine was significantly reduced from an initial design with the major radius decreased from 8.1 m (DEMO size) to 6.2 m to reduce cost. In doing so the radial thickness of the shielding was reduced to a marginal level. The lesson learned was that adequate margin should be provided at pre-final design.

One important fragility is the limitation on plasma pulse length due to residual neutron heating of the superconducting toroidal field coil, which has to be handled by the cryogenic refrigerator. A nuclear heating of 15.7 kW (instead of the 14 kW target) might require a reduction in the fusion power. Detailed calculations indicated an allowable fusion power up to 500 MW for an 1800 s repetition rate. Limited sub-cooling could allow longer operation.

Marginal shielding can lead to important fragilities and constraints on operation and is a particular lesson to be learned for low aspect ratio tokamak design.

## *7.4.4 Test Tritium Breeding Modules (TBM)*

### 7.4.4.1 Design Choices

The objective of the ITER Test TBM Program is to provide the first experimental data on their performance in a fusion environment [1, 39]. Such information would be essential to design and predict the performance of DEMO and future fusion reactors.

The main nuclear reactions involved in tritium breeding [40] are:

$n + {}^{7}Li \rightarrow n' + {}^{4}He + T - 2.5$ MeV.
$n + {}^{6}Li \rightarrow {}^{4}He + T + 4.8$ MeV.
$n + {}^{9}Be \rightarrow 2n + 2\,{}^{4}He - 1.8$ MeV.
$n + Pb \rightarrow 2n + Pb - 8$ MeV *(n,2n) and (n,3n) reactions occur in several isotopes.

Lithium acts as a tritium breeder. Beryllium and several isotopes of lead act as neutron multipliers for high energy 14 MeV (on average) neutrons. Molecular compounds of these elements are involved in several breeding blanket designs.

The plan is to test six breeding blankets. The TBM and its associated cooling and tritium extraction systems form the Test Tritium Blanket System (TBS). Breeding of tritium is not required for ITER operation because of plans to purchase it from CANDU fission reactors and recycle a sufficient inventory. Key challenges are:

- Produce tritium in the blanket and extract it at a normalized rate such that total consumption of t could be maintained with maximum breeding coverage.
- Extract heat from the blanket at high temperature to generate electricity.

The five major testing objectives are validation and demonstration of:

- theoretical predictions of the breeding blankets structural integrity under combined and relevant thermal, mechanical and electromagnetic loads;
- tritium breeding predictions;
- tritium recovery process efficiency and inventories in blanket materials;

- thermal predictions for different blanket concepts;
- integral performance of the breeding blanket systems.

The six blanket designs are:

- Helium-Cooled Lithium Lead (HCLL) by the EU [41];
- Helium-Cooled Pebble-Bed (HCPB) by the EU [41];
- Water-Cooled Ceramic Breeder (CCCB) by Japan [42];
- Helium-Cooled Ceramic Reflector (HCCR) by Korea [43];
- Helium-Cooled Ceramic Breeder (HCCB) by China [44];
- Lithium-Lead Ceramic Breeder (LLCB) by India [45].

The operation of these modules will be supported by a wide range of experimental and model simulation tools to characterize their performance for DEMO.

#### 7.4.4.2 Robustness, Fragilities and Opportunities

In addition to those above, supplementary goals include:

- keeping the tritium inventory low;
- safely containing tritium;
- engineering the breeding blanket to protect coils and leave room for plasma;
- making the breeding ratio sufficiently greater than 1.0 for non-100% coverage.

The main question is the overall fusion reactor breeding ratio, which will be extrapolated from ITER results. An analysis of various tritium breeding philosophies [46] gives breeding ratios 1.12 to 1.23 with 60% $^{6}$Li enrichment. Integrated breeding modules with neutron multipliers and cooling give ratios 1.13 to 1.15. The design of the HCPB in DEMO v2 with divertor, gives an overall machine tritium breeding ratio 1.15 [47].

It thus seems that the tritium breeding program to develop a strategy for DEMO is robust, with a sufficient margin in tritium breeding. However, the margins are not large, and a lesson to be learned is that design changes which give up breeding in the central column region could make the tritium breeding ratio less than one unless great attention is paid to mitigation.

### *7.4.5 Divertor*

#### 7.4.5.1 Design Choices

The JET divertor and its tungsten covering [48] have been extensively prototyped for ITER. Similar divertors and tungsten coatings have been studied in other machines. The Divertor Cassette Assembly is shown in Fig. 7.4.

The main functions of the divertor design are:

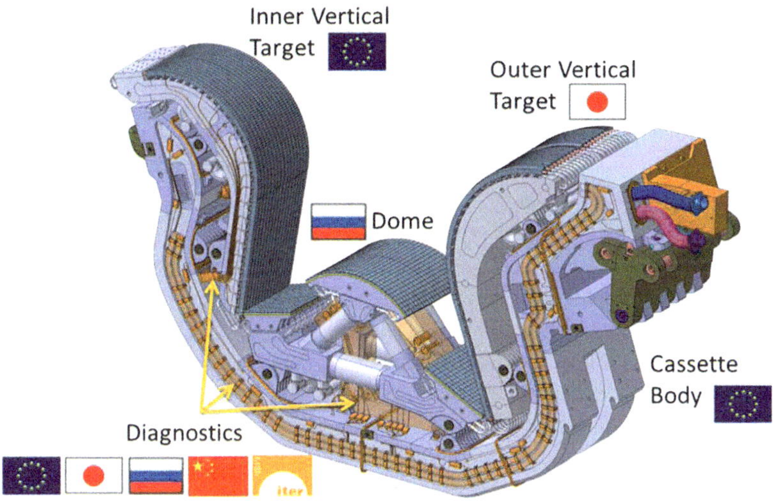


**Fig. 7.4** The Divertor Cassette Assembly and its procurement sharing. (From [48], Fig. 22)

- removing the helium and impurity content from the plasma;
- exhausting part of the plasma thermal power;
- controlling recycling.

The design consists of 54 Cassette Assemblies (CAs) which can be individually remote handled, each with a mass of 8 tonnes. One CA consists of three components: the Inner and Outer Vertical Targets, and the Dome. Tungsten is the plasma facing material with an 8 mm thickness, chosen for its high melting point, good thermal conductivity and lack of tritium retention.

In comparing the shape of the inner and outer near vertical plates of Fig. 7.4 with the last closed surface of the plasma shown in Fig. 7.2, the divertor shape closely matches that of the flux lines entering the divertor region, spreading the heat load as widely as possible, making use of the vertical plate divertor concept as prototyped in other machines.

Perhaps the most interesting part of the divertor design choices is what is not there. Elements that were often used in smaller tokamaks, which include normal conducting high-current divertor coils inside the vacuum vessel, and cryopumps inside the vacuum vessel, were not included in the design, because they are not compatible with a long-pulse high-flux of neutrons. The design does highlight the contrast between expedient solutions in smaller machines and the need for a fully integrated design in a fusion reactor.

#### 7.4.5.2 Robustness, Fragilities and Opportunities

In principle, the divertor design should be highly robust. However, there are potential fragilities in any design, and tests, prototyping and mitigation have been performed for all foreseeable cases. Nevertheless, the expected heat flux to the tungsten covered divertor is high and possible accidents are capable of delivering very high heat fluxes.

The tungsten itself can be considered a fragility. In Doublet III [49], 90% tantalum-10% tungsten was chosen for limiters instead of pure tungsten because of concern about a temperature transition. That transition played a role in a fractured limiter in Doublet IIA. This concern was studied in detail [50] where the tungsten ductile to brittle transition was assessed by combined fracture toughness and bending test analysis. Tungsten has a relatively high ductile-to brittle transition temperature (DBTT), being around 300–400 °C, which was an important concern for the designers, especially since the exploitation in the fusion environment will raise its DBTT even to a higher temperature as a result of neutron irradiation damage. The detailed tests indicated that the tungsten samples were performing as expected and within the parameters of the ITER design.

### *7.4.6 Vacuum Vessel and Disruption Mitigation*

#### 7.4.6.1 Design Choices

The vacuum vessel is a double-walled, stainless-steel, water-cooled toroidal chamber with an outside diameter of 19.5 m [1, 51]. It is extremely complicated, having to provide high vacuum [52] and support a variety of components.

From the point of view of robustness, the most important aspect is whether or not it can withstand all the various load imposed upon it [53]. Mechanical loads acting on the VV can be divided into five independent categories:

- Inertial loads: caused by accelerations due to gravity and seismic events.
- Electromagnetic (EM) loads: a strong design driver which act upon nearly all conductive structures during transient events
- Pressure loads: coolant, and incidental VV internal and external pressure.
- Thermal loads: caused by temperature gradients inside the VV structure
- Pre-tension loads: pre-tension of the divertor cassettes and pre-tension of bolts

EM loads are due to Lorentz forces when electric current crosses magnetic field lines. A 15 MA plasma disruption combined with a vertical displacement of the plasma can cause multi-MA currents to flow in the structure, along with induced eddy currents in all conducting elements within and attached to the vacuum vessel, as well as inducing multi-MA halo currents flowing on circuits involving the edge plasma and conducting structures facing the plasma. All of these extremely complicated and multi-directional loadings have been calculated, and the vessel has been engineered to withstand these events, preferably at low occurrence rates.

In view of these dangers, various disruption avoidance experiments have been carried out in existing tokamaks, and mechanisms will be put in place in ITER to avoid or mitigate such events [54]. The Disruption Mitigation System is based on the shattered pellet injection (SPI) technique for the mitigation of thermal and electromagnetic loads (TLM) and runaway electron suppression (RES). A strategy is based on avoidance of RE formation by injection of material and argon injection to dissipate the energy carried by the runaways.

#### 7.4.6.2 Robustness, Fragilities and Opportunities

The vacuum vessel has been engineered and analysed to be robust against a very wide range of load events. Nevertheless, with about 100,000 clips inside the vacuum vessel for mounting in-vessel structures, the quality control on everything needs to be extremely high to prevent any detachment and the need for repair. The fragility arises from the sheer number of components and the very high forces involved. The mitigation is that great attention has been paid to these problems.

### *7.4.7 Superconducting Magnetic Coils*

#### 7.4.7.1 Design Choices

The ITER design comprises 18 toroidal field coils using $Nb_3Sn$ superconductor to provide a toroidal magnetic field on the plasma axis of up to 5.3 T, with a maximum field of 11.8 T [1, 55]. The central solenoid also used $Nb_3Sn$ and will produce a rapidly varying field at 1.2 T/sec and up to 13 T. The superconductor NbTi was used for the 6 poloidal field coils producing maximum fields at the conductor of 6 T [56, 57] and 18 error field correction coils. Current leads for the coils were fabricated from high temperature superconductor bismuth strontium calcium copper oxide (BSCCO) 2223 tape with critical temperature 108°K [58].

Much research and innovation was required to develop these coils. They were prototyped in K-STAR, EAST and JT-60SA. These innovations included:

- high temperature superconducting leads
- cable-in-conduit conductors
- strand coating
- joint concepts
- cable compaction
- fraction of stabilizer
- cooling philosophy for stability
- mechanical support of strands

- control of fracturing of $Nb_3Sn$
- polyimide-based solid insulation
- cyanate ester insulation compatible with vacuum impregnation, mixed with diglycidylether-of-bisphenol-A (DGEBA) epoxy systems

The coils generate huge forces, and structural austenitic steels were used to support them, requiring advanced manufacturing practices. It was decided that the ripple from the 18 toroidal field coils was too high at the plasma edge and ferritic inserts are placed on the vacuum vessel to reduce the ripple from 1% to 0.3%.

Errors in the magnetic field from the main superconducting coils may result from construction errors and mechanical movement from stresses and cooling. Correction coils to reduce errors below the locked mode instability threshold [59] are inserted between the toroidal field and poloidal field magnets in three locations around the machine: six at the top, six at the bottom, and six at mid-plane.

Coils fabricated from High Temperature Superconductors (HTS) are an important design option. The advantages quoted [60, 61] are the higher operating temperature and the higher magnetic field capability than is possible with $Nb_3Sn$. ITER development showed that it took three decades to progress from the first $Nb_3Sn$ strands to a functioning high current conductor. HTS materials may allow the construction of even higher field tokamaks which might give better plasma performance. Such new HTS conductor would also bring novel problems that would have to be solved. ITER is already a high field tokamak with structural components operating at high load levels.

#### 7.4.7.2 Robustness, Fragilities and Opportunities

The coil systems are as robust as extensive engineering, testing and prototyping can make them. Nevertheless, these are huge components normally operating in rapidly pulsed fields and they are also subject to extra forces from disruptions. Since everything is first-of-a-kind, the only true test of robustness is operation.

### *7.4.8 Auxiliary Heating and Current Drive*

#### 7.4.8.1 Design Choices

ITER auxiliary heating and current drive plans are:

- 33 MW of neutral beam heating using negative ion sources of deuterium at 1 MeV [11];
- 20 MW of electron cyclotron heating and current drive for inboard first harmonic and half toroidal field second harmonic at 170 GHz [12];
- 20 MW of ion cyclotron power at 40–55 MHz [13];

- the possibility of upgrading to 130 MW total capacity of auxiliary heating, with the possibility of using 110 MW in any one scenario;
- the possibility of installation of lower hybrid current drive [14].

#### 7.4.8.2 Robustness, Fragilities and Opportunities

The robustness and fragilities of heating and current drive in ITER have been examined [62] in relation to advanced inductive scenarios. Including a wide range of scalings, they concluded:

- Inductive scenarios capable of high normalized pressure and high normalized confinement have been established by AUG, DIII-D, JET and JT-60 U.
- The parameter range where high performance is achieved is broad.

Given the variety of heating systems and scenarios and the flexibility of maintaining and upgrading systems, it seems that the ITER scenario planning is robust.

### *7.4.9 Diagnostics and Control for Plasma and Neutrons*

#### 7.4.9.1 Design Choices

The plasma diagnostic and control system for ITER [63] was designed to provide reliable operation even though its in-vessel components will be subject to neutrons and a high temperature plasma environment. ITER diagnostic design was moving towards using sophisticated methods and aiming for large spatial coverage and high signal intensities. In many cases, the in-vessel environment will result in a rapid aging of diagnostic components, so that additional measures like protection shutters or plasma based mirror cleaning were being developed. Detailed systems approaches have been developed for ITER and DEMO for plasma control [64].

For a proto-fusion reactor, the key diagnostics are for measuring neutrons and fast particles. Neutron diagnostic in ITER and on JET [65] were designed to measure key parameters and play a major role in physics measurement, machine protection and plasma optimization.

The original design for the ITER neutron camera [66] was based on the JET design and then refined [67] with the TFTR camera philosophy and designed as an integral part of all the proposed neutron diagnostics for ITER [68].

More recent designs are able to use diamond detectors for both profile measurements and accurate spectroscopy [69, 70]. The ITER Vertical Neutron Camera (VNC) and Radial Neutron Camera (RNC) designs use multichannel neutron collimators and diamond detectors to measure the time-resolved neutron/alpha emission profile for d-d and d-t plasma, fusion power density, alpha source density, neutron and alpha emissivity and ion temperature profile. Total neutron flux and fusion power measurements would also be provided by these systems.

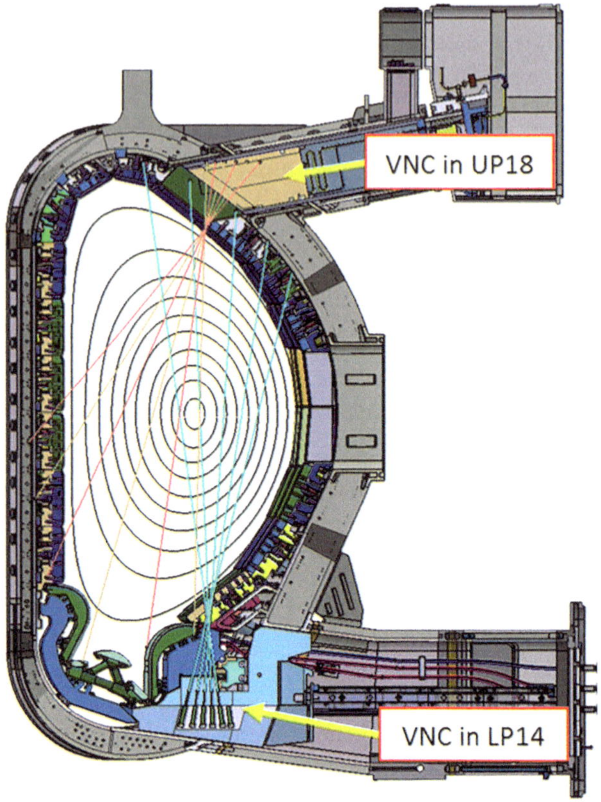


**Fig. 7.5** ITER Vertical Neutron Camera. (From [70], Fig. 6)

The VNC, shown in Fig. 7.5 is composed of two sets, each one including six detectors modules with vertical collimators alignment, one in a lower port and the other in an upper port. The detector modules included two Carbon Vapour Deposition (CVD) diamonds and two fission chambers with high purity $^{238}U$. Fission chambers are sensitive only to fast neutrons with a threshold of 1 MeV, while diamond detectors measure the neutron energy distribution, particularly the 12–16 MeV range. The cameras were designed with collimators so as to observe many lines of sight in the plasma. The expected neutron flux is estimated to be from $10^5$ up to $10^{10}$ n $cm^{-2}$ $s^{-1}$, thereby allowing either tomographic analysis or computation and comparison with analysis codes such as TRANSP.

The radial profile is measured by the RNC, see Fig. 7.6.

The RNC has two fan-shaped arrays of collimators, with detectors arranged at two locations: in the port plug for plasma edge measurement, and in the port interspace to observe the plasma core. Both systems are located in the same vacuum vessel port. There will be 20 collimators for the ex-port systems and 6 for the in-port one. Detector modules should be sensitive both d-d and d-t neutrons, with a dynamic range of four orders of magnitude and a threshold able to reject neutrons from scattering events. The RNC will be able to provide neutron energy spectra as a

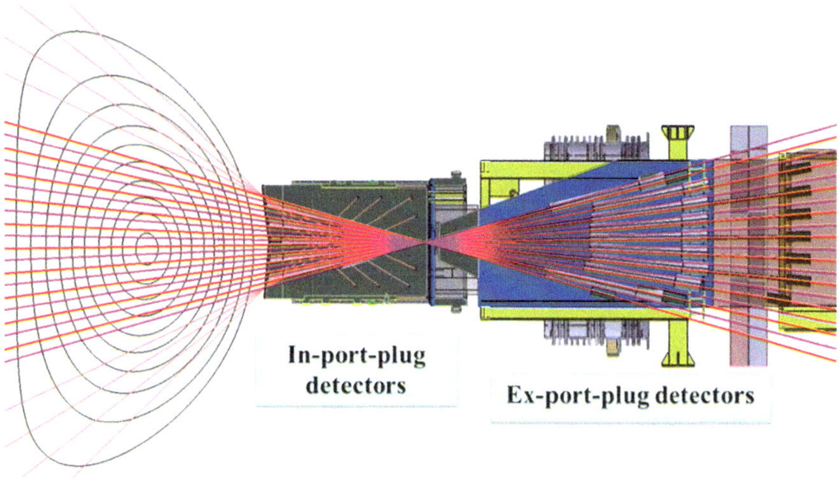


**Fig. 7.6** ITER Radial Neutron Camera. (From [70], Fig. 7)

secondary measurement. The High Resolution Neutron Spectrometer [71] will probably be used as an additional line-of-sight for the radial neutron camera, comprising a set of four spectrometers positioned after each other along a single line.

Neutron flux monitors will be installed at different locations using varying amounts of both $^{235}U$ and $^{238}U$ and appropriate moderators and shielding. Micro-fission chambers with low $^{235}U$ content will be used at the highest power levels. There are also divertor neutron flux monitors. The neutron activation system will rely on a pneumatic post system to send samples to be irradiated and then returned for integrated measurements of neutron fluence. A moveable in-vessel neutron source will be used for accurate calibration.

ITER will be equipped with a radial gamma-ray spectrometer [72] to provide time-resolved spatial measurement of gamma rays, hence the alpha particle birth profile from $D(t,\gamma)^5He$, profile of confined alpha particles, fast ion confinement effects during ICRF or NBI heating and hard X-ray profile for runaway electrons. The system will include an array of detectors facing the plasma through 6 lines of sight, and the detectors will be LaBr3 scintillator crystals and HPGe.

#### 7.4.9.2 Robustness, Fragilities and Opportunities

The large number of detectors and measurement types of neutron and gamma detectors insure that there will be a highly robust measurement system for neutrons and gamma rays from the ITER plasma, hence providing the basis for plasma measurement and control, along with the many other diagnostic systems planned.

### *7.4.10 Balance of Plant Including Power Systems, Cooling, Refrigeration and Tritium Processing*

#### 7.4.10.1 Design Choices

All of the systems discussed so far rely on a very large balance of plant capacity. All of the electrical equipment, high power and otherwise, relies on an extensive power generation and transformation system. The cryogenic and higher temperature refrigerators, the tritium processing system, the pumping systems, hot cells, computers and data processing, operator room and displays, storage, maintenance and fabrication, are all essential parts of the overall system, and will be present at the appropriate scales in all fusion reactors.

#### 7.4.10.2 Robustness, Fragilities and Opportunities

The importance of balance of plant operations should never be underestimated. Indeed, it is often the case that the most difficult technology works well because much attention is paid to its robustness and mitigation of its fragilities. In contrast, standard technology is taken for granted and then does not work well.

## 7.5 Systems Aspects of Construction and Operation

### *7.5.1 Systems Aspects of Machine Assembly and Commissioning*

The systems approaches described above [7, 8] included a careful accounting of components and their interfaces which was applied to the design and construction of ITER. The actual state of construction [2, 73, and 74] is a mixture of impressive technical achievements and the more prosaic and completely vital roles of quality control, interface control and modifications to find and correct errors and mitigate their impact on the entire system.

There have already been difficulties and complications during construction [2] which have been detected and corrected, including:

- non-compliance of steel reinforcing bars in the main tokamak hall floor, found and corrected before concrete was poured
- a design for a water cooling tower to help remove 500 MW of power was found to be too big for the site and needed redesign of the buildings

These and similar problems have led to schedule delays and cost increases. Since many countries are providing components as their share of the project, it is not possible to accurately determine the total project cost [2]. The lowest estimate is EUR 13 billion allowing for low worldwide construction costs.

### 7.5.2 Systems Integration of Construction and Operation

A detailed plan was produced for interspersing construction with operation and testing, to verify each component at the earliest stage possible, and to allow for additional experiments on other tokamaks, and additional technical progress to ensure the ultimate success of ITER. This Staged Approach [75] includes:

- Incorporation of a first divertor following First Plasma;
- Development of the Research Plan around a progressive increase in the installed Heating & Current Drive (H&CD);
- Exploration of options for studies of H-modes at lower toroidal magnetic field;
- Strengthening of the development of a disruption management program;
- A significantly more extensive program of studies on plasma-wall interactions;
- Start of commissioning of the Tritium Plant with tritium in 2033;
- Transition to experiments in D/DT plasmas in December 2035;
- Progressive transition to full DT operation during 2036;
- Development of the experimental program towards the achievement of ITER's fusion power, fusion gain and long burn duration mission goals;
- Adaptation of the evolution of the Test Blanket Module testing program.

### 7.5.3 Preparation for Assembly and Commissioning

Construction, assembly and commissioning of ITER [74] have made enormous progress despite a Covid pandemic and challenges from first-of-a-kind equipment:

- The first sets of major tokamak components were successfully completed and delivered to ITER site allowing the start of machine assembly with the cryostat, vacuum vessel sector, toroidal and poloidal field coils and correction coils.
- Assembly has begun of the first sector sub-assembly with TF Coils and thermal shielding.
- Tokamak pit supports are well advanced.
- Key infrastructure and plant systems have been completed: 400 kV connection, tokamak building, cooling plant and cryoplant plus progress on other systems.
- Research planning was progressing as foreseen.

Challenges remained ahead to complete assembly, commissioning and first plasma in order to execute the research plan, beginning with first plasma in 2025.

### 7.5.4 Robustness and Fragilities of the Construction Process

As is evident from the details of the staged approach, there remains a large number of unknowns as to how each system will function with new levels of performance not yet prototyped. To this extent, the ITER program is full of fragilities. To mitigate these fragilities, a very wide ranging programme of research and prototyping on other tokamaks [76] and within the ITER program is planned in order to ensure success.

### 7.5.5 Research Plans to Fulfil Goals

The Staged Approach foresees a Research Plan during d-t operation which will not have been fully addressed in the TFTR, DTE1 and DTE2 experiments, because of the dominance of alpha particle heating and high alpha particle pressure within the plasma. These areas of research include:

- Waves and energetic particle physics, such as induced Alfven modes;
- Fuel mixture control and helium exhaust, since deuterium injection at 1 MeV could alter the particle balance, and alpha exhaust is necessary to avoid depleting deuterium and tritium at the core;
- Self-heating and thermal stability, since self-heating as the dominant mechanism needs to be established, and near ignition, a plasma can unstably heat itself beyond the beta limit;.
- Macroscopic stability physics and control, since transport barriers inside and at the plasma edge need to be established and controlled, including ELM control, and vertical stability needs to be maintained;
- Multi-scale transport physics, since the large size and new physics of burning plasmas may alter transport;
- Physics of the plasma-boundary interface, which are highly complicated and will be in new regimes of high power management in steady state conditions;
- Integrated burning plasma scenario development and then optimized for stability and efficiency represented by Q.

Additional key physics issues for advanced modes include operation with:

- current profile control and MHD control above the no-wall $\beta$-limit;
- low normalized gyroradius;
- low collisionality pedestal coupled with a high-density divertor;
- large population of isotropic energetic particles.

### *7.5.6 Robustness and Fragilities of Physics Research Programme*

There has been very detailed planning of the ITER research programme and a huge body of experience acquired. Nevertheless, the overall fragility is that ITER is still a physics research machine at the same time as it has the goal of being a prototype for a fusion reactor. Success is therefore probable but not guaranteed, as is always the case in a fundamental research programme.

## 7.6 Conclusions and Lessons for Future Machines

### *7.6.1 Conclusions on Systems Approaches*

It now seems probable that ITER will be a success and provide a demonstration of a deuterium-tritium burning, high power, near-steady state tokamak which will serve as a basis for a future tokamak reactor. This success will be partly and crucially due to systems engineering and systems architecture practices that are guiding this project. ITER's challenges are due to large scale, high complexity, engineering and physics unknowns and planet-wide participation, collaborations and contributions. The need for a systems approach in design, interface, component control, quality control, and operation is demonstrated over and over again. A project that initially looked highly over-ambitious has turned into a model of international collaboration and project management.

The project management and control systems are now robust, and every effort is being made to overcome fragilities that have been anticipated and mitigated by systems approaches. This framework will serve as a basis for continuing innovation for ITER and especially as the basis for a future fusion reactor.

### *7.6.2 Lessons for Future Machines*

Knowledge acquired on ITER now and in the future serves as nothing less than a technological and physics basis for building a tokamak fusion reactor and for other fusion reactor concepts. ITER is the focal point for gathering and integrating the collective knowledge of the world in fusion physics leading to a reactor.

## References


1. D.J. Campbell et al., Innovations in technology and science R&D for ITER. J. Fusion Energy **38**, 11–71 (2019)
2. M. Claessens, *ITER: The Giant Fusion Reactor: Bringing a Sun to Earth* (Springer, Cham, 2020)
3. ITER Goals, ITER Goals (2021). https://www.iter.org/sci/Goals. Last accessed 17 June 2022
4. E. Joffrin et al., Overview of the JET preparation for deuterium–tritium operation with the ITER like-wall. Nucl. Fusion **59**, 112021 (2019)
5. G. Federici et al., DEMO design activity in Europe: progress and updates. Fusion Eng. Des. **138**, 729 (2018)
6. G. Federici et al., An overview of the EU breeding blanket design strategy as an integral part of the DEMO design effort. Fusion Eng. Technol. **141**, 30–42 (2019)
7. P. Barabaschi et al., System engineering and configuration management in ITER, presented by S. Chiocchio, Proceedings of 24th SOFT Conference, Warsaw (2006). https://inis.iaea.org/collection/NCLCollectionStore/_Public/38/005/38005232.pdf. Last accessed 17 June 2022
8. S. Chiocchio et al., Integrated systems engineering approach to manage the installation and start up of a fusion reactor, Semantics Scholar (2008) https://www.semanticscholar.org/paper/TOMANAGE-THE-INSTALLATION-AND-START-UP-OF-A-FUSION-Chiocchio-Kuehn/11662dc0a6a04efce1eac4891a801a2f7924c3a5?p2df. Last accessed 17 June 2022
9. ITER technical basis, Report IAEA/ITER EDA/DS/24, ITER EDA (Engineering Design Activities) Documentation Series No. 24 IAEA, Vienna (2002)
10. K. Lackner et al., Chapter 3: Equilibrium and macroscopic stability of tokamaks, in *Fusion Physics*, ed. by M. Kikuchi et al., (IAEA, Vienna, 2012), p. 225
11. R.S. Hemsworth et al., Overview of the design of the ITER heating neutral beam injectors. New J. Phys. **19**, 025005 (2017)
12. Y. Oda et al., Recent activities of ITER gyrotron development in QST, 10th International Workshop 2017 "Strong microwaves and terahertz waves: sources and applications," EPJ Web of Conferences 149, 01002. (2017)
13. P. Lamalle et al., Status of the ITER ion cyclotron heating and current drive system. AIP Conference Proceedings **1689**, 030007 (2015)
14. G.T. Hoang et al., A lower hybrid current drive system for ITER. Nucl. Fusion **49**, 075001 (2009)
15. D. Mueller, The physics of tokamak start-up. Phys. Plasmas **20**(5), 058101 (2013)
16. A.C.C. Sips et al., Progress in preparing scenarios for operation of the international thermonuclear experimental reactor. Phys. Plasmas **22**, 021804 (2015)
17. T. Rafiq et al., Fusion power production in international thermonuclear experimental reactor baseline H-mode scenarios. Phys. Plasmas **22**, 042511 (2015)
18. F.M. Poli et al., Ideal MHD stability and performance of ITER steady-state scenarios with ITBs. Nucl. Fusion **52**, 063027 (2012)
19. V. Parail et al., Self-consistent simulation of plasma scenarios for ITER using a combination of 1.5D transport codes and free boundary equilibrium codes. Nucl. Fusion **53**, 113002 (2013)
20. R.J. Hawryluk, An empirical approach to tokamak transport, in *Physics of Plasmas Close to Thermonuclear Conditions*, ed. by B. Coppi et al., vol. 1, (Commission of the European Communities, Brussels, 1980), p. 19
21. S.C. Jardin et al., Dynamic modelling of transport and positional control of tokamaks. J. Comput. Phys. **66**, 481–507 (1986)
22. W.M. Tang, Microinstability-based model for anomalous thermal confinement in tokamaks. Nucl. Fusion **26**, 1605–1618 (1986)
23. J.E. Kinsey et al., Nonlinear gyrokinetic turbulence simulations of E×B shear quenching of transport. Phys. Plasmas **12**, 062302 (2005)
24. F.D. Halpern et al., Improved model for transport driven by drift modes in tokamaks. Phys. Plasmas **15**, 012304 (2008)

25. P. Snyder et al., Pedestal stability comparison and ITER pedestal prediction. Nucl. Fusion. **49**, 085035 (2009)
26. M. Brambilla, A Full Wave Code for Ion Cyclotron Waves in Toroidal Plasmas, Report IPP 5/66 (Max-Planck-Institut fur Plasmaphysik, Garching, 1996)
27. R.J. Goldston et al., New techniques for calculating heat and particle source rates due to neutral beam injection in axisymmetric tokamaks. J. Comput. Phys. **43**, 61–78 (1981)
28. D.W. Ignat et al., Dynamic modelling of lower hybrid current drive. Nucl. Fusion **34**, 837–852 (1994)
29. A.H. Kritz et al., Ray tracing study of electron cyclotron heating in toroidal geometry, Heating in Toroidal Plasmas, Proceeding of the 3rd Joint Varenna-Grenoble International Symposium (Grenoble, France, 1982) (Brussels: Commission of the European Communities) **2**, 707 (1982)
30. J. DeLucia, S.C. Jardin, A.M.M. Todd, An iterative metric method for solving the inverse tokamak equilibrium problem. J. Comput. Phys. **37**, 183–204 (1980)
31. R.C. Grimm, J.M. Greene, J.L. Johnson, Computation of the magnetohydrodynamic spectrum in axisymmetric toroidal confinement systems. Method Comput. Phys. Adv. Res. Appl. **16**, 253 (1976)
32. J.M. Greene, M.S. Chance, The second region of stability against ballooning modes. Nucl. Fusion **21**, 453–464 (1981)
33. E. Morse, *Nuclear Fusion* (Springer, Cham, 2018)
34. D.A. Humphreys et al., Experimental vertical stability studies for ITER performance and design guidance, paper IT/2-4b, in proceedings of 22nd IAEA Fusion Energy Conference, Geneva (2008)
35. Y. Gribov et al., Plasma vertical stabilisation in ITER. Nucl. Fusion **55**, 073021 (2015)
36. P.J. Heitzenroeder et al., An overview of the ITER in-vessel coil systems, Conference: Fusion Engineering, 2009. SOFE 2009. 23rd IEEE/NPSS Symposium on Project: ITER In-Vessel Coils (2009)
37. M. Kocan et al., Steady-state magnetic sensors for ITER and beyond: development and final design. Rev. Sci. Instrum. **89**, 10J119 (2018)
38. A.R. Raffray et al., The ITER blanket system design challenge. Nucl. Fusion **54**, 033004 (2014)
39. L.M. Giancarli et al., Overview of the ITER TBM program. Fusion Eng. Des. **87**, 395–402 (2012)
40. K. Maki, Energy multiplication in high tritium breeding ratio blanket with front feeder zone for fusion reactors. J. Nucl. Sci. Technol. **25**(1), 72 (1986)
41. I. Ricapito et al., Current design of the European TBM systems and implications on DEMO breeding blanket. Fusion Eng. Des. **109–111**, 1326 (2016)
42. Y. Kawamura et al., Progress of R&D on water cooled ceramic breeder for ITER test blanket system and DEMO. Fusion Eng. Des. **109–111**, 1637 (2016)
43. S. Cho et al., Design and R&D progress of Korean HCCR TBM. Fusion Eng. Des. **89**, 1137 (2014)
44. K. Feng et al., Current progress of Chinese HCCB TBM program. Fusion Eng. Des. **109–111**, 729–735 (2016)
45. E.R. Kumar, K.N. Vyas, T. Jayakumar, Overview of LLCB TBM design and R&D activities in India. Fusion Eng. Des. **109–111**, 1522–1527 (2016)
46. F.A. Hernandez, P. Pereslavtsev, First principles review of options for tritium breeder and neutron multiplier materials for breeding blankets in fusion reactors. Fusion Eng. Des. **137**, 243–256 (2018)
47. P. Pereslavtsev et al., Neutronic analyses for the optimization of the advanced HCPB breeder blanket design for DEMO. Fusion Eng. Des. **124**, 910–914 (2017)
48. T. Hirai et al., Design optimization of the ITER tungsten divertor vertical targets. Fusion Eng. Des. **127**, 66–72 (2018)

49. M.M. Sabado, F.B. Marcus et al., Doublet III limiter performance and implications for mechanical design and material selection for future limiters, Report GA-A15633, General Atomic Company, San Diego. Symposium on Engineering Problems of Fusion Research, 8th, San Francisco, Calif., November 13–16, 1979, Proceedings. Volume 3. (A81–18901 06–75) Piscataway, N.J., Institute of Electrical and Electronics Engineers, Inc., 1979, pp. 1361–1369 (1979)
50. C. Yin et al., Ductile to brittle transition in ITER specification tungsten assessed by combined fracture toughness and bending tests analysis. Mater. Sci. Eng. A **750**, 20–30 (2019)
51. K. Iokia et al., ITER vacuum vessel: design review and start of procurement process. Fusion Eng. Des. **84**(2–6), 229 (2009)
52. R. Pearce, L. Worth, ITER Vacuum Handbook, ITER Technical Report ITR-19-004. ITER Organization (2019)
53. J.-M. Martinez, ITER Vacuum Vessel Load Specification, ITER Report ITER_D_2F52JY v3.3. ITER Organization (2013)
54. M. Lehnen, S. Maruyama, Executive Report of the ITER Disruption Mitigation Workshop 2017, ITER Technical Report ITR-18-002 (ITER Organization, St-Paul-lez-Durance, 2018)
55. ITER Design Description Document: Magnet, ITER report ITER_D_22HVSL v2.2 (2006)
56. B. Lim et al., Design of the ITER PF coils. IEEE Trans. Appl. Supercond. **21**(3), 1918 (2011)
57. A. Winkler, Transient behaviour of ITER poloidal field coils, Thesis, Karlsruher Institut fur Technologie, Fakultat fur Elektrotechnik und Informationstechnik, KIT Scientific Publishing, Karlsruhe (2010)
58. P. Bauer et al., R&D towards HTS current leads for ITER. IEEE Trans. Appl. Supercond. **19**, 1500–1503 (2009)
59. A. Foussat et al., Overview of the ITER correction coils design. IEEE Trans. Appl. Supercond. **20**(3), 402–406 (2010)
60. W.H. Fietz et al., Prospects of high temperature superconductors for fusion magnets and power applications. Fusion Eng. Des. **88**, 440–445 (2013)
61. M. Windridge, Smaller and quicker with spherical tokamaks and high-temperature superconductors. Philos. Trans. R. Soc. A. **377**, 20170438 (2019)
62. T.C. Luce et al., Development of advanced inductive scenarios for ITER. Nucl. Fusion **54**(1), 013015 (2014)
63. W. Biel et al., Diagnostics for plasma control – from ITER to DEMO. Fusion Eng. Des. **146**, 465–472 (2019)
64. Y.V. Mitrishkin et al., Plasma control in Tokamaks. Part 3.2. Simulation and realization of plasma control systems in ITER and constructions of DEMO. Adv. Syst. Sci. Appl. **03**, 136 (2020)
65. F.B. Marcus et al., Neutron emission profile measurements during the first tritium experiments at JET. Nucl. Fusion **33**(9), 1325–1344 (1993)
66. F.B. Marcus et al., A neutron camera for ITER: conceptual design, in *Diagnostics for Experimental Thermonuclear Fusion Reactors*, ed. by P. E. Stott et al., (Plenum Press, New York, 1996), p. 385
67. F.B. Marcus et al., A neutron camera for ITER (invited). Rev. Sci. Instrum. **68**, 514–519 (1997)
68. L.C. Johnson et al., Neutron diagnostics for ITER. Rev. Sci. Instrum. **68**, 569–572 (1997)
69. D. Marocco, B. Esposito, F. Moro, Neutron measurements in ITER using the radial neutron camera. J. Instrum. **7**, C03033 (2012)
70. L. Bertalot et al., Present status of ITER neutron diagnostics development. J. Fusion Energy **38**, 283–290 (2019)
71. M. Scholz et al., Conceptual design of the high resolution neutron spectrometer (HRNS) for ITER. Nucl. Fusion **59**(6), 065001 (2019)
72. M. Nocente et al., Conceptual design of the radial gamma ray spectrometer system for a particle and runaway electron measurements at ITER. Nucl. Fusion **57**, 076016 (2017)

73. B. Bigot, Progress toward ITER's first plasma. Nucl. Fusion **59**, 112001 (2019)
74. B. Bigot, Preparation for assembly and commissioning of ITER, 28th IAEA Fusion Energy Conference, Nice, France (virtual) 10-15 May 2021, presentation OV/1-1, paper IAEA–CN–286-1335, submitted to Nuclear Fusion (2021)
75. ITER Organization. ITER research plan within the staged approach,. ITER report ITR-18-003. ITER Organization (2018)
76. A. Loarte, Required R&D in existing fusion facilities to support the ITER research plan, ITER Technical Report ITR-20-008 (2020)

# Chapter 8
# Demonstration Tokamak Fusion Reactors and Their Systems Approaches

**Abstract** Many demonstration fusion reactors are being designed based on highly successful experiments. Systems approaches are being developed and used for planning and building demonstration fusion reactors. The EU DEMO is treated as a case study in how systems codes and methods are being effectively used in engineering design. Large scale demonstration reactors planned by ITER partners and based on its baseline or advanced tokamak scenarios are analysed in detail and compared. Compact fusion reactor designs have been proposed using high temperatures superconductors and unique neutron blanket and shielding concepts, which are scrutinized for robustness and fragilities and the mitigations chosen. The process of choice of goals and systems integration of tokamak fusion reactors is explored and summarized.

## 8.1 Systems Strategy for Demonstration Tokamak Reactors

This chapter synthesizes everything that has been discussed in the first seven chapters and shows not only how system approaches can improve tokamak fusion reactor design, but also that they are essential for success.

Chapter 1 introduced systems engineering, architecture and emergent properties and provided the essentials of tokamak physics. Chapter 2 demonstrated the highly interactive nature of tokamaks that make systems approaches essential from many points of view, and assembled a comprehensive design space of operational scenarios, machine components and model and analysis tools. In Chaps. 3–7, many of the relevant tokamak experiments were analysed using both systems reverse (of existing designs) and forward (designing a new machine) engineering with the aim of demonstrating ever-increasing levels of complexity. Chapter 7's discussion of ITER demonstrated that systems approaches were essential for design, management and interface control.

This Chap. 8 begins by discussing the major systems in a tokamak fusion reactor, many of which are present in ITER. It then continues with a case study of systems approaches to the EUROfusion-sponsored EU DEMO fusion reactor design [1–3]. Given the highly detailed approach summarized in these references and others,



F. B. Marcus, *Systems Approaches to Nuclear Fusion Reactors*, Springer Series in Plasma Science and Technology, https://doi.org/10.1007/978-3-031-17711-8_8

EU DEMO seems to be an excellent example of the full use of systems forward engineering and is the best documented. Checklist 1.7 from Chap. 1 provides a general outline of the systems approaches.

Following the EU DEMO case study, other demonstration and engineering fusion tokamak reactors are described and analysed. These designs fall into two categories. One group, like EU DEMO, follows the ITER route of fairly large, high field and moderate aspect ratio tokamaks, with breeding blankets and divertors also based upon ITER. They explore somewhat different optimizations of design space, but make full use of all the technological developments of ITER, of which they are partners and suppliers of breeding blanket modules.

The other group explores more compact machines with extreme regimes of very high toroidal magnetic field at the coil conductor using high temperature superconductors, both at medium aspect ratio for high field on axis, and very low aspect ratio as in spherical tokamaks. The main challenges for these groups are to manage the very high machine stresses, extreme power handling and lack of inboard space for neutron shielding and tritium breeding. As there is usually less engineering detail than for EU DEMO, their prospects are more difficult to assess at this stage.

### 8.1.1 Identify Goals and Essential Elements of a Fusion Reactor

Every commercial fusion reactor should satisfy the following:

- It must work! It needs to produce net electric power delivered to the grid, at least in the 100's of MW (electric) to justify and recover its cost.
- It must breed at least as much tritium as it consumes. The in-machine trapped inventory must be small and not present risks nor consume the initial inventory.
- It must be able to operate for decades and allow for component replacement.
- It must have a high availability to be commercial and a reliable power source.
- It must be robust against accidents and failure modes such as disruptions.
- It must be capable of being licensed as safe by nuclear regulatory authorities.

The overall desirability and the cost per MW (e) will determine the success of future fusion reactors in comparison to competing sources at the time. The age of renewable energy is emphasizing the need for base load power that a fusion reactor could provide.

In order to achieve the essential objectives, related goals need to be achieved:

- Tritium breeding blankets need a breeding ratio of somewhat greater than one to allow for areas that cannot have breeding blankets including divertors, vacuum vessel ports and possibly the inner vessel midplane wall.
- The divertor and machine walls must be able to handle both steady-state loss power and transient increases for ELMs or other instabilities.

- Plasma stabilization coils or surfaces such as those for vertical and ELM instabilities need to be highly robust and to be able to withstand high neutron flux.
- All components need to be structurally sound and withstand all forces and cycling fatigue for the lifetime of the machine. They must at least be easy to maintain or replace, especially those subject to intense neutron damage.
- Plasma and other operating systems and their scenarios need to be highly reliable. A fusion reactor is not an experiment, it has one operating scenario.
- The overall plant fusion Q must be high to produce commercial electricity.
- Even if the tokamak plant is repetitively pulsed, all components must be designed for truly steady state operation. In the case of a steady state plasma, allowances must be made for occasional plasma shut-downs and restarts.
- The design must be fully self-consistent with all interfaces accounted for and all fragilities identified and mitigated to the extent possible.

### *8.1.2 Paths to an "ITER-based" Reactor*

Several groups have risen to these challenges and have begun or made significant progress on designs of the several different options. ITER itself is a compromise between various tendencies while at the same time pushing several technologies to their limits and acting as a general prototype. ITER has the following compromise characteristics that could be the basis of one or more fusion reactor designs:

- The plasma aspect ratio is about 3, approaching that of a spherical tokamak with shield and breeding blanket.
- The toroidal field on axis at 5.3 T is already higher than most experimental tokamaks apart from dedicated high field tokamaks.
- There are two main operating scenarios:
  - The baseline scenario is inductively driven and very long pulse.
  - The advanced tokamak scenario has a high level of current drive, high normalized beta and high bootstrap current.
- Neutron shielding and tritium breeding blankets are being tested.
- Tritium machine inventory will be minimized and tritium exhaust will be processed and recycled.
- The divertor can handle the high plasma power outflow at over 100 MW.

There are several paths for achieving a fusion reactor, including:

- Use ITER experience, build a DEMO that produces electricity, leaving commercialization to the private sector.
- Build a single or multi-step Engineering Test Reactor to resolve remaining technological issues, such as high temperature tritium breeding blankets for a good breeding ratio and high efficiency heat conversion to electricity.

- Build specialized test facilities such as 14 MeV neutron generators or divertor heat test facilities.
- Choose between low or high power reactors.

## 8.1.3 Key Choices for Neutronics and Reactor Parameters

So far in the development of fusion reactors, the role of plasma physics has been the primary focus of efforts and rightly so, since unless the reactor achieves near burning conditions, there are no tritium breeding problems to worry about. Now that the focus of efforts is moving towards a commercial fusion reactor, the role of neutron shielding and the tritium breeding blanket becomes central to the design.

In a fusion reactor, there are a few key systems that have a major influence on the design. The choice of tokamak plasma, magnetic field, shaping and operating parameters place important constraints on neutron shielding design. Options include:

- ITER baseline operation scenario;
- ITER advanced steady-state operation with high bootstrap-current, beta and current drive;
- Compact spherical tokamak with very low aspect ratio tokamak;
- Compact very high axial toroidal field tokamak;
- Compact very high elongation tokamak.

All of these variants have considerable promise for achieving fusion plasma ignition conditions. However, for any or many of them to succeed, they need to have a fully integrated systems design in which all aspects are realistic or can be dealt with by innovation and testing. In particular, compact tokamaks pose extreme design difficulties on leaving enough room for coil shielding and blanket technology (especially on the inner torus wall), divertor power handling and magnetic coil strength and structural support.

### 8.1.3.1 Breeding Blanket and Shielding

Fission reactor designs are characterized mostly by their cooling scheme, including pressurized water, boiling water, liquid salt, liquid metal or high temperature gas-cooled. They can also be characterized by their moderator and breeding philosophy, such as heavy water or fast breeder reactors. Fuel possibilities include light and heavily enriched uranium, thorium and plutonium. In those fission reactors designed for breeding, fuel is often removed and taken off-site for reprocessing, whereas fusion reactors must do all reprocessing of tritium on site and in continuous operation to minimize tritium inventory.

A systems approach avoids the mistake of concentrating exclusively on the fusion containment concept and then adding the nuclear components as an afterthought.

Compared to a nuclear fission reactor, a fusion blanket has a lot more work to do. A fusion reactor's blanket, shield and vacuum vessel must accomplish the following:

- The entire machine must have an overall tritium breeding ratio greater than 1.0.
- All blanket, shield, vacuum vessel and divertor components that receive significant heat flux must run at high enough temperature for good thermodynamic efficiency for electricity production.
- The pumping power needed for circulating coolant especially across magnetic field lines must not be so large as to compromise the overall plant efficiency.
- The blanket must be designed for breeding, cooling, and separation of tritium to be recycled quickly as fuel.
- The blanket, shield and vessel trapped tritium inventory cannot be so large as to use up the initial site inventory.
- Since fission reactor-produced tritium may not be available in the future, provision may need to be made for starting the reactor with d-d fusion and breeding from the d-d and fast triton fusion product 14 MeV neutrons.
- Diffusion of tritium outside the vacuum vessel needs to be minimized.
- All sources of tritium, from the blanket, leakages and from the plasma, need to be purified and made available as fuel in real time in a continuous process.
- The overall thickness of blanket, neutron shield, vacuum vessel, thermal shield, coolant and clearances is a crucial design factor in a fusion reactor and can vary with position in the reactor. A typical thickness is 1.4 m, so less than this risks overheating coils and having a tritium breeding ratio below 1.0.
- The vacuum vessel wall facing the plasma must be thin enough to avoid compromising the high energy neutron spectrum for tritium breeding.
- Neutron damage must be carefully assessed to determine the lifetime of components in high radiation intensity locations.

When neutronics transport and breeding calculations are used for an entire tokamak fusion reactor and not just for single blanket modules, the trade-offs in blanket design become especially apparent. A comprehensive survey of the various possible designs [4] illustrated some of the trade-offs:

- Breeding blankets need neutron multipliers like lead or beryllium, and breeding material, like $^{6}$Li or $^{7}$Li. The composition and spatial distribution of these elements represent a considerable architecture and design challenge.
- The physical composition choice is important, which might be solid material, molten metal salts such as FLiBe or pebble beds, depending on coolant choice.
- There is a trade off between breeding efficiency and shielding thickness. Once a neutron enters a significant thickness of material such as tungsten, carbon, nickel or steel, the breeding of tritium from it is finished.
- The choice of coolant is crucial, whether it is water, molten metal salts, eutectic metal, liquid lithium, or helium gas, and then at what pressure and temperature. This choice affects the pumping power which can be large and the tritium recuperation method.

- If shielding is dominant in one region, such as at small major radius, then conscious choices should be made of fractions of absorbing material to minimize shield thickness or of reflecting material which could lead to a higher neutron breeding elsewhere.
- The overall assembly of first wall components, blanket, shielding, etc. has other functions than breeding, shielding, and cooling, all of which need careful design and involve trade-offs.
  - Segmentation is needed so that magnetic fields can penetrate the conducting material to reach the plasma in a finite time.
  - The plasma facing wall, which may be beryllium and a neutron multiplier, needs to not poison the plasma with impurities, and to withstand disruption.
  - All components need to be able to withstand normal thermal cycling and the mechanical, thermal and electromagnetic shock if a disruption occurs.
- Remote handling of the systems must be possible for repair and maintenance.
- The system must withstand high neutron flux for long periods, including neutron displacement damage and hydrogen isotope and helium build-up from nuclear reactions.
- Activation of all components needs to be considered and probably minimized with regard to module repair, replacement, dismantlement, and radioactive processing and storage requirements. Low activation materials have featured prominently in some reactor designs.

One of the lessons from ITER construction is that new technologies need careful design, prototyping, testing and quality control to qualify for use in a nuclear licensed plant. Newer technologies operating at extreme pressures, stresses and temperatures especially need to be carefully and demonstrably qualified.

#### 8.1.3.2 Divertor Optimization

In addition to the blanket and shielding, the other major plasma facing components is the divertor. The divertor needs to provide high performance for power loading without compromising the stability of core plasma or the fusion power generation. Divertor research is a major area of the fusion community. In the context of systems approaches, using the EU-DEMO as a case study, it was found to be necessary to fulfil two high-level requirements [5], namely: (i) the concentration of seeded impurities has to be lower than some critical value in order to not compromise the fusion plasma performance or stability and (ii) damage to the divertor plate in the case of accidental plasma reattachment must be avoided for a sufficiently long time in order to ensure safe, controlled termination of the plasma discharge.

Two figures of merit, corresponding to these criteria, have been identified and discussed. The dependence of such figures of merit on the relevant machine parameters (major radius and toroidal magnetic field) was analysed. The main conclusion was that the simultaneous fulfilment of both requirements limits the viable

EU-DEMO size both in terms of major radius and toroidal magnetic field. A strong parametric dependency on the magnetic field strength hampers the realisation of compact devices with very high fields, which would then require either the development of a technological solution to withstand extremely high heat flux during the transients or a divertor concept able to ensure that reattachment never occurs. However, large devices are limited in size by the necessary concentration of impurities to attain detachment.

## *8.1.4 Use Systems Codes to Develop Design Analysis Methods*

Planning next-step machines requires the designers to have learned lessons and knowledge acquired in previous machines. The analysis will use systems thinking to determine systems levels, elements and relations. Systems architecture methods are required to choose among variant solutions and identify knowledge gaps needing innovations.

It is also necessary to have a definite design of a fusion reactor as a basis for designing final demonstration prototypes. A major decision is the degree of ambition in the choice of plasma physics. This decision process is illustrated by the ITER double choice of an inductively driven baseline scenario, and a current drive and bootstrap current option aimed at steady state. Eventually, a fusion reactor will run on a single plasma scenario and be optimized for that. Research is planned on several existing and near term tokamaks to explore the best options.

### 8.1.4.1 General Systems Code SysML

Once the elements of a basic design are decided, there exist a number of systems tools to use to describe the system in detail and to explore variants in systems architecture. As was illustrated in Fig. 2.1 [6] of Chap. 2, the SysML language [7] can be used to extrapolate from JET operation to model an EU DEMO plasma pulse. As shown in Fig. 2.2 [8], systems architecture approaches using SysML can also be applied to the integration of the heating and current drive systems in the EU DEMO design.

### 8.1.4.2 Process

The fusion reactor systems code PROCESS [9–11] assesses the physics, engineering and economic viability of a hypothetical fusion power station using simple models of all parts of a reactor system, from the basic plasma physics to the generation of electricity. It was based on the code ETR/ITER [12] which was a national USA effort to develop a systems code, which in turn developed from the systems code for the Experimental Power Reactor [13]. PROCESS is usually used in optimisation mode, in which it finds a set of parameters that maximise (or minimise) a figure of merit chosen by the user, while being consistent with the inputs and the specified constraints. The scope of PROCESS is very wide and goes well beyond reactor physics, including conversion of heat to electricity, buildings, and costs. PROCESS has modules for many different basic fusion variants, including stellarators, inertial confinement, d-$^3$He fusion and others.

PROCESS provides time dependency and uses radial profiles of density, temperature and other parameters, with and without an edge pedestal, to calculate global quantities. Models are developed to represent the toroidal and poloidal field coils and their radial structure, and auxiliary heating access is included. Reasonable agreement in results is obtained with full one and two dimensional plasma transport codes.

Algorithms have been developed relating to the engineering aspects of the plant. The toroidal field (TF) coils and the central solenoid are assumed by default to be wound from niobium-tin superconductor with the same properties as the ITER conductors. The winding temperature and induced voltage during a quench provide a limit on the current density in the TF coils. Upper limits are placed on the stresses in the structural materials of the TF coil, using a simple two-layer model of the inboard leg of the coil. The thermal efficiency of the plant can be estimated using the maximum coolant temperature, and the capacity factor is derived from estimates of the planned and unplanned downtime, and the duty cycle if the reactor is pulsed. The need for a large central solenoid to induce most of the plasma current, together with physics assumptions that are conservative compared to some other studies, result in a large machine. Multiple constraints, working together, restrict the parameter space of the optimised model. For example, even when the ratio of operating current to critical current in the TF coils is increased by a factor of five, the total coil cross-section decreases only a little, because of the need for copper stabiliser, insulation, and structural support. The result is that the plasma major radius hardly changes. Results such as those justify the development of systems codes.

#### 8.1.4.3 SYCOMORE

The code SYCOMORE [14, 15] was developed for EU-DEMO and includes the European Integrated Tokamak Modelling framework, see Chap. 2. SYCOMORE is a modular system code which includes physics and technology models coupled to an optimizer in order to explore a large design parameter space. It was used for trade-off studies including the influence of some design-driving assumptions on the reactor performances and size, starting from a European DEMO1-like design (more than 500 MW net electric power and 2 h burn duration). The increase of the mechanical stress limits in TF and CS magnets could help reduce the reactor size, slightly more when high temperature superconductors are used in the TF coil. The tritium breeding ratio could be improved to more than 1.10 by a moderate increase of the size. Divertor coolant options were also compared, showing some differences between helium, hot and cold water scenarios at various incident divertor heat fluxes.

#### 8.1.4.4 Systems Design Efficiency and BLUEPRINT

The justification for the development of BLUEPRINT provides a very useful insight into the efficiency of systems strategy used to design EU DEMO and how it might be improved. The present strategy was to develop multiple alternative solutions for the same set of high-level plant requirements such as electricity production. The baseline determined by PROCESS was investigated in more detail from both a physics and engineering perspective as would be necessary for successful construction. The effort for more detailed investigations is very high, and the turnaround time of such studies is much longer. Therefore, the baseline design is typically frozen for a period of 2–3 years to enable a stable reactor configuration to be investigated. In parallel to the baseline activities, other reactor configurations are investigated in comparatively little detail, for example different divertor configurations and current drive scenarios. Additionally, there is a more focussed approach of examining key design issues in detail, addressing a particular design problem such as primary coolant.

The details of the general methodology are as follows:

- PROCESS and/or SYCOMORE generate a 0-D optimization, which are manually processed to create a 2-D cross section of the reactor
- These results are fed to a 2-D MHD plasma equilibrium solver, which optimizes coil positions and currents and calculates the divertor separatrix and geometry and plasma vertical stability, all taking 2 weeks.
- Next a 3-D CAD model of the tokamak is generated, taking about a month
- More detailed studies are undertaken of some components, with the main emphasis on neutronics, shielding and tritium breeding, taking 2–3 months.
- Sequential activities to create a design point can take 4–6 months.
- Detailed design of components such as superconducting magnets and breeding blankets, sometimes involving separate R&D or prototyping can take years.
- Implementation of a design is therefore long and expensive, with inconsistencies that need to be resolved, difficulties in trying new concepts.
- Note that the alternative of avoiding systems approaches would greatly increase the probability of errors and inconsistencies.

To resolve these problems, the code BLUEPRINT [16] uses a design methodology which automates and accelerates the sequence of design and analysis. The code incorporates many other code modules to integrate the design from plasma physics through to the entire power plant, incorporating Computer Aided Design (CAD) to give a fully integrated and assessed design.

The design strategy for the EU-DEMO hinges on investigating multiple reactor designs and technologies in parallel, progressively down-selecting these in the mid-2020's, in preparation for the conceptual design phase. Alternative reactor designs to the main baseline are only loosely defined. The typical activities required to generate a DEMO design point can be speeded up by four orders of magnitude –

from months to minutes – paving the way for a rigorous and broad exploration of the design space.

#### 8.1.4.5 MIRA

The systems design code MIRA was developed to use a multi-physics approach to designing a fusion power plant [17]. It includes magnetic equilibrium and plasma physics, coils engineering, reactor architecture and neutronics and reactor integration in the plant system.

The logical flow consists of two systems codes and two detailed design codes, starting with PROCESS that has a 2D reactor build, a 1D plasma and scrape-off layer physics, 0D/1D coil engineering, and 0D plant power balance. This provides input to the MIRA set of modules, which includes a 2D reactor build, 1D plasma physics, 2D/3D coil engineering, 2D free-boundary plasma equilibrium, 2D reactor neutronics and 0D plant power balance.

The results of optimizations are fed into design codes, with physics modules for 2D plasma and scrape-off layer transport, 2D plasma equilibrium and scenario development and 2D plasma vertical stabilization. These results go to detailed design engineering modules for 3D CAD layout and modelling, 3D neutronics, 3D thermo-mechanics, 3D thermal-hydraulics and 3D electromagnetics.

When applied to the DEMO 2015 baseline design, MIRA indicated that the inner wall shielding and blanket thickness could be slightly reduced, leading to a more optimized design. Future development plans involve a EUROfusion project for developing a systems code BLUEMIRA for DEMO which combines features of BLUEPRINT and MIRA.

Other systems codes have been developed such as POPCON [18], TPC [19]. HELIOS [20] and TREND [21], but are not discussed here.

## 8.2 The EU DEMO and Planning Major Steps for Designing a Fusion Reactor – Case Study

The EU DEMO concept for a demonstration tokamak fusion reactor is used here as a case study in how systems codes can be used in engineering design.

### *8.2.1 Use of Systems Approaches in the Staged Design Approach in Europe*

The EUROfusion programme developed a European roadmap towards fusion electricity [22]. This roadmap divides the quest into eight missions, for which it reviews the current status of research, identifies open issues, and proposes a research and development programme. ITER is the key facility. The roadmap is tightly connected to the ITER schedule and the vast majority of resources in fusion research are presently dedicated to ITER and its accompanying experiments. Parallel to ITER exploitation in the 2030s, the construction of the demonstration power plant EU DEMO needs to be prepared. EU DEMO should for the first time supply fusion electricity to the grid and have a self-sufficient fuel cycle. The design, construction and operation of EU DEMO will require full involvement of industry to take responsibility for commercial fusion power. The roadmap proposes an integrated way to find solutions for all challenges.

The systems engineering and architecture approaches to the staged design approach of EU DEMO have been carefully described [1, 2, 23, 24] and are summarized here. As described in [23], the progress to date on EU DEMO is summarized in a special issue of Fusion Engineering and Design with a very extensive range of articles. The pre-conceptual design activities in Europe were aimed at advancing the technical basis of EU DEMO to come into operation around the middle of this century with the main aims of demonstrating the production of few hundred MWs of net electricity, the feasibility of operation with a closed tritium fuel cycle, and maintenance systems capable of achieving adequate plant availability. EU DEMO is being designed to benefit as much as possible from the ITER experience, in terms of design, licensing, and construction. Emphasis is on an integrated design approach, based on system engineering, which provides a clear path for urgent R&D and addresses the main design integration issues.

A systems approach is required due to the main design challenges:

- large knowledge gaps in key reactor technologies;
- design dealing with uncertainties (physics/technology);
- high degree of complexity/system interdependencies;
- integration of design drivers across different systems.

The main strategy principles for the design are to use:

- modest extrapolations from ITER;
- robust design incorporating proven technologies and innovations from R&D;
- safety features and design licensability from ITER and nuclear facilities;
- a 'success orientated' DEMO design development in parallel with ITER;
- harnessing the ITER industrial base.

The main goals of DEMO are as follows:

- Conversion of fusion heat into electricity ($\sim$500 MW (e));
- Achieve tritium self-sufficiency (TBR $>$ 1);
- Reasonable availability/Several full power years;
- Minimize activation waste, no long-term storage;
- DEMO as a component test facility and pathfinder to a First-of-a-Kind Fusion Power Plant.

A major point of emphasis is that the design should be fully treated as a nuclear facility that must be licensed, with the following aspects:

- The plant design should drive R&D and not the other way round.
- Fusion is a nuclear technology and as such, will be assessed with full nuclear scrutiny by the regulator.
- There is a need for a traceable design process with a rigorous Systems Engineering approach.
- The technical solution should be based on maintaining proven design features to minimise technological risks.

ITER is planning to start operation in 2025 and reach full fusion power of 500 MW by 2035. The stages for DEMO consist of a pre-conceptual design which finished in 2020, a conceptual design, engineering design, procurement and construction, and commissioning and operations from 2050 to 2070.

Even though DEMO is an extrapolation from ITER, there are many choices to be made from the available tokamak design space. Systems codes as discussed above are an essential part of making these choices. These codes have been used to find solutions with a minimum tokamak size so as to minimize cost. The three overarching limitations preventing further reductions in size are:

- the divertor protection and power handling;
- the access to the H-mode;
- the maximum field in the conductor of the toroidal field (TF) coils and the stress in the coil casing.

### *8.2.2 Application of Systems Codes*

The initial design process of DEMO was focussed on a systems engineering and design integration approach [25] which was recognized to be essential from an early stage. Systems codes, representing the full plant by capturing the interactions between all of the important plant subsystems, were used to identify design points based on assumptions about plasma performance and technology. The systems code PROCESS was used to scope EU DEMO design studies along with another code SYCOMORE. Operating space and the consequences of choosing different global parameters were rapidly explored. Some of the point design space explored is shown in Fig. 8.1, with the lines indicating specific design options DEMO1 (pulsed ITER baseline scenario) and DEMO2 (Advanced ITER steady-state scenario), with ITER and the fusion reactor design ARIES included. The system output was then analysed with state-of-the-art tools allowing a more detailed assessment of individual aspects in several areas. This interaction is repeated until a 'stable release' is obtained for wider evaluation of the physics and engineering.

The EU DEMO simplified design model of the tokamak load assembly [26] used in the systems code PROCESS is shown in Fig. 8.2. It consists of the subsystems: (a) vacuum-vessel; (b, c) breeding blanket (d) divertor; (e, f, g) ports; (h) toroidal field coils; (i) poloidal field coils; (j) cryostat; (k) bioshield.

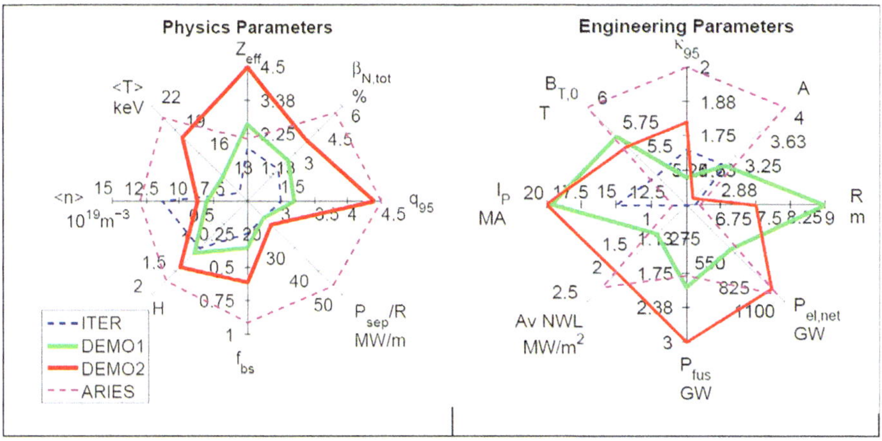


**Fig. 8.1** Physics and engineering point diagrams for inductive and steady-state EU DEMO options. (From [25], Fig. 1)

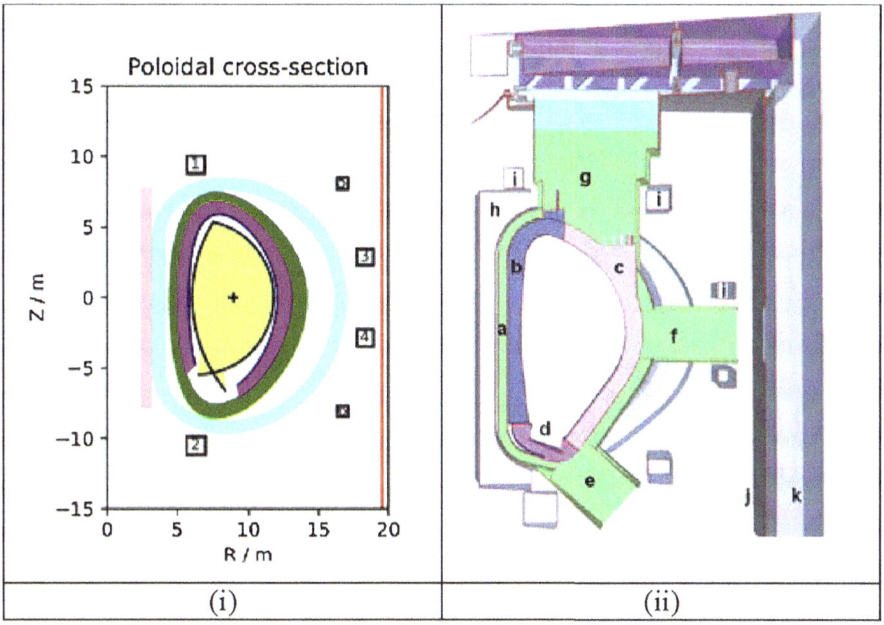


**Fig. 8.2** Systems code PROCESS model of the DEMO, (i) Elevation view of the tokamak; (ii) Tokamak radial-build: (**a**) vacuum-vessel; (**b**) breeding blanket (inboard); (**c**) breeding blanket (outboard); (**d**) divertor; (**e**) lower port; (**f**) (equatorial port; (**g**) upper port; (**h**) toroidal field coils; (**i**) poloidal field coils; (**j**) cryostat; (**k**) bioshield (From [26], Fig. 3)

Its parameters are shown in Table 8.1 for its baseline design, along with values proposed for other reactor designs. Of the six reactors shown, the first four are designed by major ITER partners and follow that design philosophy. The other two concepts explore the regimes of extremely high field and low aspect ratio.

The first thing to note is that the architecture of EU DEMO resembles ITER, only with a larger major and minor radius to accommodate the neutron shield and blanket and to produce 2 GW of fusion power. The aspect ratio, axial toroidal field, plasma current, number of poloidal coils and auxiliary heating are rather similar. However, the design team points out that there are several new challenges concerning the design and feasibility of:

- wall protection limiters during plasma transients;
- breeding blanket and ancillary systems concepts;
- advanced magnetic divertor configurations;
- breeding blanket vertical segment-based architecture;
- power conversion system options, i.e. direct or indirect;
- tokamak building concepts incl. ex-vessel maintenance;
- pumping concept based on tritium direct recirculation;
- plasma-operating scenario including support systems and diagnostics/control.

The EU DEMO approach was based on an extrapolation of the physics scenarios and technology of ITER. During their Pre-Concept Design Phase, the focus was on establishing the system requirements through a top-down systems engineering process, identifying the main technical risk/feasibility issues, and assessing the potential of a number of DEMO candidate system architectures and design concepts to meet the requirements. The supporting R&D programme aimed at responding to the critical questions and feasibility issues raised in the initial investigation of the options under study.

The minimum size of DEMO was limited by the ability to handle the divertor exhaust power for a given machine size, represented in systems code terms as $P_{sep}B/qAR_0$, where $P_{sep}$ is the power crossing the separatrix, giving a limit of 9.2 MW T m$^{-1}$, based on similarity with ITER.

A second limit on the size of DEMO was the magnet design. In the models used, the field available is principally limited by the stresses reached in the coils, rather than the superconductor performance. The forces vary with $B^2$, and since the coil cannot expand toroidally it must become radially larger rapidly limiting how small the machine can become. With an aspect ratio of 3.1, space for a breeding blanket, and stress limits of <700 MPa in the structural coil materials, targeting a field of 5 T in the plasma leads to a device with $R_0 > 7$ m without considering other limitations.

These limitations contributed to the design choice of major and minor radius of 9 m and 2.9 m, aspect ratio 3.1, a plasma current of 18 MA and toroidal field 5.9 T.

The EU DEMO design is also based on a number of conservative assumptions carried on from expected ITER baseline performance. One determining feature is that it is designed to be a pulsed machine; hence a large and high field transformer coil is required for the large plasma current and hours-long pulse duration.

**Table 8.1** DEMOs, engineering reactor and fusion reactor design parameters

| Parameter/ Tokamak | EU DEMO | JA DEMO | CFETR – DEMO | K-DEMO | ARC | ST-135 |
|---|---|---|---|---|---|---|
| Start year | 2051 | 2040 | 2030's | 2037 | ? | ? |
| Max plasma current MA | 18 | 12.3 | 11.91 | 12 | 7.8 | 7.2 |
| R(m) | 9.00 | 8.5 | 7.00 | 6.8 | 3.3 | 1.35 |
| a(m) | 2.90 | 2.42 | 2.19 | 2.1 | 1.13 | 0.75 |
| A = r/a | 3.1 | 3.5 | 3.2 | 3.2 | 2.9 | 1.8 |
| Plasma volume ($m^3$) | 2500 | 1600 | 1249 | 1150 | 141 | 40 |
| Btor (T) | 5.9 | 5.94 | 7.05 | 7.4 | 9.2 | 3.7 |
| Max pulse (s) | 7200 | Steady-state | Steady-state | Steady-state | Steady-state | ? |
| Pfusion GW | 2.01 | 1.42 | 1.446 | 2.2–3.0 | 0.525 | 0.185 |
| Power to grid GW (e) | 0.5 | 0.3 | 0.494 | 0.4–0.7 | 0.190 | ? |
| Fusion Q | 40 | 17.5 | 30.3 | Depend CD | 13.6 | 5 |
| $Q_\alpha \equiv Q/5$ | 8 | 3.5 | 6 | ? | 2.7 | 1 |
| Duty cycle | ≈30% | ? | 30–50% | High | ? | ? |
| Auxiliary heat (MW) | 120 | 84 | 48 | 80–120 | 25 ICRH, 13.6 LH | 37 |
| AuxHt/Vol ($MW/m^3$) | 0.048 | 0.052 | 0.038 | 0.070–0.104 | 0.273 | 0.925 |
| Blanket Concept (1) | Water Cooled lithium Lead PbLi (WCLL), PW | Water Cooled solid breeding (WCSB), PW, pebbles | Water cooled ceramic with $Be_{12}Ti$ and $Li_2TiO_3$, PW | Water cooled ceramic pebbles with $Be_{12}Ti$ and $Li_4SiO_4$, PW | Molten salt FLiBe as breeder and coolant in tank, | ? Inner shield made of tungsten carbide with water cooling |
| Blanket Concept (2) | He cooled pebble bed be. $Li_4SiO_4$ | × | He cooled ceramic with be and $Li_4SiO_4$ | × | × | |
| Breeding comment | WCLL shields better, breeds worse than HCPB | Pebble ratio $Be_{12}Ti$ to $Li_2TiO_3$ = 7/3 best breeding | Helium cooled is first choice with 500 °C outlet. TBR = 1.2 | Helium has too high a pumping power. | Low pumping power, TBR = 1.1 | TBR less than one? |

The divertor power handling capability is also similar to that of ITER, as well as the on-axis toroidal field. A current-drive machine has also been extensively examined as DEMO2.

### 8.2.3 *Systems Design Point Studies, Sensitivities and Trade-Offs*

EU DEMO trade-off studies were conducted for the aspect ratio, breeding blanket thickness, the number of TF coils and a double null divertor. A sensitivity analysis was also carried out on electrical output and pulse duration by varying key physics parameters by $\pm 10\%$. The strongest sensitivities are found for the plasma elongation, confinement and density. Detailed studies were also carried out in particular technology areas.

#### 8.2.3.1 Peak Toroidal Field at the Coils

In a superconducting tokamak magnet, the superconductor itself takes up relatively little volume in the winding pack. Higher field windings generate higher forces, and the stresses in the inboard leg of the TF coil casing quickly reach the maximum allowable stress for a given geometry. This provides an effective limit to the practically-achievable field. The solutions require either higher-strength structural materials than currently available, or an increase in machine size to provide space in the radial build for enlarged magnets. The stress varies as $B^2/A$. A 12.5 T peak field gives an inboard leg width of around 1.3 m. Assuming an HTS conductor which could perform at 20 T instead, the resulting forces would require a doubling in the radial thickness of the inner TF limb, plus a small increase required in the winding pack size due to quench protection.

#### 8.2.3.2 Wall-Plug Efficiency of Heating and Current Drive for Steady-State

For the pulsed DEMO-1, the auxiliary heating power requirements during the flat-top are only about 30 MW for MHD control plus some tens of MW for burn control when needed, so about 50 MW in total. The plant electrical output to the grid of 500 MW assumes this auxiliary power is delivered at 40% efficiency.

For scenarios at high non-inductive current fraction like flexi-DEMO, auxiliary power for current drive is required at much higher levels than DEMO-1 also because the same control functions as for DEMO-1 have to be provided as well. Thus, achieving a high value of wall plug efficiency is crucial to ensure an acceptable electricity output. With only 6 MA of driven current, the current drive efficiency needs to be above 60 kA/MW. A proposal for LHCD [27] for ITER claimed it could drive an 8.5 MA plasma current with 20 MW of power, giving 425 kA/MW.

#### 8.2.3.3 Plasma Elongation and Aspect Ratio

Increasing plasma elongation could greatly increase the fusion power output. Vertical stability of the plasma could be improved by reducing the aspect ratio, but this increases the plasma minor radius and decreases the achievable field in the plasma, generally driving up the device size. Toroidally conducting inserts into the blanket in key locations could help, but at the possible cost of tritium breeding and a significant increase in engineering complexity. The final choice of plasma parameter values must respect what is reasonably achievable within whole-machine engineering constraints.

### *8.2.4 Systems Integration and Key Design Issues*

Certain design, physics or technology choices are so integral to the plant architecture, that they have significant implications on a large number of systems that must be integrated into the plant. A thorough examination of system integration aspects is essential to ensure that the total view of the plant is maintained from the very beginning. All factors affected by the numerous design choices to be made must be identified, evaluated, and properly weighted to allow overall design convergence. These key design issues include:

- Power handling capability of the first wall facing the plasma;
- Coolant choice for divertor and breeding blankets.
- Choice of divertor configuration.
- Remote maintenance philosophy.
- Balance of plant issues, especially for a pulsed reactor.
- Plant building layouts.
- Fuel cycle efficiency to minimize tritium inventory.
- Plasma operating scenario.

As a particular example, we focus on blanket design.

### *8.2.5 Fusion Blanket and Shield Thickness for DEMO*

The most critical difference between a fusion reactor and existing experiments is the need to shield the magnetic field coils from neutron damage and heating and to breed tritium. An important and critical impact on fusion reactor design is therefore the physical thickness needed. Since 14 MeV is the centre of mass energy of a d-t produced neutron, lab frame energies can reach up to 18 MeV in the case of a 1 MeV deuteron fusing with a thermal triton.

The blanket shielding and tritium breeding in the context of the EU DEMO design is examined in detail [28]. Considerations of tritium breeding and shielding requirements constrain the design of the test tritium breeding modules for ITER.

The systems engineering codes called PROCESS and BLUEPRINT [29] were used to model the DEMO design and in particular, the blanket and shielding. These codes were developed to cover the middle of the design cycle. PROCESS was expanded to include expanded sensitivity tools and high temperature superconductor REBCO modelling. Key features of BLUEPRINT include a free-boundary plasma equilibrium solver, 3-D CAD generation and a tritium cycle model.

The utilization of the space on the inner side of the torus represents a crucial design aspect in tokamak design and deserves some further consideration, since it is necessary to minimize the thickness of the region occupied. This distance crucially relates the maximum toroidal field at the superconducting coil to that at the plasma centre, thus determining the costs of and stresses on the toroidal field coil.

These factors provide a strong incentive to reduce this distance. However, satisfying the energy conversion and tritium breeding requirements in the blanket and providing the radiation attenuation in the blanket/ shield necessary for magnet protection favours a relatively large distance. In the current DEMO design, the space utilization on the inner side of the torus and the required fractional coverage the breeding blanket needed to achieve tritium self-sufficiency has been set on the basis of extensive neutronics calculations [30]. The blanket modules are part of the basic geometric configuration, and in particular, part of the radial reactor build. The main adopted design guidelines and criteria are:

- Tritium breeding requirements;
  - an overall tritium breeding ratio $\geq$1.05;
  - thin plasma facing components;
  - limited penetrations, about 85% coverage by breeding blanket;
  - constraints on occupied divertor space;
- Neutron shielding requirements;
  - max displacement damage in vacuum vessel (2.75 dpa);
  - cutting/re-welding location in vessel component cooling pipes;
  - helium production 1 appm;
  - total neutron fluence to epoxy insulator $10^{22}$ m$^{-2}$;
  - fast neutron fluence to the $Nb_3Sn$ $10^{22}$ m$^{-2}$;
  - neutron fluence to Cu stabilizer between TF coil warm ups 1–2·$10^{21}$ m$^{-2}$;
  - volumetric nuclear heating in winding pack 50 W m$^{-3}$;
  - port interspace: Shutdown dose rate 12 days after shutdown ~500μSv/h (target);
  - port cells (occasional access) 100μSv/h (target);
  - in-cryostat area, 100μSv/h (target);

- tokamak building areas beyond port cells requiring frequent access;
- shutdown dose rate 1 day after shutdown 10μSv/h (target).

A generic CAD neutronics model was generated from the DEMO Configuration Geometry Management (CGM) model which included the toroidal field coil, vacuum vessel, divertor, blanket segment box, vessel ports, and plasma chamber, represented with "envelopes" without internal structure. The model converted to analysis for MCNP/TRIPOLI-4 (3D Monte Carlo neutron radiation transport codes/neutron and photon transport and electron-photon shower) using the McCad conversion software, then used for the four blanket types.

The bottom line of these various calculations and different blankets is that the distance from high field side of the superconducting coil to the first wall facing the plasma is 1.4 m in order to provide sufficient shielding, both in terms of material damage and vacuum vessel nuclear heating.

The distance of 1.4 m, composed of 60 cm vacuum vessel, shielding and gaps plus 80 cm breeding blanket, is a crucial factor in any fusion reactor design. It could be reduced by using only shielding to protect the coil, and rely on breeding elsewhere than the inner wall, but then a high tritium breeding ratio elsewhere would need to be demonstrated. Analysis for the entire DEMO machine of four types of blanket being tested in ITER, being HCPB, HCLL, DCLL, WCLL, found overall tritium breeding ratios of 1.09–1.15. Two specialized designs were retained for DEMO.

- WCLL, with water cooling under fission Pressurized Water Reactor (PWR) conditions;
- HCPB breeding blanket with Be and $Li_4SiO_4$ pebbles.

Of the two, a WCLL shields better but breeds tritium worse than HCPB, hence they could be used in different parts of DEMO.

Due to the numerous penetrations, neutron streaming on the outboard also represents a serious design issue. A biological shield is necessary to reduce the radiation biological dose outside the reactor to the maximum permissible dose for occupationally exposed individuals. The walls of the reactor building can serve the dual purpose of providing necessary containment and biological shielding.

On the opposite side of design space, spherical tokamak designers are exploring the options of a very small central column area with a high temperature superconducting toroidal field core made of yttrium barium copper oxide. The design uses a neutron shield of only 32 cm thickness, composed of ceramic tungsten carbide [31] which combines the neutron reflection of tungsten and the neutron moderating properties of carbon. There is a further 6 cm gap to the plasma.

### 8.2.6 Systems Management Structures

The EU DEMO design team completed the pre-conceptual design phase in 2020, entered the conceptual design phase in 2021, and identified the systems engineering management structure necessary to carry forward the project [2]. In such a large and complicated project, particular issues arise:

- There is a high degree of complex system interdependence.
- There are multiple design drivers across different systems.
- The design has to deal with multiple physics and engineering uncertainties.
- There are inefficiencies in iterating the design (as discussed for BLUEPRINT).
- A shortage of systems engineering skills hampers design integration.
- Design teams need better coordination.

The case is then made for a close-knit structure that addresses these needs. This management structure was designed to achieve:

- integration of lessons learned, especially from ITER;
- better articulation between design and R&D activities;
- a central team that achieves improved coordination and project organization;
- project documentation including baseline and knowledge management;
- quality assurance planning and structure;
- strengthening project governance.

### 8.2.7 Robustness and Fragilities

One of the central goals of the systems approach to EU DEMO is the continuous attempt to build robustness into the system by a combination of:

- Reliance on ITER R&D inputs and construction experience
- Use of the full knowledge base of the physics and engineering communities
- Identifying potential fragilities in the system and performing specific studies and optimizations to mitigate or remove the fragilities, for example:
  - Excessive stresses in the toroidal field coil;
  - Damaging heat loads to the thin first wall and the divertor;
  - Risk of low tritium breeding ratio;
  - Premature use of insufficiently tested technology;
  - The need for full integration of the entire plant at the design stage.

  If fragilities in systems do occur, a key factor in repairing damage is the remote maintenance capability of a fusion reactor [32], involving the need to:

  - recognize maintenance technical risks;
  - develop integrated maintenance systems;
  - address key maintenance risks;

- conduct tests on proof-of-principle designs, including pipe connections, in-vessel component handling, and port access systems, which includes pipe module handling and contamination control doors.

Given the high complexity and the need for very high levels of safety, reliability, accuracy and speed of DEMO remote handling operations [1], it is vital that the control systems are tested using realistic sensors, movers and payloads to demonstrate feasibility during the concept design phase. A scheme has been proposed for a maintenance test facility to mitigate risks.

Looking at the entire system, the fragilities can be roughly divided into local and overall types. The local ones are in a system that can be repaired, and the others can cause complete systems failure. It would seem that the philosophy of systems approaches, modularity and the possibility of repair minimizes most foreseeable fragilities.

There remain the fragilities from a huge extrapolation from existing d-t operation JET plasmas with either the beginnings of self alpha heating or the beginnings of sustained steady state operation. ITER will provide essential intermediate operation, but the results will be available only in the 2030's. When comparing DEMO to JET, it has much more volume, heating power, stored energy, fusion power and pulse length. It also has a lower heating power per unit volume. Understanding of plasma and technology scaling laws has vastly improved and extrapolations may be reliable, but it remains to be demonstrated.

In summary, the positive results from JET d-t experiments, the currently successful construction of ITER, world-wide collaboration and supporting physics and technology projects, give the current DEMO design a good probability of leading to a successful fusion reactor.

## 8.3 Japan Demo and Fusion Reactor Designs and Their Variations

### *8.3.1 Current Japan DEMO Design*

A number of fusion Demonstration Reactor (DEMO) designs have been developed over the years in Japan [33] including the current design [34, 35] together with progress on design issues [36]. A path and strategy toward a demonstration plant (DEMO) to achieve net electric power generation and to establish various technologies for successive commercial plants was based on the Japan's action plan for fusion energy development. Required innovations to attain the commercial plant as a final target of the DEMO plant were considered and a variety of novel innovations introduced in previous reactor designs in Japan were reviewed in order to demonstrate a possible path to the fusion power plants.

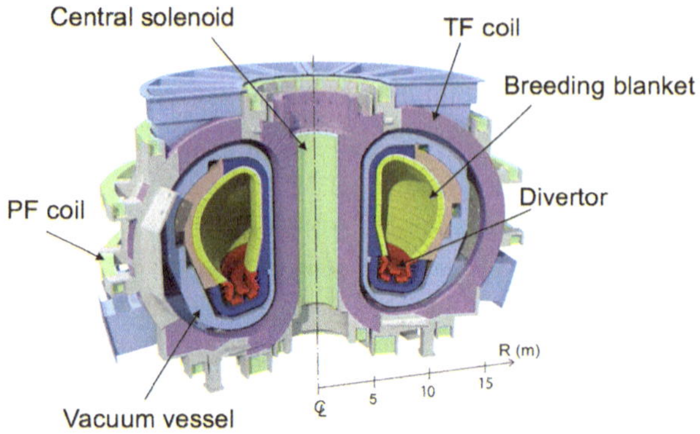


**Fig. 8.3** Conceptual view of Japan's fusion reactor JA DEMO. (From [35], Fig. 2)

In preparation for DEMO construction, some intermediate engineering facilities need planning [33]:

- A fusion neutron source for material irradiation (FNS)
- A cold test facility for divertor heat load by the middle 2020's
- A large-scale facility for handling tritium by 2030
- A facility for maintenance technology at a realistic scale
- Experiments on JT-60 SA to demonstrate steady-state with heating and current drive by 2030

The basic operation mode of Japan's DEMO is steady-state, in contrast to EU's DEMO in which pulsed mode has been assumed for the basic operation mode. The role of JT-60SA is therefore crucial. Relevant current drive and bootstrap current experiments in Japan and elsewhere have been reviewed [37].

A schematic of the Japan JA DEMO is shown in Fig. 8.3 and its parameters are shown in Table 8.1. In comparison to the EU Demo, the JA Demo is smaller, has lower plasma current and is designed for steady state operation.

Tritium breeding and heat removal is based on a water-cooled solid breeding (WCSB) with pebbles blanket with pressurized water reactor (PWR) fission reactor conditions of 290 °C–325 °C temperature water at 15.5 MPa. The best breeding ratio was found with 70% pebbles of $Be_{12}Ti$ and 30% pebbles of $Li_2TiO_3$.

An interesting plan for starting operations is to begin without an initial tritium inventory and operate in deuterium only until enough tritium is generated by d-d fusion and by 2.5 MeV neutrons breeding tritium in the blanket [38] leading to full d-t operation in less than 100 days.

The current baseline design of the divertor is based on:

- a conventional single null divertor (SND) with a divertor leg longer than ITER
- plasma detachment with impurity seeding
- an ITER-like tungsten monoblock heat sink cooled by pressurized water

The design settled on the use of $Nb_3Sn$ and NbTi as superconductors as in ITER. Other superconductors were considered, but had only small advantages, whereas superconductors for ITER have had decades of testing and development.

Current drive is essential for the success of a steady-state DEMO, but there is a major concern that current drive efficiencies are not high enough for a comfortably large energy gain, hence considerable development is still needed. Such considerations led the EU DEMO team to choose pulsed operation.

### 8.3.2 Robustness of Japan DEMO Current Design

In order to increase the robustness of the design and to protect against fragilities, a number of engineering issues were examined in detail [36]:

- Countermeasures against a loss of coolant accident inside blanket
  - In the "honeycomb-shape" blanket, little retention of tritium was found by the flow analysis of the purge gas.
  - A simple concept for cylindrical structures was designed to meet the target tritium breeding ratio and keeping pressure tightness.
- Steady power generation
  - The total power consumption of in-plant systems was 386 MW, giving a net electric output of 254 MW (e).
  - Consistency between cooling system and tritium concentration control was confirmed for safety management.
- Safety
  - Accident sequences and mitigation systems were designed.
  - A mitigation system of countermeasures was confirmed. In summary, safety at several levels and hence robustness was being incorporated into the JA DEMO design right at the start of the process.

### 8.3.3 Japan DEMO Designs and Options in Systems Architecture

Over the years, Japan has explored many possible variants for DEMO [33].

SSTR's (Steady State Tokamak Reactor) features [39] included a very high bootstrap current ratio of 70% allowed by high $\beta_p \geq 2.0$ and q of 4–5 with $\varepsilon\beta_p < 0.5$ and $\beta_N = 3.5$. The maximum toroidal field at the coils of 16.5 T was generated by $Nb_3Al$ superconductor. The remaining plasma current of the total of 12 MA was driven by 70 MW of 1 MeV neutral beams, with an overall fusion power production of 2.5 GW.

IDLT (Inductively operated Day Long Tokamak) [40] was based on the idea of being as close to ITER as possible without any major jumps in technology. It was planned to use inductive technology with a large central solenoid in this R = 10 m machine with an aspect ratio of 5.35 that could maintain a plasma for 12 h. The maximum toroidal field at the coil is 12 T, and 8.08 T on the plasma axis giving a plasma current of 13.9 MA, qualifying this as a high field tokamak. The pulse length was planned at 12–13 h and dwell time of 300 s. Neutral beam current drive would have been used for current profile control and aiding the rather small bootstrap current fraction of 35%. Auxiliary heating power is about 80 MW. The plant was designed for a net electric power output of 1.026 GW, at 31% plant efficiency. A continuous electric power output is maintained during the plasma dwell time by burning hydrogen produced by electrolysis during the pulse.

CREST (Compact REversed Shear Tokamak) [41] was designed with a major and minor radius of 5.4 m and 1.6 m, giving an aspect ratio of 3.4, closer to that of ITER, an elongation of 2 and a high normalized beta of 5.5 based upon a reversed shear plasma, stabilized by a closed conducting shell made of zirconium. Current drive was provided by neutral beam injection with the idea that high injected toroidal rotation will stabilize resistive wall modes. The zirconium would help to maximize tritium breeding and minimize activation. A unique energy conversion feature was to use high temperature superheated steam at 1000 °C to produce hydrogen at 90% efficiency. External load balancing could be achieved by power trading with electrolysis. The goal was to produce 1.16 GW of electricity at a competitive cost.

DEMO-CREST (Demonstration fusion power plant – CREST) [42] involved a two stage approach, starting as a minimum extension from ITER in plasma parameters, giving moderate plasma performance of $\beta_N = 1.9$ for break-even in net electric output. The next step would increase up to $\beta_N = 2.5$ for a net electric power of 500 MW, which would require a maximum magnetic field at the coils of 16 T, a thermal efficiency of 30%, NBI system efficiency of 50% and 200 MW of current drive power. A two stage blanket design was planned going to higher thermal efficiency in the second blanket which could produce 1000 MW of electricity.

VECTOR (VEry Compact TOkamak Reactor) [43] was a futuristic, idealized and economic design with a very low aspect ratio of 2.0, vertical elongation of 2 and major radius 3.75 m characteristic of spherical tokamaks. The maximum toroidal field at the coils was 19.6 T and the plasma current was 18.3 MA, achieving $\beta_N = 3.75$ with a fusion power of 1.8 GW. A neutron shield was able to protect the inner toroidal field coil. There was no central solenoid and no breeding blanket at small major radius in the design to allow the smallest aspect ratio possible, thereby relying entirely on current drive. The material SiC-SiC, a ceramic matrix composite, was used as the structural material to minimize neutron activation and long term radioactivity. A high temperature helium gas coolant is used to maximize the thermodynamic efficiency of conversion of heat to electricity.

SlimCS (Slim Central Solenoid) [44] evolved from the VECTOR design and was based on less optimistic assumptions. A low aspect ratio of 2.6 was used as before, but a central solenoid was added to the design which is used for plasma shaping rather than current drive. The planned normalized beta was reduced to 4.2 and the density limited to the Greenwald limit.

### 8.3.4 Summary of Previous Design Variants

The evolution of Japan's DEMO designs provides an illustration of the extremes of choices available for systems architecting:

- SSTR – high aspect ratio and high bootstrap current for steady-state operation
- IDLT – ITER like inductively driven tokamak with large central solenoid
- CREST – an advanced tokamak with reversed shear operation and current drive with a low activation zirconium first wall and high temperature.
- DEMO-CREST – an advanced tokamak with very high thermal efficiency blankets
- VECTOR – a compact spherical tokamak reactor with low activation materials
- SlimCS – a modified compact spherical tokamak reactor with a small central solenoid for plasma shaping
- Japan DEMO – a steady-state current driven tokamak with moderate aspect ratio and ITER-like technology

It is interesting to note that after exploring all these variations, the main Japan DEMO design is fully based on ITER and is following on from its baseline alternative scenario of steady-state operation with current drive.

## 8.4 China Fusion Engineering Test Reactor (CFETR)

### 8.4.1 Two Phase Design

A conceptual design was developed for the CFETR [45, 46]. Several operating modes are planned, and the one corresponding to DEMO operation is shown in Table 8.1. With the EAST superconducting tokamak, China is playing a key role in advanced steady-state experiments. The CFETR is designed for fusion powers in the range of 200 MW to over 1 GW. The CFETR engineering design was approved, and the project started on December 2017. Two phases are planned:

- Steady-state operation and tritium self-sustainment will be the two key issues for the first phase with a modest fusion power up to 200 MW.
- The second phase aims for DEMO validation with a fusion power over 1 GW.

### 8.4.2 Tritium Fuel Cycle Studies for CFETR

Tritium breeding and recycling is a main study goal of CFETR [47]. In order to attain tritium self-sufficiency as well as ensure safe handling of tritium, it is important to establish a closed tritium fuel cycle. Therefore, a tritium plant which consists of various tritium processing systems and tritium safety systems is indispensable for the successful operation of CFETR.

Two types of breeding blankets were planned for CFETR:

- Water cooled ceramic with $Be_{12}Ti$ and $Li_2TiO_3$, PWR conditions
- Helium cooled ceramic with Be and $Li_4SiO_4$

The helium cooled blanket was their first choice with 500 °C outlet and therefore higher efficiency for electricity generation. Both blanket designs can reach tritium breeding ratios of up to 1.2 under appropriate conditions.

Detailed simulations of tritium to deuterium ratios in the plasma have also been carried out [48]. They found that there are important trade-offs between fuelling depth, d-t ratio in the plasma, total fusion power and fractional burn up, which in turns effects the rate at which tritium needs to be reprocessed.

### 8.4.3 *Toroidal Field Coils*

Prototype toroidal field coils are being built, using 3 types of conductors spread over 12 winding levels [49]. High performance $Nb_3Sn$ generates the highest field at 14.7 T, the intermediate ITER-like $Nb_3Sn$ generates up to 11.1 T, and NbTi generated the remaining field up to 6.6 T. The conductors are placed in rectangular CICCs to withstand the electromagnetic loads.

The central solenoid might use the high critical temperature superconductor Bismuth–Strontium–Calcium–Copper–Oxide (BSCCO) 2212 cable-in-conduit-conductor (CICC) which could greatly exceed the capabilities of $Nb_3Sn$. BSCCO is already being used in ITER for cable connectors. By using fairly well known technology and relying on ITER experience and EAST simulations, a relatively early schedule is planned for CFETR construction and DEMO-level performance.

## 8.5 Korea K-DEMO

### 8.5.1 *A High Field Reactor*

K-DEMO [50] is aimed at near term implementation. Much of the design philosophy is similar to CEFTR, but with important differences such as the preferred choice of pressurized water for cooling the blanket and more open choices about the best approach for heating and current drive. Its design parameters are given in Table 8.1 and its device core features are shown in Fig. 8.4. Attention has been paid to access for remote maintenance, vertical access and coolant supply. The size and high toroidal field qualifies the K-DEMO as a relatively high field and compact machine, compared to EU-DEMO.

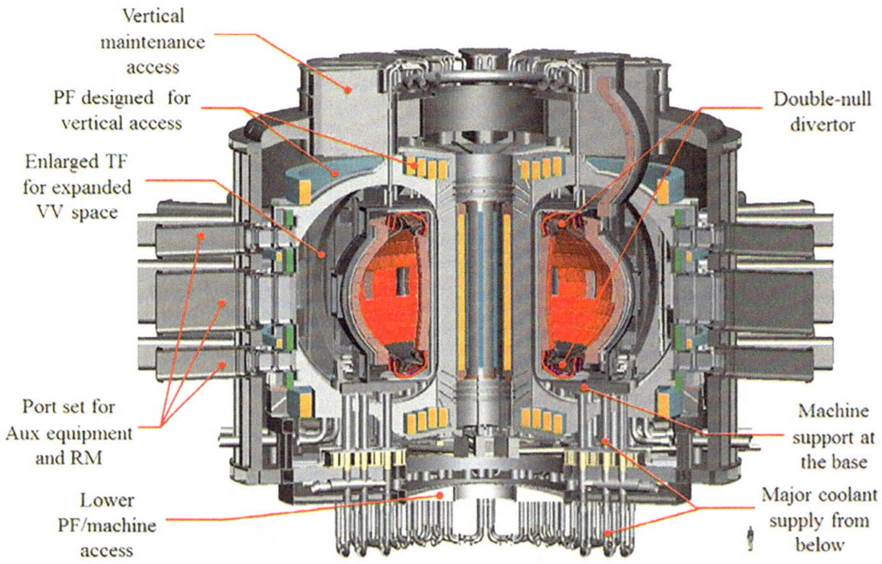


**Fig. 8.4** K-DEMO device core design features. (From [50], Fig. 1)

### 8.5.2 *Fragility – Cyclotron/Synchrotron Radiation*

One fragility is that high toroidal field and electron temperatures result in a significant cyclotron/synchrotron radiation loss of 42 MW even with a 90% reflectivity assumed for the first wall. Such a number illustrates that a low auxiliary heating power might not be enough to drive a high field reactor to near ignition conditions. The auxiliary heating power is therefore high as well as being needed to maintain current drive.

## 8.6 USA DEMO and Fusion Reactors

### 8.6.1 *A High Elongation Fusion Reactor*

The ARIES series [51] of reactor designs included the ARIES-RS and the ARIES-AT [52]. This continuing series of reactor designs explored the various physics concepts reflected in the DEMO designs already discussed in this chapter.

The physics basis of the ARIES-AT design has been explored [53] by using the TSC code. ARIES-AT makes good use of a higher than usual elongation of 2.2, which can have important economic value in increasing the plasma current for a given major and minor radius. ARIES-AT had plasma current 12.8 MA, axial toroidal field of 5.8 T, major and minor radius of 5.2 and 1.3 m, an aspect ratio of

4.0, a triangularity of 0.90, a toroidal beta of $\beta = 9.1\%$ and a normalized beta $\beta_N$ of 5.4, giving a total fusion power of 2 GW. These beta values were chosen to be 10% below the ideal MHD stability limit. The bootstrap current fraction was 0.91.

The effects of H-mode edge gradients and the stability were analysed. The current drive was provided by fast wave ICRF for on-axis current drive and a LH system for off-axis drive.

### 8.6.2 Water or Liquid Nitrogen Cooled Toroidal Field Coil Reactors

Although water-cooled toroidal field coils are no longer being pursued as a design feature for tokamak fusion reactors, it is useful to have a historical look at some design attempts [54]. For high beta operation of about 10%, the coil resistive power can be low enough for net power gain which could be especially relevant in radiation test facilities or fusion-fission hybrid blanket operation. Room-temperature copper coils would offer design advantages.

There are a wide range of possible normal conductors besides water cooled copper, including super pure aluminium or copper cooled by liquid nitrogen or even liquid hydrogen [55]. The ARIES-ST (spherical tokamak) design with a toroidal beta of 54% was considered with normal conductors. Despite the magnet resistive power, the reactor could still produce a net electric power of 1 GW.

This is not to say that resistive coils should be an option in tokamak reactor design space. Nevertheless, there are applications with poloidal field coils near the plasma where actively cooled normal conductors play an important role.

Over the years, there were proposals in the USA for high-field copper devices such as BPX (Burning Plasma eXperiment) [56], FIRE (Fusion Ignition Research Experiment) [57] and IGNITOR (IGNIted TOroidal Reactor) [58]. These proposed machines used water cooled coils to minimize the technological steps to observe the physics of burning plasmas.

### 8.6.3 Advanced Tokamak Reactors

Currently, the USA DEMO and reactor thinking is being developed along lines following the three major tokamak experiments, specifically:

- Advanced tokamaks following DIII-D;
- High field tokamaks following ALCATOR C-MOD;
- Spherical tokamaks following NSTX-U.

A future programme on DIII-D [59] for the development of advanced tokamak physics for reactor design was proposed [60]. The DIII-D view of an advanced tokamak involves high pressure operation, favourable stability and transport properties and a self-driven ('bootstrap') plasma current to sustain operation efficiently and without disruptions.

DIII-D has demonstrated high $\beta$ access, off-axis current drive, profile control, and stability and ELM control in reactor relevant physics regimes. Radiative cooling techniques have been used to reduce power to the divertor while maintaining high performance pedestals. This physics development is underpinned by model creation and validation which can be used for reactor design. DIII-D has recently upgraded its capabilities in auxiliary heating power and current drive along with 3-D field capabilities to further develop advanced tokamak reactor concepts.

## 8.7 Large Scale DEMOs – Robustness and Fragilities

### *8.7.1 Common Features of Large Scale DEMOs*

The first major example of a proto-reactor is ITER itself. ITER is the focus of much of the world's effort on technological and physics development towards a fusion reactor. Although not certain, it has every chance of demonstrating fusion conditions near ignition, alpha particle self-heating, steady state operation with current drive and neutron shielding and tritium breeding technology.

With an aspect ratio of 3.1, ITER already reaches a small aspect ratio when the need for fully shielding 14 MeV neutrons is taken into account. The axial toroidal field of 5.3 T is higher than most existing machines, and the maximum magnetic field at the central solenoid is 13 T, and 11.8 T at the toroidal field coil, so it is a high field as well as a low aspect ratio tokamak. Some of the coil stresses are near their limits, and require special steels as structural elements. Reaching these limits has required much development and innovation [61].

DEMO designs have been developed as EU-DEMO, Japan-DEMO, China CFETR and Korea K-DEMO as summarized in Table 8.1. India is also conducting studies on large-scale demonstration fusion reactors [62]. The EU-DEMO is the largest and they all have a major radius between 6.8 and 9.0 m, aspect ratios of between 3.1 and 3.5, axial toroidal fields of 5.9 up to 7.4 T, and plasma currents between 12 and 18 MA. They can all be characterized as having medium aspect ratio and high axial toroidal field.

The other important similarity is that all these groups are testing shield and blanket modules on ITER. Even though the blanket modules are different in important ways, they all share similar philosophies about total thickness of shield and breeding blanket and the need to fully shield superconducting coils from neutrons, making it difficult to have a much more compact reactor.

The ITER participating countries and design groups have developed similar philosophies and design spaces in the development of ITER and have realized the very extensive innovation necessary and the design choices made in order to solve the problems of building a large scale fusion reactor that produces net electric power and breeds tritium.

There are important variations between the various DEMO designs. The history of DEMO designs such as the Japan DEMO and the USA ARIES series has included a wider range of design. Most of the recent designs have reached a general consensus on the broad design envelope for a future tokamak fusion reactor.

On the question of very high magnetic field, ITER is already using high-temperature superconductor BSSCO for leads for the coils. They considered that even though high temperature superconductors are available for large magnets, $Nb_3Sn$ and NbTi would be used for the toroidal and poloidal field coils because there was decades of development for these conductors and because of stress limits in the coil supports at higher fields.

The China CEFTR designed a toroidal field coil with $Nb_3Sn$ operating at 14.7 T at the coil and was planning to use the high temperature superconductor BSCCO for its central solenoid for an even higher field.

### *8.7.2 Robustness Summary for Large DEMO Machines*

- The physics scenarios have been prototyped in smaller machines and an intensive programme worldwide is coordinated to verify remaining physics and engineering questions.
- The large plasma size and high plasma current are very probably enough to obtain near ignition conditions which will be verified in ITER.
- The careful and detailed systems engineering and systems architecting approaches applied to ITER, EU-DEMO and possibly other machines give a high probability that the machines can be successfully designed, built to specification and operated.
- The DEMO designs are testing much of their technology on ITER, and will have extensive prototyping and operation experience to support their machines.
- Existing nuclear reactor technologies such as PWR cooling are being applied to the tritium breeding blankets and shielding.

Fragilities are being carefully identified and mitigated as best as possible. Some of these fragilities include

- Even though the designs are considered conservative by the standards of very high field tokamak and spherical tokamak reactors, the designs are still pushing the edges of technology. There may be unexpected surprises.
- DEMOs involve nuclear engineering with radioactive materials and heat loading. Maintenance and duty cycle become challenging.
- Large scale reactors and balance of plant are fundamentally complicated, hence expensive. Every effort is being made to reduce costs in designs, and commercial production costs should be much less than these innovative prototypes.

- The large plasma volumes of the DEMO machines have relatively low heating power density and are subject to volumetric power loss by impurity radiation. As an example, the amount of auxiliary heating power necessary for ignition has been optimized by the "(J.G.) Cordey Pass" method [63]. This approach determines the point in operating density and temperature which minimizes the auxiliary power needed to heat a plasma to controlled ignition.

In summary, there is now a high probability that a successful fusion reactor can be constructed which produces a large net electricity output.

## 8.8 Very High Magnetic Field Reactors

The ALCATOR series of high field experiments by MIT and joint work with Commonwealth Fusion has led to the design of a high field fusion machine SPARC [64] and the ARC fusion pilot plant [65, 66]. A high field reactor roadmap [67] and a general commentary [68] were produced.

### *8.8.1 SPARC (Soonest/Smallest Private-Funded Affordable Robust Compact)*

SPARC is planned to be a superconducting coil tokamak producing up to 140 MW of fusion energy in a compact, short current flattop (10 s) machine. It is of comparable size to DIII-D, Asdex Upgrade, EAST and KSTAR. By using rare-earth (Yttrium or Gadolinium) barium copper oxide (REBCO) superconductors, its magnetic field on axis of 12.2 T is three to four times higher than machines such as JET. It includes a neutron shield to protect the coils, but does not breed tritium. Its planned parameters are $R = 1.85$ m, $a = 0.57$ m, $B_T = 12.2$ T, $I_p = 8.7$ MA, vertical elongation = 1.97, triangularity = 0.54, double-null divertor configuration, auxiliary heating power of 25 MW ICRH at 120 MHz, transformer flux = 42 WB, and fusion power gain $Q = 11$. The operation space was initially scoped out using the systems code POPCON (Plasma Operation Contours) [18]. SPARC is shown in Fig. 8.5.

The main goals are to:

- act as a technological prototype for a compact very high field reactor;
- show that a compact machine can achieve reactor level parameters;
- demonstrate alpha particle heating.

The SPARC project has two phases:

- Phase 1 lasted 3 years (2018–2021) involving two major milestones:
  - construction and operation of an HTS-based toroidal field coil
  - completion of the SPARC design

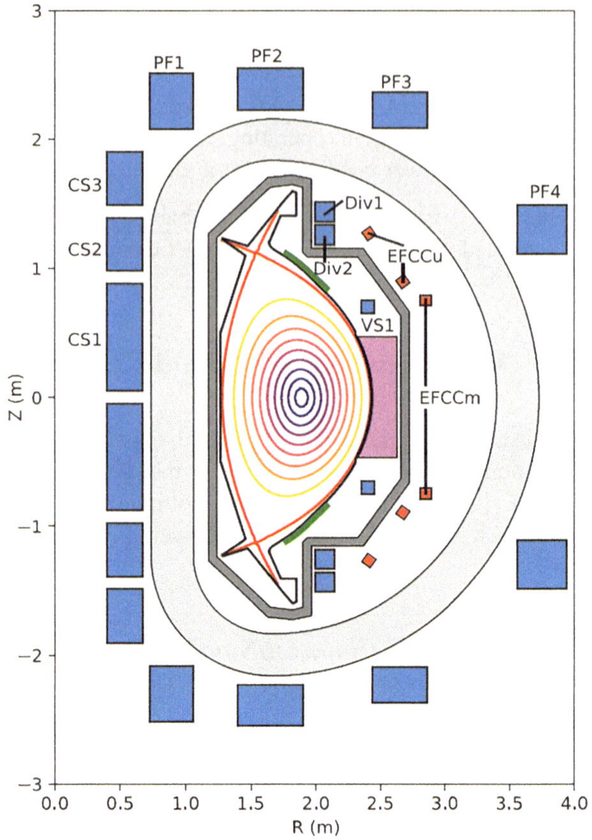


**Fig. 8.5** SPARC tokamak. (From [64], Fig. 2)

- Phase 2 started in 2021 and was planned to last 4 years for final design, construction and commissioning, then moving rapidly to full d-t operation.

The toroidal field coils plus all the poloidal field coils on the outside use REBCO superconductor, now available and sufficiently tested. The inner poloidal field coils are labelled CS for the central solenoid and outer as PF. A full-sized prototype toroidal field coil was successfully tested to 20 T at the coil.

Inside the toroidal field coils, there are two pairs of copper coils used to produce the divertor field labelled Div plus three sets of error-field correction coils. There are a pair of vertical stability coils inside the vessel labelled VS1.

The vacuum vessel design is double-walled, with space in-between the walls for gas heating and cooling. Roughly half of the space between the vacuum vessel walls, as well as the space between the vacuum vessel and the toroidal field coils, is filled with neutron shielding material. The vacuum vessel has three ports at each toroidal location. Upper and lower passive stability plates are located between the vacuum vessel and the plasma to improve vertical stability and allow operation at high elongation.

The SPARC plasma pulse with a 10 s current flattop was simulated with the Tokamak Simulation Code [69] verifying the plasma control system and the vertical stability of this approximately 2/1 vertically elongated plasma. Extensive simulations and use of scaling formulas were used to justify plasma performance. An overview of the SPARC physics basis [70] predicted 140 MW of fusion power with a plasma gain of $Q \approx 11$ with:

- ICRF heating with $^3$He minority at 11.1 MW with 25 MW installed;
- H-mode with a minimum L-H threshold of 21 MW at $1.9 \times 10^{20}$ m$^{-3}$;
- Peeling-limited pedestal (EPED) and ITG-dominated core (TGLF, CGYRO);
- ~78% of ICRF and ~ 23% of alpha power goes to heat bulk ions;
- Core exhausts ~40% of the power by radiation;
- $T_e \approx T_i$;
- Divertor with 1 Hz sweep and 50% of heat removed volumetrically;
- 3.1% of alpha power lost to walls.

Preparing the SPARC site for construction started in the second quarter of 2021, with first plasma planned for 2025.

### 8.8.2 ARC (Affordable Robust Compact)

The ARC pilot fusion reactor is aimed at producing long pulse fusion power of 525 MW and involves several technological steps beyond the SPARC prototype. ARC is shown in Fig. 8.6 [65, 66].

The ARC reactor is about the size of JET or JT-60SA, but has a toroidal magnetic field of 9.2 T at a major radius of 3.3 m. Novel design features include:

- Demountable high temperature REBCO superconductor toroidal field (TF) magnets with resistive joints [71];
- Poloidal magnetic field coils located inside the TF;
- Vacuum vessel immersed in a molten salt FLiBe blanket which shields the coils and has an overall tritium breeding ratio of 1.08;
- A low FLiBe pumping power that is less that 1% of total fusion power;
- Vacuum vessel that can stand high neutron flux and be replaced once a year by vertical removal;
- Steady state operation with bootstrap current plus highly-efficient lower-hybrid current drive on the high-field side;
- A long-legged advanced divertor able to handle the high power heat load and surrounded by the blanket;
- Robustness due to operation at steady-state and far from operational boundaries.

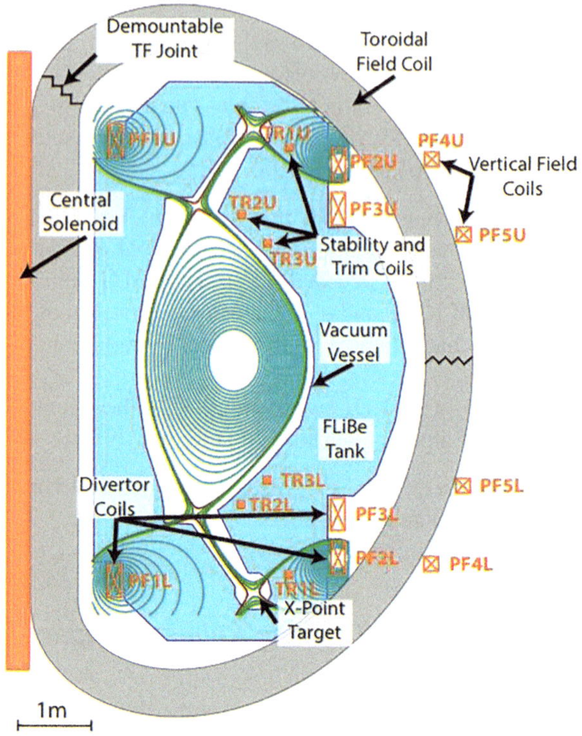


**Fig. 8.6** ARC fusion reactor cross section. (From [66], Fig. 1)

The suggested advantages of this high field approach to fusion reactors [67] include high temperature superconductor of the REBCO type that has critical temperature, critical magnetic field and critical current densities an order of magnitude higher than $Nb_3Sn$. REBCO-wound solenoids have operated at up to 40 T. This superconductor can be deposited as a thin layer on a steel substrate resulting in tapes that are ideal for magnets. Also revolutionary is the higher specific heat and larger operating margin at higher temperatures, which allows construction with demountable joints and gives far more flexibility in construction.

The use of FLiBe as a bath of molten salt for the breeding blanket and shield greatly reduces the amount of material exposed to high neutron flux. It can be drained and thereby allows easy maintenance of the vacuum vessel and related components inside the toroidal field coil. Demountable coils offer great flexibility for modification and maintenance, including the possibility of replacing the vacuum vessel or the entire power core. The FLiBe would operate at ≥630 °C, allowing high thermodynamic efficiency for electricity production. It has high electrical resistivity, greatly reducing the pumping power compared to liquid lithium. FLiBe has a long history going back to MSRE [72].

### 8.8.3 Innovations Required for Very High Field Reactor

A high field fusion reactor would be much less expensive than lower field machines. However, its roadmap requires development and demonstration of some extremely challenging technical accomplishments:

- High-field HTS magnets – High temperature superconductors such as REBCO are now available as tape on substrate and has been demonstrated to very high field. With its high critical temperature, it not only reaches much higher fields, it has better quench protection against pulsed magnetic fields and stresses, and the energy cost of the cryogenic refrigerator is less when additional heating is present, for example from neutron leakage. It could prove far more reliable than $Nb_3Sn$ or NbTi superconductor. However, it needs demonstrating that it can be constructed at large size and that it and its structure can survive the very high forces and stresses.
- Demountable superconducting magnets via normal conducting joints – Superconducting joints in REBCO have been demonstrated and can completely revolutionize tokamak reactor design. Being able to lower complete toroidal components into the demounted coil space make construction and repair much easier. Poloidal coils can be close to the plasma, recovering the advantages of such machines as DIII-D and TCV for shaping and creating divertor plasmas. It remains to be demonstrated that such joints can withstand stresses at very high field and can be opened and put back in place on a regular basis. It is encouraging that demountable joints have been successfully used in the test coil for SPARC [73].
- A tank with a molten salt FLiBe blanket and shield – The tank of FLiBe represents a major saving of space inside the toroidal field coils, especially with the unique feature of surrounding the divertor. It is the key innovation that allows a highly compact machine. It combines breeding and shielding with a method of high efficiency heat transfer and making tritium bred in the blanket accessible to a tritium removal and processing plant. However, molten salts can be a very difficult technology to handle in a complicated environment.
- High-field side LHCD – Lower hybrid microwave launchers can be compact. A high-field-side launch offers the prospect of high efficiency and easy entry into a compact reactor. At very high plasma densities of compact reactors, it offers a way of indirectly heating the ions efficiently via electron-ion heat transfer.
- High divertor power load – The vertical face divertor solution offers a way of handling very high divertor heat load and uses a similar philosophy to ITER heat handling, although the power loading is necessarily higher in a more compact reactor.
- High fusion Q – All of the current knowledge about plasma scaling supports that the proposed H-mode plasma in ARC should be able to reach ignition conditions. This conclusion is supported by detailed plasma simulations and will be prototyped in SPARC.

- High neutron and plasma particle and radiation loading – The FLiBe blanket provides a method of removing high heat levels from the vacuum vessel at good thermodynamic efficiency. If the wall loading proves too high for the material, the first wall can be replaced. However, such an operation could be extremely complicated and remains to be demonstrated.
- Reactor maintenance – The demountable toroidal field coils provide a simple way of removing, repairing or replacing the vacuum vessel and other components, also greatly reducing down time. Nevertheless, remote handling, especially involving piping and electrical connections can be extremely time-consuming and complicated.

The very-high magnetic field SPARC experiment and the ARC reactor design represent important and innovative combinations of design space choices that in principle allow a much more compact design than other DEMOs. It may or may not be successful, but it is a design option which might result in a much less expensive reactor core, and therefore worth exploring. SPARC especially might provide early physics information on dominant alpha particle heating of plasmas.

How much less expensive will be the balance of plant will be remains to be seen, but having a relatively inexpensive reactor core greatly aids the plant economics.

### 8.8.4 *Fragilities of Very High Field Tokamaks*

A major difference between the design progression of JET to ITER to DEMO and the very high field route is that ARC has a number of completely different technologies than SPARC, which uses a FLiBe blanket and demountable superconducting coils. The roadmap listed for ARC provides a fairly comprehensive list of the fragilities of the design which require mitigation, as follow:

- High-field HTS magnets – Although REBCO and other HTS materials are more widely available, they are still far behind the level of demonstration of $Nb_3Sn$ and NbTi. ARC relies on the use of REBCO to achieve higher magnetic fields. The ARC reactor has a major radius less than half of the other DEMOs listed in Table 8.1 though still larger than the spherical tokamak ST-135. System studies in the larger DEMO reactors indicate great difficulties in withstanding magnetic field stress in coils and supports at such high fields.
- Demountable superconducting magnets via normal conducting joints – Demountable superconducting coils represent a truly major challenge. The technology for making joints with superconducting tapes needs development. Superconducting to normal connectors have proven to be sites of high fragility.
- A tank with a molten salt FLiBe blanket and shield – The major design choice that allows a smaller machine is the use of the molten salt FLiBe in a tank surrounding the vacuum vessel and immersing the divertor. It is a highly interesting design,

but most certainly needs to be demonstrated. Questions of repair in case of a fault remain open. If the FLiBe cools down because of a malfunction, it might be difficult to reheat and remove.

- High-field side LHCD – The efficiency of high-field side LHCD and its feasibility in the crowded space and in a neutron flux remains to be demonstrated.
- High divertor power load – The very high divertor power loading in a compact machine may be extremely difficult to handle. Impurity radiation from material from the divertor could lead to high radiation losses in the very high density plasma in addition to cyclotron radiation.
- High fusion Q – The axial toroidal field of ARC is 9.2 T, compared to that of K-DEMO with 7.4 T, so the main difference is not there. The ARC plasma current is 7.8 MA, not very different from JET's maximum current of 7 MA before divertors were installed. It remains to be demonstrated that high Q can be achieved in a machine the size of JET with similar plasma current.
- High neutron and plasma particle and radiation loading – The very high power density of neutrons and plasma losses to the walls could be too high to handle and sublimate or sputter material into the plasma. The vessel wall facing the plasma still needs to be thin to allow successful tritium breeding.
- Reactor maintenance – The idea of simply draining out the FLiBe from the tank, decoupling the toroidal field, replacing or repairing the tokamak core in a short period of time that does not compromise the reactor duty cycle seems an extremely ambitious goal, especially in a highly radioactive environment.
- An integrated very high field reactor – The reactor core itself already represents a major set of challenges. When integrated with the balance of plant, which includes pulsed power supplies, auxiliary heating systems, a high power cryogenic refrigerator, FLiBe circulation, heat transfer, power generation, tritium extraction and recycling, access to a compact machine, the costs may no longer be small, and the plant not so easy to operate and maintain.

Without a fully detailed reactor design, it is difficult to assess the robustness and fragility of this very high field system. Obviously such advances as a jointed very high field REBCO superconducting magnet needs further demonstration, but with resources such as MIT and its Francis Bitter National Magnet Laboratory, such a design is certainly worthy of consideration. The wall and divertor loadings are very high, and immersing everything in a high temperature molten salt is obviously a technological challenge.

In summary, if the SPARC prototype is successful, it would represent a major advance in fusion physics. The ARC type of fusion reactor would even so represent a major technological step from SPARC, and a full engineering design would be needed before its probable success can be evaluated.

## 8.9 Compact Spherical Tokamak Fusion Reactors

### *8.9.1 ST-135 and STEP (Spherical Tokamak for Energy Production)*

The spherical tokamak is a fascinating invention by Martin Peng and Dennis Strickler [74] which was initially developed in START [75]. It represents the tokamak physicist's dream of realizing very high beta and plasma current at low toroidal field and size.

Spherical tokamak reactor concepts have been developed by the Tokamak Energy Company Ltd. [76–78]. In compact reactors, the fusion gain Q [76] depends mainly on the absolute level of the fusion power and the energy confinement, but weakly on the device size. Small low-cost pilot plants may be feasible.

In a compact spherical tokamak (ST) power plant [78], the elongated shape leads to significant reductions in major radius and/or field for comparable fusion performance. A compact spherical tokamak might achieve a high fusion gain at a modest toroidal field [77]. A physics level design for ST135 ($R_0 = 135$ cm), a successor to ST-40, was developed having $a = 75$ cm, aspect ratio 1.8, central column current = 25 MA, $R_{cc} = 0.25$ m, plasma current = 7.2 MA, $B_{T0} = 3.7$ T, elongation $\kappa = 2.64$, peak field magnet = 20 T and fusion power = 185 MW at Q of 5.

Development and demonstration of feasibility is essential in at least four areas:

- the central column with high stresses on the high field REBCO superconductor, which in addition would require demountable joints;
- the divertor with high power loads;
- the inboard shielding to protect and reduce heating on the central coil;
- access, repair, maintenance and compatibility with nuclear licensing standards.

A neutronics analysis of ST135 [31] used the geometry shown in Fig. 8.7. The design incorporated tungsten carbide and its composites as an attractive class of neutron-attenuating materials. The fundamental materials properties of monolithic tungsten carbide, including thermal transport, mechanical properties and plasma interaction, were found to be generally favourable in comparison to metallic tungsten shields. A new material WC-FeCr was also investigated for its accident safety, thermo-mechanical properties, and irradiation behaviour. Further study of irradiation damage and hydrogen trapping will be required.

The central column only has neutron shielding and no tritium breeding capability which must be carried out by a breeding blanket elsewhere. The design leaves only 0.50 m between the outer edge of the central superconductor and the inner edge of the plasma to allow for structure, cooling, the vacuum vessel, gap to plasma edge, and neutron shielding. With a shield thickness of 0.31 m, the neutron heating of the central column is 100 kW, requiring a cryogenic refrigerator with a wall plug power of 3.0 MW, an acceptable level, but neutron damage will also occur.

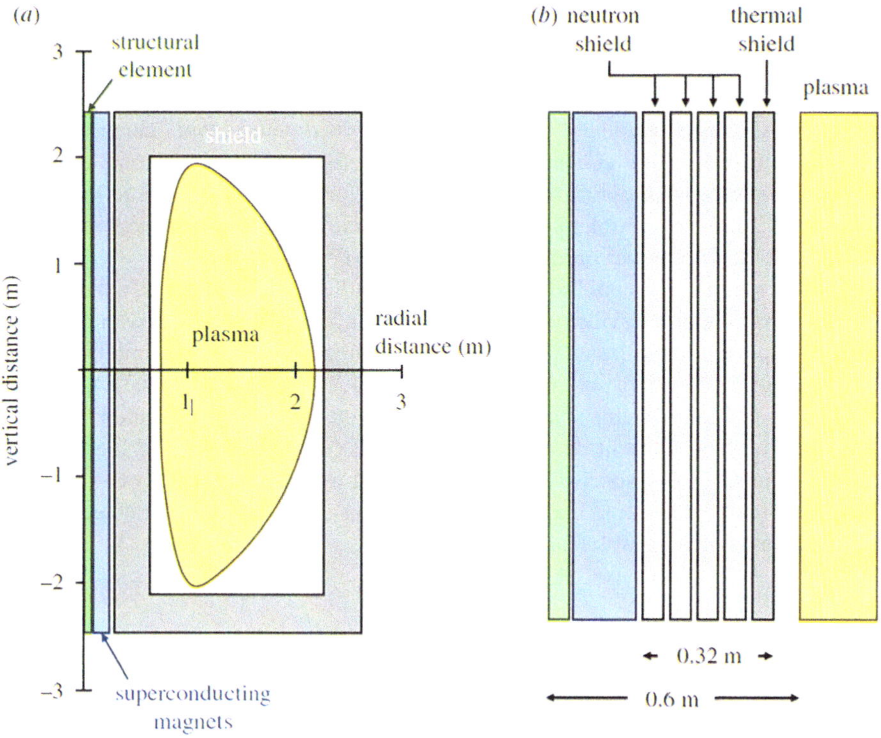


**Fig. 8.7** (**a**) SST-135 spherical tokamak geometry and (**b**) close up of central column with neutron shield and water cooling channels. (From [31], Fig. 1)

For the next step in building a modular reactor beyond ST135, a higher fusion Q would be needed up to 20, and the tritium breeding ratio would need to be greater than 1. The reactor would be larger than ST135 but still relatively small compared to ITER-like DEMO designs. The energy confinement, shielding thickness, mechanical stresses and divertor loads will be crucial design elements.

In parallel to ST-135, the UK government was sufficiently impressed with the success of MAST and MAST-U that they are committing over £200 million to design STEP (Spherical Tokamak for Energy Production [79–81] which is planned to be a 100 MW net-electric power-producing reactor costing £2 billion.

There does not yet seem to be a fully engineering design of these proposed reactors published in the literature, hence it is difficult to evaluate the feasibility of such machines which are so radically different from the larger DEMOs.

### 8.9.2 USA Sustained High-Power Density (SHPD) Tokamak Facility

In the USA, a fusion nuclear-science facility combined with a component test facility (FNSF/CTF) was being considered [82] for developing fusion materials and components. An extensive optimization showed that for R = 3 m, magnetic field at the TF coil of 20 T, no central solenoid for plasma start-up allowing aspect ratios near 2, then net electric power production up to 300 MW is possible.

Further studies [83] concluded that a new facility is needed to access a fusion pilot plant regime. Such a device would require a high fraction of self-driven current, a high core plasma pressure and high surface-average and divertor-parallel heat flux. The parameters of such a tokamak would be $R = 1.4 \pm 0.2$ m, B = 4-6 T, A = 2–2.5, $P_{H\&CD} = 50$ MW. Systems studies indicated that the performance would be that of a fusion pilot plant. Current designs integrate high current density and high toroidal field magnets, a low aspect ratio, strong shaping and liquid metal systems for the divertor, first wall and blankets. Engineering calculations indicated that a pre-conceptual design was feasible.

### 8.9.3 Robustness of Spherical Tokamak Reactors

The spherical tokamak reactors ST-135 and STEP are difficult to evaluate because they represent the biggest move from standard design space. There is only minimal shielding at small major radius, leaving the rest of the machine to provide a tritium breeding blanket. In comparison, the VECTOR DEMO design with a major radius of 3.75 m was a spherical tokamak design, but one that still maintained adequate shielding everywhere.

Robustness aspects of spherical tokamaks include:

- High current and high beta – Spherical tokamaks have great promise, since their small aspect ratio and naturally stable vertical elongation allow high current and high beta in very compact and inexpensive machines.
- Extensive experimental experience – There are MA-level spherical tokamak facilities around the world which have demonstrated good plasma performance and very high beta.
- A committed government funder – The UK has proposed contributing large sums to design STEP to achieve economical fusion power in a short time.
- A clear design vision – Even though there are challenges requiring innovation and ingenuity, there may be pathways to success.
- Economically promising – If successful, a spherical tokamak reactor could be a highly economical approach to a fusion reactor.
- Synergies with other compact tokamak designs – The design can share innovation studies with reactors such as ARC, with high temperature REBCO superconductors, normal conducting joints for demountable coils, FLiBe breeding blanket, safety and maintenance philosophy.

### 8.9.4 Fragilities of Spherical Tokamak Reactors

- Shielding thickness on inner core – Designs such as ST-135 with a major radius of 1.35 m leave only a tiny space of 1/3 of a meter for shielding, and the design allows a significant amount of neutron heating and damage of the central toroidal coil.
- Plasma physics at low aspect ratio and high beta – In a spherical tokamak, the plasma beta can be tens of percents and the diamagnetic effect can reduce the axial toroidal field considerably. There is also a very large ratio between the toroidal field at large and small minor radius. Therefore, plasma and alpha particle physics could be very different, as has already been observed.
- Systems architecture of poloidal coils – There is a big jump in machine design to a reactor from existing spherical tokamaks which have poloidal field coils not only inside the toroidal field coils, but also inside the vacuum vessel. Such a design led to major coil failure on NSTX-U that required a total redesign.
- Space from plasma to superconductor – In the attempt to minimize the plasma to coil distance, there is a risk of compromising safety margins on components with very high temperature and neutron deposition gradients.
- Tritium breeding – The challenge is to transform machines with toroidal and poloidal coils very close to the plasma into a d-t fusion reactor with shielded coils and an overall machine tritium breeding ratio greater than one. A very simple model of a rectangular vacuum vessel gives the ratio of the inner wall to the total area as $0.5(1\text{-}A^{-1})/(1 + \kappa^{-1})$. Using the geometry for ST-135, $A = 3$ and $\kappa = 3$, the ratio is 19% of neutrons lost for breeding. Allowing for more losses in the divertor and ports, it will be challenging to reach a tritium breed ratio $\geq 1$.

All of these fragilities can in principle be overcome, but the spherical tokamak reactor pushes so many technologies to the limit that a full engineering design needs to be done and components proven before any external evaluation. If successful, the economic advantages of compact spherical tokamaks would be large.

### 8.9.5 A Pulsed Reactor Concept Using Coaxial Helicity Injection

A concept for a pulsed fusion reactor using Magnetized Target Fusion and coaxial helicity injection to create and rapidly compress a spherical tokamak was proposed by General Fusion [84]. The concept is to create a liquid metal flux conserver with a liquid metal shaft running down the centre. A current flows through the shaft producing a toroidal field. A spherical tokamak plasma is then formed by coaxial helicity injectors, and the liquid metal is rapidly compressed by pistons, maintaining the spherical tokamak configuration. The idea is to compress the plasma in 40 ms by a linear factor of 10 from a major radius of the flux conserver of 2 m to 0.2 m, thereby producing a thousand-fold increase in the plasma density. The plasma would have a temperature of 12 keV, a density of $2 \times 10^{23}$ m$^{-3}$ and a plasma beta of 40%. A cycle time of 1 s would produce a net electric power of 40 MW. Several problems remain which are under investigation.

## 8.10 Summary

In systems approaches to fusion reactor design, the first steps are to define goals and constraints, and then to examine variants at various systems levels on the way to a design. The survey of tokamak fusion reactor designs in this chapter have illustrated that designers can either regard constraints, carefully arrived at by years of prototyping and research, as appropriate boundaries for designs, or rather as challenges, to be overcome by determination, ingenuity and innovation.

For example, the major partners in the ITER have built prototyping machines and are performing ITER-relevant experiments. Their reactor designs are solidly based on these machines and experiments plus ITER R&D and construction experience, and therefore have comparable designs. These designs still involve a great deal of innovation and the parameters of the machine are already reaching design limits in high toroidal field, low aspect ratio and very advanced technology. ITER is pursuing parallel scenarios of their baseline H-mode scenario and advanced tokamak scenarios which offer a wide range of possibilities of making a reactor that is more compact and economical.

The very high field tokamak designs SPARC and ARC are characteristic of the MIT ALCATOR series, which has produced excellent physics results. The teams have chosen the route of compact, economic and near term reactor development.

The compact size of a spherical tokamak holds out the prospect of compact and inexpensive reactors. It may be that a comprehensive design such as the 4 year effort underway for STEP will be successful.

However much optimism is present, machines are still subject to the real constraints of physics and engineering. It is only a machine designed to satisfy all these constraints that has a chance of working successfully.

The first eight chapters of this book have demonstrated the progress made towards tokamak reactors and the increased probability that they will be built and operated successfully. The systems approaches described here have allowed a detailed understanding of these machines and provided a methodology for analyzing existing ones and planning and building new experiments and reactors. ITER and DEMO teams are using systems engineering and architecture tools that are considered essential for designing, building and operating large proto-reactors and reactors. Such tools are especially necessary since there is no consensus on the ideal form of a tokamak fusion reactor; hence a careful use of a design space is essential to find the optimum paths to reactor concepts. The actual process of designing a tokamak fusion reactor, given its size and complexity, will certainly make full use of systems approaches.

In conclusion, there are now many nations and companies and groups actively designing DEMOs and fusion reactors, based on high successful physics and engineering experiments. The prospects for tokamak fusion reactors are now more real and exciting than at any time in the past. This book has so far demonstrated that these advances require systems approaches.

## References


1. G. Federici et al., Overview of the DEMO staged design approach in Europe. Nucl. Fusion **59**, 066013 (2019)
2. G. Federici et al., The plan forward for EU DEMO. Fusion Eng. Des. **173**, 112960 (2021)
3. G. Federici et al., The EU DEMO staged design approach in the pre-concept design phase. Fusion Eng. Des. **173**, 112969 (2021)
4. M. Abdou et al., Blanket/first wall challenges and required R&D on the pathway to DEMO. Fusion Eng. Des. **100**, 2 (2015)
5. M. Siccinio et al., Figure of merit for divertor protection in the preliminary design of the EU-DEMO reactor. Nucl. Fusion **59**, 106026 (2019)
6. I. Jenkins et al., A SysML model of the tokamak subsystems involved in a DEMO pulse, Report EUROFUSION WPPMI-CP(16) 15445. Proceedings of 29th Symposium on Fusion Technology (SOFT) (2016)
7. OMG systems Modelling Language, An OMG systems modelling language publication, Document formal/19-11-01 (2019)
8. G. Grossettia et al., Systems engineering perspective to the integration of the heating and current drive system in the EU DEMO: analysis of requirements and functions. Fusion Eng. Des. **136A**, 53 (2018)
9. M. Kovari et al., "PROCESS": a systems code for fusion power plants – part 1: physics. Fusion Eng. Des. **89**, 3054 (2014)
10. M. Kovari et al., "PROCESS": a systems code for fusion power plants – part 2: engineering. Fusion Eng. Des. **104**, 9 (2016)
11. Knight PJ, Kovari MD. A user guide to the PROCESS fusion reactor systems code, Culham Centre for Fusion Energy, 12 January 2016 (2022). https://ccfe.ukaea.uk/wp-content/uploads/2019/11/A-User-Guide-to-the-PROCESS-systems-code.pdf. Accessed 28 Jan 2022
12. R. L. Reod (ed.), *ETR/ITER systems code, report ORNL/FEDC-87/7* (Oak Ridge, Oak Ridge National Laboratory, 1988)
13. D. McAlees, F. B. Marcus (eds.), *Oak Ridge Tokamak Experimental Power Reactor Study-1976-Part 2 – Plasma Engineering in a Deuterium-Tritium Fueled Tokamak*, Report ORNL-TM-5573 (Oak Ridge National Laboratory, Oak Ridge, 1976)
14. C. Reux et al., DEMO reactor design using the new modular system code SYCOMORE. Nucl. Fusion **55**, 073011 (2015)
15. C. Reux et al., DEMO design using the SYCOMORE systems code: influence of technological constraints on the reactor performance. Fusion Eng. Des. **136**, 1572 (2018)
16. M. Coleman, S. McIntosh, BLUEPRINT: a novel approach to fusion reactor design. Fusion Eng. Des. **139**, 26 (2019)
17. F. Franza et al., MIRA: a multiphysics approach to designing a fusion power plant, 28th IAEA Fusion Energy Conference, Nice, France (Virtual) 10–15 May 2021, Presentation TECH/2-3Rb, Paper IAEA–CN–286-0666, Submitted to Nuclear Fusion (2021)
18. W.A. Houlberg, S.E. Attenberger, L.M. Hively, Contour analysis of fusion reactor plasma performance. Nucl. Fusion **22**(7), 935 (1982)
19. M. Nakamura et al., Efforts towards improvement of systems codes for the broader approach DEMO design. Fusion Eng. Des. **87**(5–6), 864 (2012)

20. J. Johner, Helios: a zero-dimensional tool for next step and reactor studies. Fusion Sci. Technol. **59**, 308 (2011)
21. T. Hartmann, Development of a systems code to assess omplications of physics assumptions on the design of a demonstration fusion power plant (DEMO) (Dissertation), Technische Universität München (2013)
22. A.J.H. Donné, The European roadmap towards fusion electricity. Philos. Trans. R. Soc. A **377**, 20170432 (2019)
23. G. Federici, F. Maviglia, J. Holden, Special issue on European programme towards DEMO: outcome of the pre-concept design phase, Fusion Eng. Design (2022). https://www.sciencedirect.com/journal/fusion-engineering-and-design/special-issue/10RRZQ6LW4H. Accessed 28 July 2022
24. R. Kembleton et al., EU-DEMO design space exploration and design drivers, Fusion Eng. Design. https://www.sciencedirect.com/journal/fusion-engineering-and-design/specialissue/10RRZQ6LW4H. Accessed 28 July 2022
25. G. Federici et al., Overview of the design approach and prioritization of R&D activities towards an EU DEMO. Fusion Eng. Des. **109–111**, 1464 (2016)
26. G. Federici et al., DEMO design activities in Europe: progress and updates. Fusion Eng. Des. **136**, 729 (2018)
27. G.T. Hoang et al., A lower hybrid current drive system for ITER. 22nd IAEA Fusion Energy Conference, Geneva; 13–18 October 2008
28. G. Federici et al., An overview of the EU breeding blanket design strategy as an integral part of the DEMO design effort. Fusion Eng. Des. **141**, 30 (2019)
29. J. Morris et al., Preparing the systems code Process for EU-DEMO conceptual design. 28th IAEA Fusion Energy Conference, Nice, France (Virtual) 10–15 May 2021, Presentation TECH/2-3Ra, Paper IAEA–CN–286-0806, Submitted to Nuclear Fusion (2021)
30. U. Fischer et al., Neutronic performance issues of the breeding blanket options for the European DEMO fusion power plant. Fusion Eng. Des. **109–111**, 1458 (2016)
31. S.A. Humphry-Baker, G.D.W. Smith, Shielding materials in the compact spherical tokamak. Philos. Trans. R. Soc. A **377**, 20170443 (2019)
32. O. Croft et al., Overview of progress on the European DEMO remote maintenance strategy. Fusion Eng. Des. **109–11**, 1392 (2016)
33. K. Okano, Review of strategy toward DEMO in Japan and required innovations. J. Fusion Energy **38**, 138 (2019)
34. K. Tobita et al., Japan's efforts to develop the concept of JA DEMO during the past decade. Fusion Sci. Technol. **75**(5), 372 (2019)
35. K. Tobita et al., Conceptual design of Japan's fusion DEMO reactor (JA DEMO) and superconducting coil issues. J. Phys. Conf. Ser. **1293**, 012078 (2019)
36. Y. Someya et al., Progress in design and engineering issues on JA DEMO. 28th IAEA Fusion Energy Conference, Nice, France (Virtual) 10–15 May 2021, Presentation TECH/2-1, Paper IAEA–CN–286-0802, Submitted to Nuclear Fusion (2021)
37. M. Kikuchi, A review of fusion and tokamak research towards steady-state operation: a JAEA contribution. Energies. **3**, 1741 (2010)
38. S. Konishi et al., Possible scenario to start up DT fusion plant without initial loading of tritium. J. Plasma Fusion Res. **76**, 1309 (2000)
39. M. Kikuchi, Steady state tokamak reactor based on the bootstrap current. Nucl. Fusion **30**, 265 (1990)
40. N. Inoue et al., Feasibility study for an inductively operated daylong tokamak reactor. Proceedings of 14th IAEA Conference on Controlled Fusion and Plasma Physics Research, vol. 3 (1993), p. 347
41. K. Okano et al., Compact reversed shear tokamak reactor with a superheated steam cycle. Nucl. Fusion **40**, 635 (2000)
42. R. Hiwatari et al., Demonstration tokamak fusion power plant for early realization of net electric power generation. Nucl. Fusion **45**, 96 (2005)

43. S. Nishio, K. Tobita, S. Konishi et al., Tight aspect ratio tokamak power reactor with superconducting TF coils. 19th IAEA Fusion Energy Conference, Lyon, IAEA-CN-FT/P1–21 (2002)
44. K. Tobita et al., SlimCS – compact low aspect ratio DEMO reactor with reduced size central solenoid. Nucl. Fusion **47**, 829 (2007)
45. Y.T. Song et al., Concept design of CFETR tokamak machine. IEEE Trans. Plasma Sci. **42**(3), 503 (2014)
46. J. Li, Y. Wan, Present state of Chinese magnetic fusion development and future plans. J. Fusion Energy **38**, 113 (2019)
47. X. Wang et al., Current progress of tritium fuel cycle technology for CFETR. J. Fusion Energy **38**, 125 (2019)
48. W.S. Lin et al., Burning fraction, radial transport, and steady state profiles of multispecies particles in CFETR burning plasmas. Nucl. Fusion **60**, 076007 (2020)
49. X. Liu et al., Progress in the conceptual design of the CFETR toroidal field coil with rectangular conductors. Nucl. Fusion **60**, 046032 (2020)
50. K. Kim et al., Design concept of K-DEMO for near-term implementation. Nucl. Fusion **55**, 053027 (2015)
51. F. Najmadabi et al., The ARIES tokamak fusion reactor study. IEEE Thirteenth Symposium on Fusion Engineering, Knoxville, TN, USA, vol. 2 (1989), p. 1021
52. F. Najmabadi et al., The ARIES-AT advanced tokamak, advanced technology fusion power plant. Fusion Eng. Des. **80**, 3 (2006)
53. S.C. Jardin, N. Pomphrey, J. Delucia, Dynamic modelling of transport and positional control of tokamaks. J. Comput. Phys. **66**, 481 (1986)
54. J. Kalnavarns, D.L. Jassby, Resistive toroidal-field coils for tokamak reactors, Princeton Plasma Physics Laboratory Report PPPL-1685, Princeton, USA (1980)
55. R. Woolley, P. Heitzenroeder, Resistive magnets in magnetic fusion energy research, Snowmass 1999 fusion summer study (Technology Issues Working Group, 1999)
56. R.J. Goldston, Burning plasma experiment physics design description. Fusion Technol. **21** (3P1), 1050 (1992)
57. D.M. Meade, FIRE, a next step option for magnetic fusion. Fusion Eng. Des. **63–64**, 531 (2002)
58. B. Coppi et al., Optimal regimes for ignition and the Ignitor experiment. Nucl. Fusion **41**(9), 1253 (2001)
59. J.L. Luxon, A design retrospective of the DIII-D tokamak. Nucl. Fusion **42**(5), 614 (2002)
60. R.J. Buttery et al., DIII-D research to prepare for steady state advanced tokamak power plants. J Fusion Energy **38**, 72 (2019)
61. D.J. Campbell et al., Innovations in technology and science R&D for ITER. J Fusion Energy **38**, 11 (2019)
62. R. Srinivasan, Conceptual study of Indian fusion power plant. Proceedings of 2nd Technical Meeting on First Generation of Fusion Power Plants: Design and Technology (IAEA Publications, Vienna, 2008)
63. G.O. Ludwig et al., Physics performance analysis of low-power tokamak reactors. Nucl. Fusion **49**, 085026 (2009)
64. A.J. Creely et al., Overview of the SPARC tokamak. J. Plasma Phys. **86**, 865860502 (2020)
65. B.N. Sorbom et al., The engineering design of ARC: a compact, high field, fusion nuclear science facility and demonstration power plant. Austin: IEEE Symposium of Fusion Energy (2015)
66. A.Q. Kuang et al., Conceptual design study for heat exhaust management in the ARC fusion pilot plant. Fusion Eng. Des. **137**, 221 (2018)
67. M. Greenwald et al., The high-field path to practical fusion energy, plasma science and fusion center report PSFC/RR-18-2. Cambridge: Massachusetts Institute of Technology; 2018.
68. H. Zohm, On the use of high magnetic field in reactor grade tokamaks. J Fusion Energy **38**, 3 (2019)

69. S.C. Jardin et al., Physics basis for the advanced tokamak fusion power plant, ARIES-AT. Fusion Eng. Des. **80**, 25 (2006)
70. P. Rodriquez Fernandez et al., Overview of the SPARC physics basis towards the exploration of burning-plasma regimes in high-field, compact tokamaks. 28th IAEA Fusion Energy Conference, Nice, France (Virtual) 10–15 May 2021, Presentation OV/P-4, Paper IAEA–CN–286-0856, Submitted to Nuclear Fusion (2021)
71. N. Yanagi et al., Design and development of high-temperature superconducting magnet system with joint-winding for the helical fusion reactor. Nucl. Fusion **55**, 053021 (2015)
72. Molten-Salt Reactor Experiment, Wikipedia. 2022. https://en.wikipedia.org/wiki/Molten-Salt_Reactor_Experiment, see also film https://youtu.be/tyDbq5HRs0o. Accessed 17 June 2022
73. D. Whyte, Achieving a 20 tesla superconductor magnet for the SPARC fusion device, IAEA webinar on Building Stars: Breakthroughs in Fusion R&D, 24 March 2022; 2022. https://youtu.be/5KkVGASnPJ0. Accessed 20 July 2022
74. Y-K.M. Peng, D.J. Strickler, Features of spherical torus plasmas. Nucl. Fusion **26**, 769 (1986)
75. A. Sykes et al., High-β performance of the START tokamak. Plasma Phys. Control. Fusion **38**, B247–B260 (1997)
76. A.E. Costley et al., On the power and size of tokamak fusion pilot plants and reactors. Nucl. Fusion **55**, 033001 (2015)
77. A. Sykes et al., Compact fusion energy based on the spherical tokamak. Nucl. Fusion **58**, 016039 (2018)
78. A.E. Costley, Towards a compact spherical tokamak fusion pilot plant. Philos. Trans. R. Soc. A. **377**, 20170439 (2019)
79. I.T. Chapman, A.W., Morris on behalf of the UKAEA team and collaborators, UKAEA capabilities to address the challenges on the path to delivering fusion power. Philos. Trans. R. Soc. **A377**, 20170436 (2019)
80. A. Vaughn, UK takes step towards world's first nuclear fusion power station. New Scientist 3312; 2 December 2020
81. STEP website, UK Atomic Energy Authority, Culham Centre Fusion Energy, STEP (Spherical Tokamak for Energy Production), 2021. https://ccfe.ukaea.uk/research/step/. Accessed 21 February 2021
82. J.E. Menard, Compact steady-state tokamak performance dependence on magnet and core physics limits. Philos. Trans. R. Soc. **A377**, 20170440 (2019)
83. J. Menard et al., Mission and configuration studies for a U.S. sustained high-power density tokamak facility. 28th IAEA Fusion Energy Conference, Nice, France (Virtual) 10–15 May 2021, Presentation TECH/2–4, Paper IAEA–CN–286-1013, Submitted to Nuclear Fusion (2021)
84. M. Laberge, Magnetized target fusion with a spherical tokamak. J. Fusion Energy **38**, 199 (2019)

# Part IV
# Helical, Linear and Inertial Fusion Reactor Concepts

# Chapter 9
# Helical Fusion Reactor Concepts

**Abstract** Helical coil systems, with the stellarator, heliotron/torsatron, heliac and helias as members, were designed as steady-state toroidal confinement devices by providing a twisted field from external coils instead of plasma current. A design space is established and a case study is made of the superconducting helias W7-X. Previous members of this pioneering Wendelstein series are described. The goals, design choices and operational results are analysed for the superconducting LHD, which is the largest of the Heliotron series. Other stellarators are examined to catalogue the choices made in their helical design space. The successful W7-X and LHD results have led to fusion reactor designs, the HELIAS-5B and the FFHR-d1, which are examined for their robustness and fragility properties.

## 9.1 Introduction to the Stellarator and Heliotron Families

Helical coil systems of the stellarator family were designed as steady-state toroidal confinement devices by using external coils instead of plasma current. Early stellarators were difficult to construct and to operate successfully because of complex coils leading to small magnetic field errors which induced a major loss of plasma confinement. Partly as a result, tokamaks took the lead in $n\tau T$ results. Modern superconducting stellarators such as the largest helias (helical advanced stellarator) Wendelstein 7-X (W7-X) and the heliotron / torsatron Large Helical Device (LHD) have established that high performance and steady-state plasmas can be obtained. Designs for fusion reactors grew from these successes.

### 9.1.1 Systems Analysis Strategy

First a brief discussion of stellarator physics and construction issues are presented in order to build a stellarator design space, followed by systems analysis methods. Experimental results of existing stellarators are reviewed and the systems aspects of designing a stellarator fusion reactor are addressed. A case study in systems forward



F. B. Marcus, *Systems Approaches to Nuclear Fusion Reactors*, Springer Series in Plasma Science and Technology, https://doi.org/10.1007/978-3-031-17711-8_9

engineering, i.e. the conception and design, is made of the superconducting helias W7-X as a successor to the pioneering Wendelstein series. Reverse systems engineering, i.e. an examination of the goals, design choices and operational results, is performed for the superconducting LHD which evolved from the Heliotron series. Detailed presentations of the physics of stellarators are available [1–7].

## 9.1.2 *Magnetic and Coil Configuration*

It was soon recognized that toroidal confinement fusion reactors would best operate in steady-state, but intrinsic particle vertical drift needed to be counteracted. Tokamaks and reversed field pinches countered the drift by a current flowing toroidally in the plasma, thereby creating a poloidal magnetic field. Stellarators instead built twisting coils outside the plasma that created a rotational transform in the vacuum magnetic field. The coil winding for a traditional stellarator such as CLEO [8] is shown in Fig. 9.1 [9].

At first appearance the configuration is similar to a tokamak with poloidal coils, except the coils are twisted. The plasma cross section is non-circular, because adjacent coils have currents going in opposite directions producing multipole fields. Due to equal and opposite helical coil currents, their total is zero.

By contrast, in a heliotron, there are no toroidal field coils and the helical coils all carry currents in the same direction to also produce a toroidal field. As the coils twist around the vacuum vessel and plasma, they reconnect with themselves again after

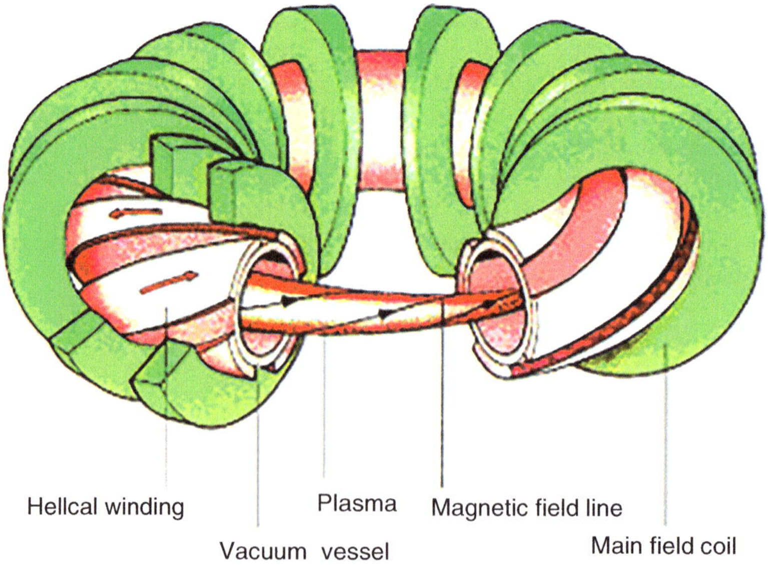


**Fig. 9.1** Magnetic field of a stellarator. (From [9])

one or more complete toroidal turns. The periodicity in the poloidal plane is characterized by '$\ell$', where there are $2 \times \ell$ coils, and 'm', which is the periodicity in the toroidal direction. As well as cancelling particle drifts, the coils also provide plasma equilibrium and stability. Since the field strength increases outwards towards the coils, a magnetic well can be produced that stabilizes most instabilities. The coils can be designed to suppress bootstrap induced-currents flowing in the plasma, hence avoiding current-driven instabilities.

In designing a stellarator, there is extreme flexibility in configuration choice because toroidal, helical with uni- or bi-directional currents and combined coils can create both toroidal and poloidal fields of high complexity.

### 9.1.3 *Early History of Stellarator Development*

Early stellarator results were discouraging. At the Princeton Plasma Physics Laboratory where the stellarator was invented and developed by Lyman Spitzer [10, 11], their early experiments culminated in the large Model-C. Its results implied that stellarator transport was dominated by a Bohm scaling law where the confinement time decreases as the plasma temperature increases.

Progressive advances were typified by the early Wendelstein series. Wendelstein stellarator experiments [12] started at the IPP (Institute for Plasma Physics) at Garching near Munich after the 1958 Geneva conference and were named after the local Bavarian mountain as a modest imitation of the American Project Matterhorn at Princeton. (The stellarator W7-X was built at Greifswald.)

The first devices, Wendelstein 1-A (W 1-A) and W 1-B were racetrack stellarators. The toroidally circular $\ell = 2$ stellarator W2-A operated a steady-state barium plasma which was dominated by collisional diffusion, thus contradicting the hypothesis of Bohm diffusion according to the Princeton C-Stellarator results.

After other stellarators such as Wendelstein 2-A and 2-B found much better scaling, it was determined that poor confinement in the C-stellarator was due to errors in coil windings, causing particles to drift out along field lines.

To demonstrate that a stellarator could be constructed without error fields, the table-top CLASP was built with very high precision engineering. It demonstrated that single particles could be very well confined for $10^7$ orbits [13].

Ohmic and auxiliary heating played a major role in stellarator development. In W2-B, mainly ohmic heating was used to reach higher plasma conditions. In W7-A, the predecessor to W7-AS, the plasma was initially produced by ohmic heating. The first net-current-free plasma was achieved in 1980. Electron cyclotron heating gave $T_e \geq 2$ keV. Ion temperatures of 1 keV and $n_e = 10^{20}$ m$^{-3}$ were reached. In the 1980s, stellarators had demonstrated that they had similar confinement properties to tokamaks. The stage was thus set for W7-AS (Advanced Stellarator) which operated from 1988 to 2002 and which is described below.

### *9.1.4 Required Properties of Stellarator Fields*

Stellarator magnetic fields need to accomplish a number of important tasks.

- Confine single particle orbits in a vacuum magnetic field using externally created helical coils.
- Provide MHD equilibrium for finite pressure plasmas.
- Adjust the pressure-induced bootstrap current.
- Provide MHD stability against plasma perturbations at fluid and micro levels.
- Allow for high plasma density.
- Allow external heating access to reach fusion temperatures.
- Provide plasma energy confinement to reach fusion conditions.
- Provide external field structures to allow for divertor installation.

When optimized as discussed at the end of this section, especially with regard to "symmetry," the fields can achieve many desirable properties [14]. These fields can be described analytically, for example by Boozer coordinates [15]. The vacuum field can be altered by induced currents from finite plasma pressure, such as the bootstrap current [16]. In tokamaks, it can reduce the need for external current drive, but in stellarators it can be a source of instability. It is sometimes desirable to remove it from a machine by field design.

Given these general principles, let us briefly examine the history of stellarator development from the confining field point of view.

### *9.1.5 Main Stellarator Configurations and Design Space Options*

The main types of configuration are:

- Traditional stellarator with separate toroidal field and helical coils such as CLEO or Wendelstein 7A, see Fig. 9.1;
- Helias, such as Wendelstein 7AS or Wendelstein 7X advanced stellarators, where the toroidal and poloidal coils have been combined into a set of toroidal field coils, each of which has out-of-plane distortions to create the stellarator field. Modular field coils and correction coils can be added to vary fields or to correct error fields, see Figs. 9.3 and 9.4;
- Heliotrons/Torsatrons such as the Large Helical Device (LHD), consisting of only the helical windings, but with all currents going in the same direction to create a toroidal as well as stellarator field, see Fig. 9.6;
- Heliac, such as the TJ-II stellarator, where the plasma winds around a central conductor as well as having external helical coils, see Fig. 9.7;
- HSX-type stellarator, which is compact and generates a large bootstrap current in the plasma [17];
- Tokamak-stellarator hybrid, such as CTH and the designed but not built NCSX, also CHS-qa and ESTEL [18].

### 9.1.6 MHD Equilibrium and Stability

The general principle of stellarator 3-D equilibrium [1] is that the pressure of a finite beta plasma is balanced by distortion of the vacuum magnetic field. The magnetic field lines lie along nested equilibrium surfaces which are non-circular and variable in the toroidal direction. Equilibria are found by minimizing the Hamiltonian of the plasma energy minus the vacuum energy. An early solution to the Grad-Shafranov equation used for finding plasma equilibria was developed by Palumbo [19], which reduced the 3-D problem to two coupled ordinary differential equations. This allowed the numerical solution for "isodynamic" (see below) magnetic surfaces where the modulus of the magnetic field is constant on magnetic surfaces.

An early code was the BETA equilibrium, stability and transport code [20]. Calculations of equilibria were made for heliotron/torsatron large experiments [21] such as LHD in Japan. Other equilibrium solvers include VMEC, NESTOR [22] and PIES [23]. The high beta limits to equilibrium in a classical stellarator have been investigated [24] and the results are shown in Fig. 9.2 for the cases of a zero net current stellarator and a fixed iota stellarator. Instabilities have been investigated [25] with codes such as CAS3D. The computational Alfvén and sound continua were identified with the corresponding ones in cylinder theory.

### 9.1.7 Particle and Energy Transport

Because of their 3-D geometry, stellarators have additional loss mechanisms [1]. Collisionless losses can be reduced by optimized designs. Optimized energy transport rates have been determined by the International Stellarator Database, giving ISS04 scaling [26], where ι-bar is the rotational transform divided by 2π.

$$\tau_E^{ISS04} = 0.134 a^{2.28} R^{0.64} P^{-0.61} \bar{n}_e^{0.54} B^{0.84} \left(\frac{\iota}{2\pi}\right)_{2/3}^{0.41} \tag{9.1}$$

A density limit initially based on W7-AS [27, 28] was found to be:

$$n_c = 1.46 \pm 0.07 \left(\frac{P_{abs}}{V_p}\right)^{0.48 \pm 0.03} B^{0.54 \pm 0.05} \left[10^{20} m^{-3}\right] \tag{9.2}$$

It is essential in stellarator design to minimize the loss of fast ions for neutral beam injection and of alpha particles in fusion reactors. Accelerated methods for calculating these losses [29] were developed.

In following up on the fundamental neoclassical transport, a thorough benchmarking of the various numerical methods used to calculate mono-energetic neoclassical transport coefficients in the complex 3-D magnetic-field topologies of planned and existing stellarators has been performed. It was carried out within the

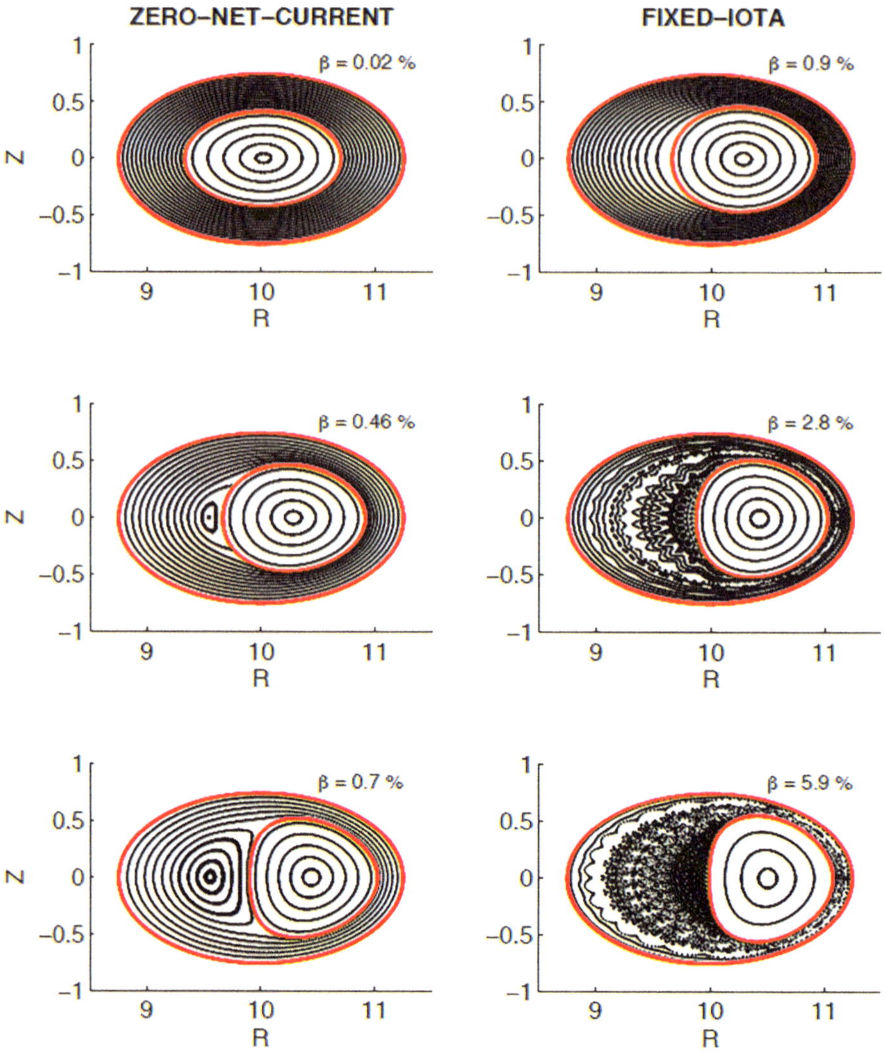


**Fig. 9.2** High beta MHD stellarator equilibrium. (From [24], Fig. 2)

IEA Implementing Agreement for Cooperation in Development of the Stellarator Concept [30], which involved Australia, the European Union, Japan, Russia, Ukraine, and the United States.

### 9.1.8 Symmetry and Stellarator Design Space Optimization Criteria

The various design criteria of the magnetic field to be optimized [31] include:

- magnetic surfaces
- error fields
- shear and rotational transform profile
- MHD-stability
- neoclassical transport
- bootstrap current
- confinement of α-particles
- technically feasible coils
- plasma creation mechanism
- auxiliary heating
- divertor concept

The starting point for optimisation is the fundamental concept of "symmetry" [14]. In a circular torus of major radius R, the minor radius, poloidal angle and toroidal angle are orthogonal co-ordinates that can describe any position inside the torus. In a tokamak with an internal current creating the poloidal magnetic field, all field properties are identical as one moves around the toroidal angle co-ordinate, and hence it is an "axisymmetric" geometry with the property that single particle drifts are contained in the magnetic field.

In a stellarator, the fields inside the plasma are created by external coils with a helical component and thus true axisymmetry is not available. In this case it is important to note that the collisionless orbit loss is prevented by an additional constant of motion arising if the confinement domain exhibits continuous symmetry. Axisymmetry is sufficient but not necessary. Symmetry in the magnitude of magnetic field B is sufficient for the loss cone to be removed. A stellarator configuration being symmetric in the absolute value of B can be produced with no need for any net toroidal current. Therefore the name "quasi-symmetry" was used.

There are three types of symmetries:

- With "quasi-helical symmetry," mirror reflected particles drift helically instead of being lost; however a bootstrap current will develop which reduces the rotational transform. This is counterproductive to achieving good plasma equilibrium and confinement. The construction of HSX used this configuration.
- With "quasi-axisymmetric" configurations, the reflected particles drift toroidally. However, the pitch of constant magnetic field is smaller than the pitch of the field lines, so that the bootstrap current now increases the rotational transform, thereby introducing stellarator fields into tokamak-like configurations.
- With "quasi-isodynamic" configurations, particles reflected within one field period drift poloidally instead of becoming lost. The bootstrap current can be made zero not only in total but also for each flux surface separately. This configuration allows for nearly shearless rotational transform profiles so that major resonances can be avoided and was the basis of W7-X design.

The quasi-isodynamic configuration allowed optimisation with respect to the following criteria:

- high quality of vacuum-field magnetic surfaces (regular boundary, avoidance of resonances, adjustment of the shear, small thickness of islands);
- good finite-ß equilibrium properties, small change of rotational transform;
- good MHD stability properties;
- small neoclassical transport;
- small bootstrap current;
- good collisionless particle containment;
- good modular coil feasibility.

The methods for optimization of stellarator fields have been explored [32, 33]. Drift trajectories of trapped particles in many older stellarator configurations were not generally confined, leading to large losses. A goal of stellarator design is to minimize the average radial guiding-centre drifts or the density of helically trapped particles. A magnetic field is termed omnigenous, which is the same concept as isodynamicity, if the radial drift of all particles averages to zero. Omnigeneity is guaranteed for a field which is quasi-symmetric, meaning that the magnetic field varies on a flux surface only through a fixed linear combination of the Boozer angles [15]. A field can never be exactly quasi-symmetric, but fields can be found which are approximately so. However, requiring even approximate quasi-symmetry places a severe constraint on stellarator optimization. It is therefore desirable to examine non-quasi-symmetric fields which are still nearly omnigenous. Fields can be constructed which are nearly omnigenous and yet are far from being quasi-symmetric. Omnigenous fields have many properties which aid plasma confinement.

An important subset of omnigeneous fields is the set of fields which are quasi-isodynamic when it is omnigenous and the constant- B contours close poloidally. The vacuum field of the W7-X stellarator is designed to be approximately quasi-isodynamic, and approximately remains so at high beta due to a deeper magnetic well which helps to confine fast particles. In a quasi-isodynamic plasma, the part of the long-mean-free-path (banana) regime distribution function determined by the collisional constraint is found from an equation which is identical in form to that for an axisymmetric plasma.

The bootstrap current can occur in both stellarators and tokamaks [16] at finite plasma pressure. In tokamaks, it can reduce the need for external current drive, but in stellarators, it can be either positive or negative and also be a source of instability. It is sometimes desirable to remove it from a machine by field design.

### 9.1.9 Coil Errors and Tolerances

Early stellarators suffered poor confinement due to error fields from coil winding imperfections, hence avoiding these is crucial to success. Therefore, it is necessary to both minimize coil fabrication costs while maintaining sufficient tolerances

[34]. The Hessian matrix method was used to rapidly identify important coil deviations. The method was applied to the design of the Chinese First Quasi-axisymmetric Stellarator configuration. Relaxing unnecessary coil tolerance could lead to substantial reduction in time and cost. New methods of designing stellarator coils by allowing variable winding surfaces [35] also hold the prospect of easier coil design for future machines.

## 9.2 Wendelstein7-AS (W7-AS)

### *9.2.1 W7-AS Systems Architecture and Technical Choices*

The design details and operational results of W7-AS were summarized [36, 37]. It is schematically shown in Fig. 9.3, including the planar and non-planar coils and the plasma equilibrium poloidal cross section at two toroidal locations separated by half a stellarator period.

The parameters of W7-AS and its superconducting upgrade W7-X, along with the largest of the Heliotron series and their superconducting upgrade LHD, are shown in Table 9.1. W7-AS was designed as a flexible system to produce a very wide range of stellarator fields, especially in comparison to classical stellarators.

In terms of robustness, the wide range of field shapes were designed to give a much greater possibility of obtaining a good equilibrium shape and particle and

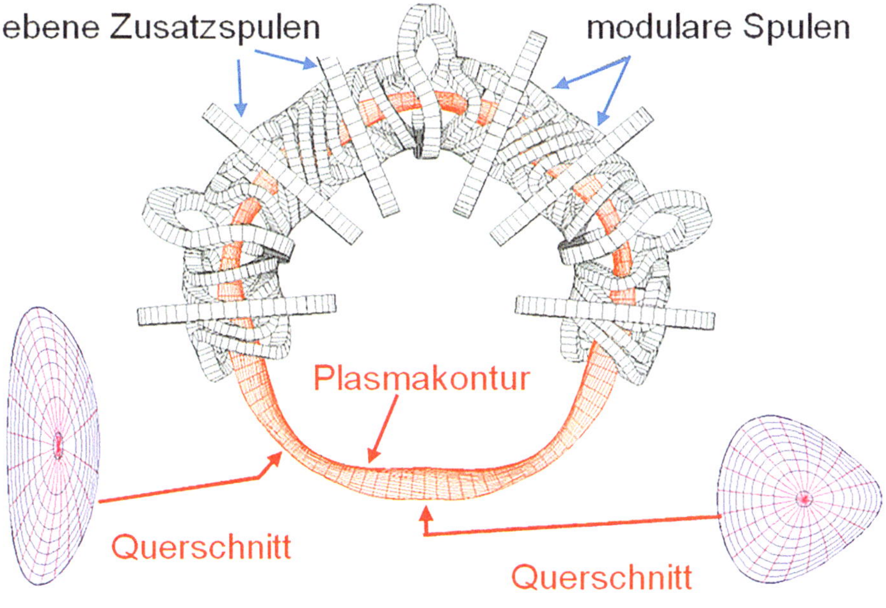


**Fig. 9.3** Schematic of Wendelstein 7-AS. (From [37])

**Table 9.1** Parameters of Wendelstein 7-AS and 7X, and Heliotron-E. Large Helical Device and Heliotron-J with Reactor Concepts

| Parameters\machine | W7-AS | W7-X | HELIAS-5B | Heliotron-E | Heliotron-J | LHD | FFHR-d1A |
|---|---|---|---|---|---|---|---|
| Major radius (m) | 2.0 | 5.5 | 22.0 | 2.2 | 1.2 | 3.6–4.0 | 15.6 |
| Minor radius (m) | 0.13–0.18 | 0.50 | 1.8 | 0.20 | 0.20 | 0.60 | 3.744 |
| Divertor (yes or no) | Yes | Yes | Yes | Yes | Yes | Yes | Yes |
| Superconducting coils (yes or no) | No | Yes | Yes | No | No | Yes | Yes |
| Axial toroidal field (T) | 2.6 | 2.5 | 5.9 | 2.0 | 1.5 | 3.0 | 4.7 |
| Configuration type | Helias | Helias | Helias | Heliotron | Helical axis | Heliotron | Heliotron |
| $\ell$ poloidal periods | 2 and 3 | 2 and 3 | 2 and 3 | 2 | 1 | 2 | 2 |
| m toroidal periods | 5 | 5 | 5 | 19 | 4 | 10 | 10 |
| Rotational transform $\iota$ | 0.24–0.58 | 5/6...5/4 | Around 1 | 2.5 | 0.4–0.65 | 1.0 | 1.5 |
| ECRH (MW) | 2.4 | 10–15 | Yes | 0.72 | 0.4 | 5.5 | Yes |
| ICRH (MW) | 1.0 | 3.5 | Yes | 2.0 | 1.4 | 2.0 | Yes |
| NBI (MW) H or D | 2.8 | 14 or 20 | Yes | 4 | 0.8 | 34 | Yes |

plasma confinement. In particular, the in-vessel control coils could be used for error field correction and for alterations to magnetic islands at the edge of the plasma. It was designed to overcome the difficulties of previous stellarators, including magnetic error fields in the C-Stellarator and the lack of optimization of quasi-symmetry properties. The complicated coil shapes were a source of risks in fabrication, assembly and operation, which were mitigated by very careful quality control.

W7-AS had the following characteristics:

- 45 modular coils produced both toroidal and poloidal field components with dominant $\ell = 2$ and $\ell = 3$ poloidal components.
- An additional set of toroidal field coils enabled change of rotational transform.
- Five separate modular coils at the corners of the pentagon-like shaped system with $m = 5$ toroidal periodicity allowed variation of the toroidal mirror ratio.
- The plasma position could be changed with vertical field coils.
- An ohmic system was used to modify rotational transform, to study current driven instabilities and to compensate the bootstrap current.
- Five in-vessel control coil pairs changed the $B_{5,m}$ field spectrum to vary the edge island size and connection length for divertor operation.
- W7-AS pioneered the island divertor for the line of low-shear Wendelstein devices. Five divertor module pairs (top-bottom) were installed at the 5 toroidal positions with elliptical plasma cross-section.
- Inserts provided the target and baffle plates and gave room for fairly closed divertor chambers. Diversion occurred with the edge islands, preferentially via the 5/m ($m = 8, 9, 10$) island chains.
- The strike points were defined by the intersection of the outer island separatrix with the target plates.
- The divertor gave access to plasmas with high density at quasi-steady-state conditions, good energy and low impurity confinement properties, large radiation from the plasma edge, and partial detachment. The scrape-off layer and the discrete divertor modules established a 3-D edge geometry, which was modelled by the 3-D edge transport code EMC3/EIRENE.
- Plasma heating was accomplished with
  - ECRH (70 GHz, 140 GHz, $< 2.4$ MW absorbed power);
  - NBI $< 2.8$ MW absorbed power; in the last phase with all-co (or other) orientation;
  - ICRH (up to 1 MW antenna power);
  - OH power which could be neglected in all heating scenarios.

### *9.2.2 W7-AS Stellarator Field Optimization*

W7-AS was the first stellarator designed with modular coils in the helias configuration, to test this reactor-relevant concept that would avoid the necessity of interwound coils. The geometry of W7-AS was only partially optimised due to lack of computer power during its design years. Even so, the parallel currents were reduced in W7-AS in comparison to an equivalent classical $\ell = 2$ stellarator and the equilibrium was improved for high ß due to the reduced displacement of the magnetic axis, confirmed experimentally for beta up to 3.4%.

The impact of the partial optimisation on neoclassical effects was more difficult to demonstrate. The measured (tokamak-like) bootstrap current was in agreement with the neo-classical calculations. Neo-classical transport, which governs the plasma core in the long-mean-free-path regime, is strongly affected by the ambipolar radial electric field. A positive electric field gave rise to low core transport and thus high electron temperatures, consistent with the profile characteristics of internal transport barriers.

Wendelstein stellarators were all designed with low shear in the vacuum field in order to avoid low-order rational surfaces inside the plasma, which was observed to result in good confinement in the proximity of low-order rational surfaces. The confinement therefore strongly depended upon the rotational transform $\iota$. In these $\iota$ windows, transitions into improved regimes like the H-NBI regime (close to $\iota = 1/3$) or the H* mode (around $\iota = 1/2$) were observed to occur.

## *9.2.3 W7-AS Operational Scenarios*

### 9.2.3.1 Fast Particle Confinement Improvement with Mirror Ratio

Energetic fast ions, such as beam injected ions or fusion products, when in trapped particle orbits, can experience rapid losses in many stellarator configurations. One way of reducing these losses, incorporated also in the W7-X design, was to increase the magnetic fields strength in regions of large curvature. In addition, high beta plasmas can increase the depth of a magnetic well.

A radially oriented neutral beam injector was used to create trapped fast particles, and the loss rate of these ions was measured as a function of the mirror ratio which was varied by changing the currents in different coil sets. It was found that the particle loss dropped very rapidly as the mirror ratio was increased above 1.0.

### 9.2.3.2 Successful Island Divertor Operation and High Density H-mode (HDH)

An island divertor system allowed the achievement of the High Density H-mode [36, 38]. The island divertor concept was used in W7-AS and is used in W7-X. Depending on the rotational transform $\iota$, the plasma was bounded either by smooth flux surfaces or by a separatrix formed from naturally occurring magnetic islands (at edge iota values 5/m with m = 8, 9, 10...). The additional control coils acted to enhance or reduce the radial extent of the edge islands, thereby altering the plasma radius as well as influencing the field line pitch inside the islands. The modules had a length of about 700 mm, covering only a fraction of the toroidal circumference.

The divertor enabled very high line-averaged plasma densities up to $4 \times 10^{20}\ m^{-3}$ in quasi-stationary conditions with separatrix edge and scrape-off densities up to $8 \times 10^{19} m^{-3}$ and fractional radiation levels up to 90%, permitting divertor detachment.

#### 9.2.3.3 Turbulent Transport and Confinement

The outer plasma zone of W7-AS is governed by turbulent transport. It was found [36] that the energy confinement was comparable to that of tokamaks by converting the ι, B and geometry into an equivalent plasma current that produces a comparable rotational transform. There were some differences:

- No isotope effect was observed.
- At high densities, a transition to HDH mode occurred with a strong increase in energy confinement times and a strong drop in impurity confinement times.

#### 9.2.3.4 Improved Confinement Regimes

There are three distinct improved confinement regimes:

- H-mode
- High density H-mode (HDH) regime
- H-NBI-mode

These regimes depend on the density, heating method, divertor characteristics and rotational transform. The main point is that improved confinement is available over a wide range of operation.

#### 9.2.3.5 Operational Limits: Density, Temperature, Confinement, Beta

High values of the fusion triple product in a relatively small machine and good performance demonstrated the excellent confinement properties of stellarators. Using up to 2 MW of ECRH and 2.8 MW of NBI, the following parameters were separately obtained: $T_e = 6.8$ keV, $T_i = 1.8$ keV, $n_e = 4 \times 10^{20} m^{-3}$, beta=3.4%, $\tau_E = 0.06$ s. and the simultaneous triple product $n_e T_i \tau_E = 5 \times 10^{18} m^{-3}$ keV s.

The maximum beta values of 3.4% on average and 7% core value of W7-AS were obtained at low toroidal field of about 1 T [39, 40].

### *9.2.4 W7-AS Goals Achieved, Robustness and Fragilities*

W7-AS fulfilled its major goals:

- to test the modular coil concept;
- to develop the island divertor;
- to test the effectiveness of the first steps in stellarator optimisation and thus to contribute to the scientific basis for W7-X;
- to contribute to the physics of toroidal confinement of stellarators.

W7-AS achieved the following operational successes:

- Divertor operation was possible with good plasma confinement under quasi-steady-state conditions in an ELM-free variant of the H-mode.
- The H-mode power threshold was low.
- Close to operational limits, the plasma performance was found to be free of violent instabilities.
- The ß-values limited by the heating available were high.
- Operation at high densities was possible.
- The plasma performance did not noticeably decline close to limits.
- W7-AS established the Wendelstein concept as a potential fusion reactor.

## 9.3 Wendelstein 7-X: A Case Study in Systems Forward Engineering of a Superconducting Helias

### *9.3.1 Systems Analysis Strategy for the Superconducting W7-X Helias as a Case Study*

#### 9.3.1.1 Case Study Strategy

This section examines the Wendelstein 7-X (W7-X) as a case study in designing, obtaining funds, building and commissioning a new fusion experimental machine. Checklist 1.3A from Chap. 1 gives general systems principles for forward engineering and Checklist 1.3B outlines detailed steps to follow using these principles. W7-X is suitable for this type of case study, being the largest stellarator that is also well documented [41, 42].

#### 9.3.1.2 W7-X Capabilities

Figure 9.4 shows a schematic diagram of W7-X [43]. There are 50 non planar coils and 20 planar coils with superconducting NbTi, which are grouped in 7 sets of 10 coils wired in series with 7 independent power supplies plus 5 water cooled trim coils (not shown) to compensate for error fields. The non planar coils create a helias

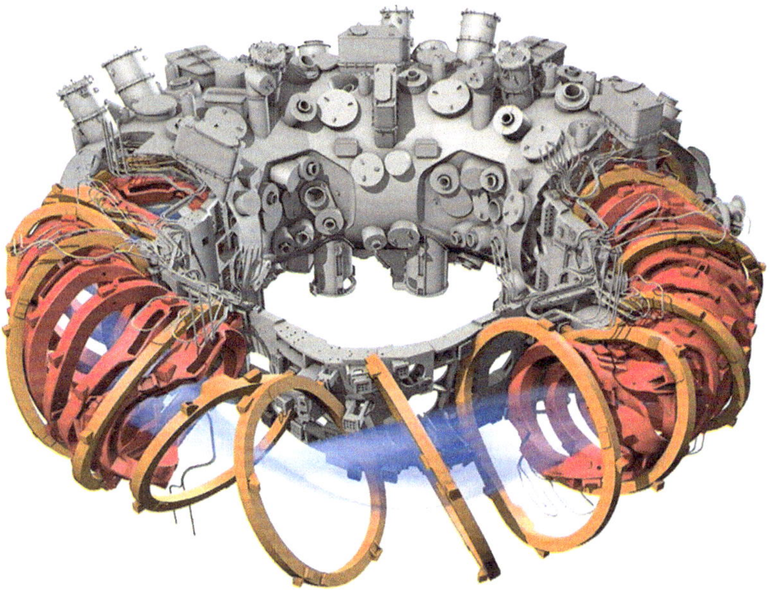

**Fig. 9.4** Schematic diagram of the superconducting W7-X. (From [43], Fig. 1, permission from ©Euratom)

field. The planar coils can be operated to shift the high-iota rotational transform and low-shear profile. The various available profiles of iota included standard, high mirror ratio, high iota, and low iota. The parameters of W7-X with maximum heating power and installed divertors are given in Table 9.1.

#### 9.3.1.3 Systems Forward Engineering: Getting Started

The starting point for the forward system engineering of a new fusion machine is Checklist 1.3B, beginning with the 'getting started' section and setting goals, reviewing other machines, examine and choose machine parameters.

### *9.3.2 Systems Approaches to the Design and First Goals of W7-X*

The physics and engineering design for W7-X included detailed calculations of stellarator equilibrium, stability and operation [44, 45]. The timeline of the project was Phase 1 Euratom approval in 1991, Phase 2 approval in 1995, project start in 1996, move to Greifswald in 2000 [41] and project operation start in 2015.

A major source of delay in the project was the decision to build W7-X at a greenfield site in the former East German and Hanseatic city of Greifswald because of the long scientific traditions of the Ernst-Moritz-Arndt University, founded in 1456 with expertise in plasma physics [46]. A political goal was to contribute to the development of research infrastructure in the New Federal States.

W7-X as a continuation of W7-AS had the overall goals of reaching reactor relevant parameter regimes and to demonstrate the reactor capability of modular stellarators. For this purpose the size of Wendelstein VII-X was chosen large enough to allow powerful heating methods and quasi-steady state operation.

The initial goals of W7-X were to:

- achieve quasi-steady state operation in a reactor relevant parameter regime with temperatures above 5 keV and densities above $10^{20}$ $m^{-3}$;
- demonstrate stable plasma equilibrium with beta = 5%;
- confine a plasma over a sufficiently long time, allowing extrapolation to reactor parameters;
- control plasma density and impurity content;
- operate the experiment in steady state with a field generated by modular and superconducting coils.

The logic of choosing these goals was as follows:

- W7-X did not aim at ignition; hence provisions for handling radioactive materials were not needed.
- The parameter regime with ion temperature above 5 keV was chosen because, in a reactor, this regime has to be reached with auxiliary heating only and without assistance of alpha-particle heating, which when added in a reactor, leads to near ignition conditions.
- The aim of beta = 5% resulted from economic requirements of nuclear fusion and not so much from physics requirements.
- Since plasma confinement and stability depend mainly on properties of the magnetic field such as rotational transform, shear, magnetic well, local mirrors and the magnitude of Pfirsch-Schlüter (similar to bootstrap) currents, it required a careful optimisation of the magnetic field and the modular coils.

The following design principles were taken into account in developing the detailed systems architecture:

- High quality of the magnetic surfaces;
- Good finite-equilibrium properties;
- Good MHD-stability properties;
- Reduced neoclassical transport;
- Small bootstrap current in the long-mean-free-path regime;
- Good modular coil feasibility.

Generally, strong geometrical shaping of the plasma boundary improved confinement and stability properties while adversely affecting coil feasibility. Thus, additional conditions on the shaping parameters at the plasma boundary were used to ensure coil feasibility. Important issues for selecting the coil geometry were:

- the minimum distance between coils and plasma and the minimum bending radius of the coils;
- a sufficiently broad accessible range of variable magnetic field parameters, like field strength, rotational transform and axis position.

## 9.3.3 *Review and Prioritize Goals and Evaluate Variants for W7X*

### 9.3.3.1 Systems Review Process in Early Stages

The processes for reviewing and prioritizing goals and evaluating variants for W7-X were similar to those of Checklist 1.3B of reviewing goals, scope major variants, and produce report with design specifications.

### 9.3.3.2 Prioritizing Goals and Making Choices of Systems Architecture

The magnetic field configuration of W7-X differed from W7-AS mainly in the shape of the magnetic surfaces and the existence of a helical magnetic axis. The W7-X helias design had much more computing power available and used it to minimize bootstrap currents while keeping MHD-stability by the creation of a magnetic well. Helias configurations were investigated for the architectural variants of 4, 5 or 6 field periods. They were characterized by very low outward axial shift at high plasma pressures and there was negligible change of the rotational transform with finite $\beta$. Configurations were found that were stable up to $\beta$ of 9% against resistive interchange modes. In the special case of quasi-helically symmetric configurations, neoclassical transport could become very low.

A helias option of 5 field periods and number of coils was selected for W7-X for the following reasons:

- MHD-ballooning instability modes led to $\beta$-limits below or equal to 3% in systems with 4 field periods.
- The 4-period case would have been preferable due to less complex coil geometry, but high beta was felt a necessary goal for reactor relevance (with the superconductors available at that time).
- The number of modular coils per period was chosen as a compromise between maximizing the distance between adjacent coils for heating and diagnostics access and minimizing the magnetic field ripple arising from these gaps.
- A compromise was found with 14 coils per period.
- The magnetic field on axis was 3 T while using superconducting coils for steady state operation, determined by the available space for the modular coils, the maximum current density in the coils and the mechanical stresses.

As in W7-AS, several groups of similar type coils were separately connected in series with independent power supplies to vary the magnetic configurations. The following design options were considered:

- The rotational transform of 5-period Helias configurations can be adjusted in the range between 0.7 and 1.4 with a shear less than 0.2.
- A design goal was to be able to experimentally vary the rotational transform. Options included:
  - dividing the modular coils into an upper and lower winding pack and feeding each one separately with current or
  - adding a separate coil system which superimposes a toroidal field parallel or antiparallel to the field of the modular coils.
- By choosing the second option, the rotational transform could be varied by ±20%. Separate coils would also allow shifting the magnetic axis by about 1% of the major radius.
- Vertical field components were introduced by appropriate current polarities.

The dimensions of W7-X were determined by the following considerations:

- A heating power of at least 20 MW was considered necessary to reach reactor relevant parameters.
- The geometric dimensions of the device had to be chosen to give access to NBI and allow for divertors.
- The distance between plasma and wall had to be large enough to keep wall loading and impurity influx low.

Due to these considerations, a major radius of 6.5 m and an average plasma radius of 0.65 m were chosen.

#### 9.3.3.3 Superconducting Coil Considerations

The axial toroidal field of W7-X design option of 3 T led to a field of 6 T at the superconducting coils, so NbTi winding could be used. Working with industry, potential methods of producing and winding the modular coils were developed. The mechanical stresses in and between the coils were found to be very large, so the coils would need to be enclosed in casings and fixed to a structural shell, with everything enclosed in a cryostat with heat shields.

#### 9.3.3.4 Basic Systems Parameters for Further Design

Having settled on the intermediate goals for W7-X, the next steps would have been similar to those from Checklist 1.3B, which were to scope the highest system level of major variants, develop or find models for primary level systems, estimate performance, and then produce a first report and discuss and present with wider group and laboratory or project management.

The primary systems to be considered are the plasma, vacuum vessel, plasma facing components, divertor, planar coils, non-planar coils and error field correction coils, electrical power supplies, diagnostics, machine and plasma control, gas pumping and injection, auxiliary plasma heating, gas injection, coil cooling, cryostat engineering, heat shields and machine structure.

Of these systems, the most important consideration was to scope the highly interactive systems of the plasma and helias coils, assuming that the toroidal field could be altered by the planar coils, and then to determine the maximum plasma performance.

The next step from Checklist 1.3B is to produce definitive proposals and engineering design for Phase I proposal. It is necessary to choose one or more most promising variants and begin detailed design of level 1 and level 2 subsystems of key areas. A Phase I proposal needs a specific design involving appropriate physics and engineering modelling, cost and time estimates and whether the proposal fit into the European fusion strategy. Detailed systems architecture is required.

#### 9.3.3.5 Phase I (1991) and Phase II (1995) Proposals and Project Start

After an initial design and choice of magnetic field configuration, the W7-X proposal was able to obtain phase 1 Euratom approval for the project in 1991, and after a very detailed physics and engineering design, phase 2 Euratom approval was obtained in 1995 [41]. The official project start was in 1996, with the move to Greifswald in 2000.

### *9.3.4 W7-X Revised Proposal, Robustness and Tradeoffs*

Checklist 1.3B notes that important activities are to assess robustness and mitigate fragilities, then to refine the design. Even at the beginning of the systems architecture process, the designers were well aware of the trade-offs between robustness of plasma performance. They were trying to design a highly optimized magnetic field yet flexible magnet configuration. The trade-off was with the difficulties in manufacture and construction of such highly distorted non-planar coils in several different types to obtain flexibility, compounded by extreme forces in a superconducting coil environment. The trade-offs were influenced by the history of error fields ruining plasma confinement in stellarators and the need to demonstrate that reactor-level plasma confinement was possible.

The designers chose to weigh the design heavily in favour of a highly optimized magnetic field and a magnetic coil system that could be varied to explore different configurations to find optimum performance. The design was also optimized to allow for auxiliary heating and a divertor.

The construction was also staged so that the best technological solutions in auxiliary heating could be added when the machine was ready, for example with high power and high frequency gyrotrons for ECRH.

The high performance designed into the system for physics robustness led to strong fragilities in the manufacture, construction and assembly of the components. These fragilities were overcome with excellent engineering, quality control and lots of time and money, but the resulting system was a complete success.

### 9.3.5 W7-X Final Design for Construction

The final design of W7-X and initial construction details [47] is summarized in Table 9.1 and parameters for machine construction.

- Use 50 non-planar coils to create the helias magnetic field configuration.
- Use 20 planar coils to vary the axial toroidal field and to vary the rotational transform ι around the value 1, corresponding to the tokamak $q = 1$, in the range between 5/6 and 5/4.
- Add 5 error and island production coils that allow the machine to function even in the presence of very small coil winding errors.
- Produce a magnetic well depth of typically 1% to provide stability and to minimize the axial displacement of high beta plasma equilibrium.
- Stage the installation of auxiliary heating power, starting at 15 MW and later increasing to 30 MW from a mixture of ECRH, NBI and ICRH.
- Install ECRH at 140 GHz, which is resonant at the second harmonic of the EC frequency at 2.5 T. The axial magnetic field was designed to be increased from the nominal field up to 3 T so as to be resonant with 140 GHz ECRH and thus allow off-axis heating.
- Attach neutral beam injectors with up to 130 kV to allow beam penetration at high plasma density.
- Provide for pulse length of up to 30 min with full cooling and island divertor operation.
- Choose superconducting coils wound from a cable-in-conduit conductor composed of 243 strands of NbTi enclosed by an aluminium jacket.
- Specify coil winding form accuracies of a few tenths of a millimeter.

### 9.3.6 Procurement, Construction, and Initial Testing of W7-X

Checklist 1.3b recommends that once the detailed engineering design was finalized, and construction was ready to begin, then detailed management and construction structures and procedures are required. The difficulties of the long construction process also led to the heroic efforts of overcoming these difficulties [41]. The construction process difficulties included the insolvency of a coil manufacturer in

2002 along with external audits of the project. The original promised completion date at Phase 1 was 1999, and the actual completion date was at the end of 2015, while the project costs had doubled to over a billion Euros, including building costs and salaries. The hardware costs were 400 million Euros. There was a first restructuring of the project in 2003, and the device assembly was interrupted due to the need for coil repair in 2004.

The major problems were with the coil construction and a very wide range of issues caused delay. They were overcome by careful project management. An analysis concluded that the major issues could have occurred in the construction of any large scale fusion project, especially with superconducting coils.

The biggest issue was required manufacturing tolerances, which are an essential part of fabricating a stellarator. Other problems included:

- deviations and damages of SC strands;
- voids in cast steel coil casings;
- geometrical deviations in coil casings;
- residuals of Cu-stainless steel soldering flux;
- Al and stainless steel welds to be requalified;
- quench detection cable damage;
- insulation faults in the coil header;
- danger of shorts in the coil header.

## 9.3.7 W7-X Operational Scenarios

W7-X started operation at the end of 2015 and results were reported 2 years later [48]. W7-X operated as a fully integrated part of the EUROfusion programme. Engineering systems such as the operation of the superconducting magnets [49] and trim coils [50] were documented. It was the excellence of the operation of the superconducting coils and the error correction capability of the trim coils that allowed highly successful operation. W7-X high performance [43] and divertor [51] operation were also highly successful. Initial plasma heating was with 8 MW of 140 GHz ECRH at the second harmonic from long pulse gyrotrons. Some NBI had already been installed and was being increased in power, with ICRH coming later. All plasma-facing components will eventually be water cooled.

### 9.3.7.1 Wall Conditioning for High Performance

After baking the vacuum vessel at 150 °C, glow discharge and ECRH conditioning discharges in helium were important for density and edge radiation control. The plasma-facing components covering about half of the surface area were made of graphite and steel.

Routine access to high plasma densities was made possible with boronization of the first wall. After boronization, the oxygen impurity content was reduced by a factor of 10 and the carbon impurity content by a factor of 5. The reduced (edge) plasma radiation level allowed higher densities without radiation collapse. The plasma had well above $1 \times 10^{20}$ m$^{-2}$ line-integrated density and $T_e = T_i = 2$ keV central temperatures at moderate ECRH power.

#### 9.3.7.2 High Density and High Stellarator Triple Product with ECRH

Using a hydrogen plasma with pellet injection at a central density of $8 \times 10^{19}$ m$^{-3}$, the ions were indirectly heated and a temperature of 3.4 keV with $T_i$ approximately equal to $T_e$ was obtained with second harmonic ECR heating in X-polarization. A stellarator record was obtained for the triple product $n_e T_i(0)\tau_E = 0.64 \times 10^{20}$ m$^{-3}$ keV s, about 1/5th of the highest steady-state value in tokamaks achieved by JT-60 U. The peak diamagnetic energy was 1.1 MJ and volume-averaged normalized plasma pressure was $\beta = 1.5\%$. With a combination of ECRH and NBI, the central plasma density reached the cut-off of O-mode second harmonic heating of $2 \times 10^{20}$ m$^{-3}$. The record values of triple product provide evidence for the W7-X goal of obtaining reduced neoclassical energy transport [52].

#### 9.3.7.3 Stellarator Optimization

Various scenarios demonstrated successful operation of W7-X and showed that the variable coil system could be used to optimize discharges. One of the optimizations was to minimize the bootstrap current in the plasma, which was measured time-dependently for different stellarator configurations produced with different coil currents, for the following cases:

- low iota
- standard
- high mirror
- high iota

This optimization allowed the demonstration of good agreement with transport codes and the minimization of the bootstrap current in high iota discharges.

#### 9.3.7.4 Island Divertor Operation

The magnetic island divertor was first tested in the predecessor experiment W7-AS. The divertor was able to function because W7-X has a rotational transform of about 1 and low magnetic shear at the plasma boundary. Therefore, large magnetic islands can form, and divertor plates can be located on field lines locally, as is shown in Fig. 9.5 [51]. Because of the intrinsically low error fields and the possibility of further compensation with the trim coils, these island divertor modules have proved very effective in handling exhaust power which was very evenly distributed.

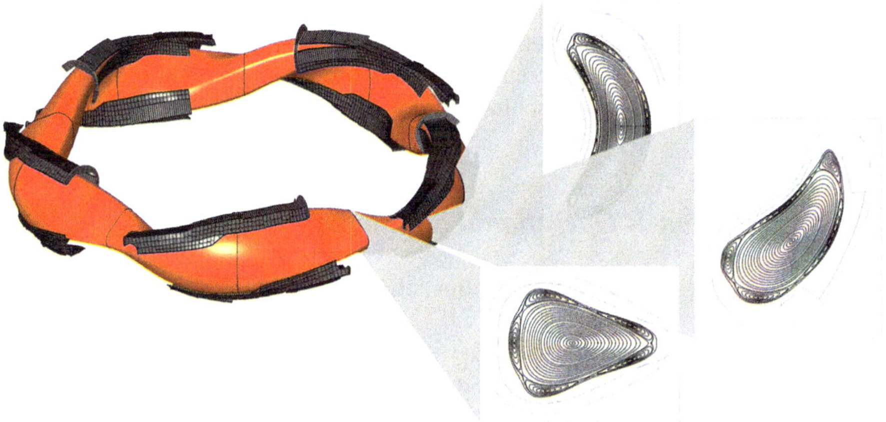

**Fig. 9.5** W7-X island divertor schematic including exhaust targets and vacuum field with islands at three poloidal locations, one without a divertor, two with. (From [51], Fig. 3)

Further results on divertor operation from W7-X and their implications for steady state operation were discussed [53, 54]. The performance of turbulence-suppressed discharges demonstrated that successful neoclassical optimization had been achieved and indicated the existence of a "stability valley" between ITG and TEM instabilities.

Island divertor performance was remarkably good, with record-length detachment and simultaneous efficient particle exhaust, benign impurity level, and good main plasma confinement for discharges lasting nearly 30 s.

### *9.3.8 W7-X Metrics and Performance*

W7-X had already approached many of its initial goals by demonstrating high beta operation, high temperatures and high density, with discharges lasting up to 30 s. To achieve these together and in steady-state is within reach of W7-X operation, given the plans to increase heating power and to fully adapt the machine for steady state operation. W7-X can already be classified as a success, but it took a very long time and much effort and money to bring it to this point.

## 9.4 Heliotron-E

### *9.4.1 Heliotron-E Systems Architecture and Technical Choices*

The Heliotron (also known as torsatron) series of experiments, founded by Koji Uo at Kyoto University, started with Heliotron A in 1959 and culminated in the Heliotron-E experiment, which started operation in 1980 and produced highly successful results [55]. These led to the superconducting Large Helical Device and the advanced helical concept Heliotron-J, see Table 9.1. Heliotron E (H-E) experiments for its first 8 years of operation were summarized [56] when it had provided enough data to justify future machines based on its success.

## *9.4.2 Heliotron-E Operational Scenarios*

### 9.4.2.1 High Density and High Temperature Plasmas

H-E demonstrated that ECR heated plasmas are able to produce target plasmas for NBI or ICRF so as to obtain high density and high temperature currentless plasmas. The highest electron or ion temperatures were both about 1.5 keV depending on heating method, obtained in the low density regime of less than $10^{19}$ $m^{-3}$.

### 9.4.2.2 Currentless Plasmas Without Disruptions

It was demonstrated that currentless plasmas have no major disruptions. Quasi-steady plasmas can be produced by controlling impurity ions with titanium gettering and carbon coating.

### 9.4.2.3 High Beta Operation

At low toroidal field of 0.94 T, high volume averaged beta of 2% was demonstrated, with average density of $8 \times 10^{19}$ $m^{-3}$ and equal ion and electron temperatures of 350 eV. Pressure-driven instabilities were observed for peaked pressure profiles and sometimes relaxation oscillations were observed that were similar to tokamak sawteeth.

In ECRH plasmas, neoclassical transport was found to be dominant in the region inside the half radius, but the overall energy confinement time scaled differently, being proportional to $n^{0.66}P_{heat}{}^{-0.53}$.

# 9.5 Heliotron-J, a Helical Axis Configuration

## *9.5.1 Heliotron-J Systems Architecture and Technical Choices*

During LHD construction elsewhere, the University of Kyoto continued with Heliotron J. The theoretical basis [57, 58] involved a helical axis stellarator with an aspect ratio of 11 based on a modulated $\ell = 1$ helical coil. The bumpiness of the magnetic field was comparable to the $\ell = 1$ helical component with the opposite sign in order to reduce neoclassical diffusivity. The bumpy field was also designed to control the bootstrap current.

The optimized configuration had an almost flat rotational transform of 0.7 and a magnetic well at zero beta. A volume averaged beta up to 4.5% was thought to be possible. With the increase of beta, confinement of high energetic trapped particles was expected to improve. A divertor configuration was also explored.

The construction of Heliotron-J was completed impressively rapidly [59]. Its design and construction principles and results have been summarized [60]. The helical field is provided by a single $\ell = 1$ coil and the vacuum vessel (containing the plasma) wrapping around each other as they twist along in the toroidal direction. There are also two types of toroidal coil and three pairs of poloidal coils to provide flexibility of the magnetic configuration. The machine is smaller and more compact than Heliotron-E.

The heating systems available are 0.4 MW of ECRH, 1.6 MW of NBI and 0.8 MW of ICRH. Plasma parameters of up to a density of $1 \times 10^{20} m^{-3}$, electron temperature of 3 keV and ion temperature of 0.4 keV have been achieved.

### *9.5.2 Heliotron-J Operational Scenarios*

The operational scenarios of Heliotron-J have explored a number of areas:

- An H-mode transition is triggered by strong gas puffing during NBI.
- An enhanced confinement regime is observed during plasma current ramp up with ECH resulting in the formation of an electron Internal Transport Barrier.
- Turbulent properties were studied by varying isotope ratio and zonal flow [61]
- Effect of magnetic configuration was determined on energy confinement and energetic particle driven MHD modes [62]

## 9.6 Superconducting Large Helical Device (LHD)

Superconducting stellarators/helias such as W7-X and heliotrons such as LHD are meant to explore technologies that can be used in fusion reactors and also to study the physics of very long pulse operation. Reverse systems engineering is used on LHD to examine its goals, design choices and operational results.

### 9.6.1 LHD Goals, Systems Architecture and Technical Choices

The parameters of LHD are shown in Table 9.1, and a schematic of the coil system is presented in Fig. 9.6 [63], showing the superconducting helical coils.

The original goals and design of LHD [64, 65] were based upon initial theoretical studies [66]. The Japanese National Institute for Fusion Science (NIFS) in Toki was set up to build the LHD and to support fusion related activities [65, 67].

The design of LHD was begun in 1986, and in 1988, the plan was to finish to basic design and detailed design phase I and phase II, then begin construction in 1991 and start plasma operation in 1995. LHD in fact started in 1998, not very delayed for a project of this complexity, especially when setting up an entirely new institute. A large number of papers on impressive physics results were published in the 2000 IAEA fusion energy conference.

The original design was meant as a bridge between Heliotron-E and a fusion reactor and had considered the use of either normal or superconducting coils [64]. They settled on superconducting coils which would allow reactor-relevant quasi-steady-state operation. The main LHD goals were to:

- carry out transport study in a wide range of plasma conditions and to achieve high triple product $n\tau T$ plasma conditions extrapolable to reactor plasmas.
- realize high-beta plasmas with average beta value of 5%, which is required for reactor plasmas, and to understand the related plasma physics.
- attain quasi-steady state operation and related database using helical divertors.
- study the behaviour of high-energy particles in this nonaxisymmetric system, and to simulate alpha-particle behaviour.

**Fig. 9.6** LHD superconducting helical coils. (From [63])

- increase the comprehensive understanding of toroidally confined plasmas by carrying out studies complementary to those in tokamaks.

The original design parameters before detailed engineering were as follows: R = 4 m, minor radius within coils = 0.96, $B_0$ = 4 T, B at windings = 8 T, m = 10 toroidal periods, rotational transform at centre of 0.5 and edge 1.0, with 20 MW of auxiliary heading power.

As is usual in Japan, the project chose a major industrial company Hitachi [68] to collaborate on the final design and to build the machine. The cost of LHD was about 50 billion yen, roughly half a billion dollars, hence similar to the original W7-X budget.

The final design of the machine used a two stage approach to the toroidal field. It would be 3 T initially, with the possibility of 4 T later and a much higher level of auxiliary heating, especially NBI.

Three types of plasmas were set as goals, considering known transport laws:

- CASE 1 (high $n\tau T$), $B = 4$ T, $\langle T_i \rangle = 3\text{–}4$ keV, $\langle n \rangle = 10^{20}$ $m^{-3}$, $\tau_E = 0.1\text{–}0.3$ s
- CASE 2 (high $T_i$), $B = 4$ T, $T_i(0) = 10$ keV, $\langle n \rangle = 2 \times 10^{19}$ $m^{-3}$, $\tau_E = 0.05\text{–}0.1$ s
- CASE 3 (high $\beta$), $B = 1\text{–}2$ T, $\langle \beta \rangle = 5\%$

This set of parameters was felt to be sufficiently close to reactor levels to be able to study fusion relevant plasmas.

An in-depth analysis of the theoretical aspects of the LHD was carried out [66] on this $\ell = 2$ and m = 10 period configuration. They considered many different physics and engineering effects in optimizing the heliotron design:

- Helical period m = 14 option: There was a size limitation on the pitch parameter $\gamma = a_c/r\ (m/l)$ to be less than 1.3, otherwise the outermost flux surface would have touched the coil. If $\gamma$ was greater than 1.3, the rotational transform at the edge would have exceeded 1.0, creating a severe limit on beta. Therefore m = 14 was rejected.
- m = 12 option: The need for leaving space for a divertor led to the rejection of the m = 12 option.
- m = 10 option: With 10 periods, there was no contact of the outermost flux surface with coils, and there was room for a helical divertor. An equilibrium axial beta up to 15% was found to be possible. The ratio of quadrupole components of the field and axial position were varied to find the optimum stability conditions. Configurations to minimize fast particle loss were also optimized, as well as energy confinement.

LHD was equipped with two different divertor systems:

- The Helical Divertor (HD) had a shape similar to the double-null open divertor in tokamaks except for its helically twisted structure.

- The Local Island Divertor (LID) used an m/n = 1/1 (where m is azimuthal wave number) magnetic island produced by an externally applied resonant field to guide particle and heat fluxes to the locally inserted divertor plates.

## 9.6.2 *LHD Operation Scenarios*

### 9.6.2.1 Initial Operation

Over the more than two decades of successful operation, the LHD has explored a wide range of operating scenarios, starting with a summary of the first 2 years of operation [69]. With a toroidal field of 2.89 T, it has used heating powers of 4.2 MW of NBI, 1.3 MW of ICRH and 0.9 MW of ECRH. Results included parameter values which were separately achieved of $T_e = 4.4$ keV, $\tau_E = 0.3$ s, and stored energy of up to 880 kJ. The confinement was 1.6 times the International Stellarator Scaling 95, and confinement was comparable to ELMy H-mode in tokamaks. A high beta of 2.4% was achieved at 1.3 T without confinement degradation. A plasma with 200 kJ was sustained for 5 s by ICRH. An NBI plus ICRH heated pulse was sustained for 1 min. LHD continued to produce excellent results in the following years:

### 9.6.2.2 Improved Particle and Energy Confinement Regimes

One important improvement of energy confinement resulted from an inward shift of the plasma axis [70], although this may be due to a very broad heating profile due to NBI beam deposition. The achievement of an improved particle and energy confinement regime was reported [71]. With hydrogen ice pellet injection plus NBI heating, a peaked axial density of $1.1 \times 10^{21} m^{-3}$ was obtained. A high ion temperature of 5.6 keV was separately obtained by NBI at an electron density of $1.6 \times 10^{19} m^{-3}$. An internal ion transport barrier was observed.

### 9.6.2.3 Design Parameters Achieved in Hydrogen Plasmas

LHD had achieved many of the project goals [67]:

- $T_i = 10$ keV with an Ion ITB and impurity hole on axis;
- $T_e = 20$ keV with an Electron ITB;
- $n_e = 1.2 \times 10^{21}$ $m^{-3}$ with a super dense core;
- $\beta$ average = 5.1% at 0.425 T and 4.1% at 1 T, with current free plasmas;
- plasma duration up to 54 min, 28 s with dynamic wall retention.

A theoretical study of the beta limit in LHD [72] indicated that the achievable beta limit could be even higher and would be accessible with more heating power.

#### 9.6.2.4 High Power Deuterium Plasmas and Fast Particle Studies

The LHD started deuterium experiments in 2017 with the aim of exploring fast particle physics and reactor relevant regimes [73]. For the first time in LHD, 1.0 MeV triton burnup could be measured by 14 MeV neutrons produced from fusion during slowing down. The triton has the same orbit characteristics as a 3.5 MeV alpha particle. A burnup ratio of up to 0.45% was measured with the Sci-Fi detector [74]. Total 2.5 MeV neutron rate reached $3.3 \times 10^{15}$/s. The vertical neutron camera measured shifts in the peak position of the neutron emission. Using a full range of neutron diagnostics [75], a study of triton burnup concluded that LHD has comparable triton confinement and alpha confinement to that of tokamaks.

#### 9.6.2.5 H-mode and Divertor Detachment

The role of edge magnetic field structure and topology on detachment and compatibility with core plasma confinement was investigated in the LHD [76]. A sharp boundary between the confinement region and the stochastic layer could be introduced with Resonant Magnetic Perturbations (RMP) with an $m/n = 1/1$ helical structure. A confinement mode transition occurred with the application of RMP during the detached phase when working in deuterium. The divertor heat load decreased during the detached phase by around 70%. During constant neutral beam heating power, RMP led to detachment, the formation of an ETB (electron temperature barrier) and a stored energy increase of about 50%.

### *9.6.3 LHD Robustness and Fragilities*

The LHD was a highly robust machine, having been constructed and operated rapidly and successfully. It met all of its goals and demonstrated equivalent confinement properties including that of fast particles to tokamaks.

The fragilities were in the nature of the heliotron configuration itself. In the design of LHD, the difficulty of obtaining a large gap between the helical coils and the plasma was apparent. Multipole fields fall off rapidly as distance is increased from the coils, as will be necessary for a breeding blanket in a fusion reactor.

The other fragilities are the highly intertwined nature of the helical coils, vacuum vessel and other components and the very high tolerances needed. The ability to repair a failed coil or other component in a highly radioactive environment remains to be demonstrated.

Thus the evaluation of fragility will depend on the success of fusion reactor designs of this concept and of any intermediate machine between LHD and a reactor.

## 9.7 Alternative Helical Configuration Systems Architecture and Operational Scenarios

### *9.7.1 TJ-II Heliac*

TJ-II was designed in Spain as a low magnetic shear Heliac, in which the plasma winds around a central conductor. It is shown in Fig. 9.7 [77], and its parameters, such as R = 1.5 m, a = 0.22 m and 4 toroidal periods, are shown in Table 9.2. The magnetic field of 1.2 T is generated by a system of poloidal, toroidal and vertical field coils. The coil configuration allowed the rotational transform and magnetic well depth to be varied over a wide range. It was constructed between 1991 and 1997, and has been operating successfully ever since, with main characteristics:

- strong helical variation of its magnetic axis;
- very favourable MHD characteristics with potential for high beta operation;
- operational flexibility;
- bean shaped plasma cross section.

The existence of closed and nested magnetic surfaces, in good agreement with calculations, was demonstrated by magnetic surface measurements. Plasma heating was provided by ECRH from gyrotrons producing up to 600 kW at 53.2 GHz and NBI up to 1.2 MW. Transitions of L-mode to high confinement H-mode were observed TJ-II [78]. The shear of the negative radial electric field increased at the transition by an amount that depended on the magnetic configuration and heating power. The results indicated suppression of turbulence.

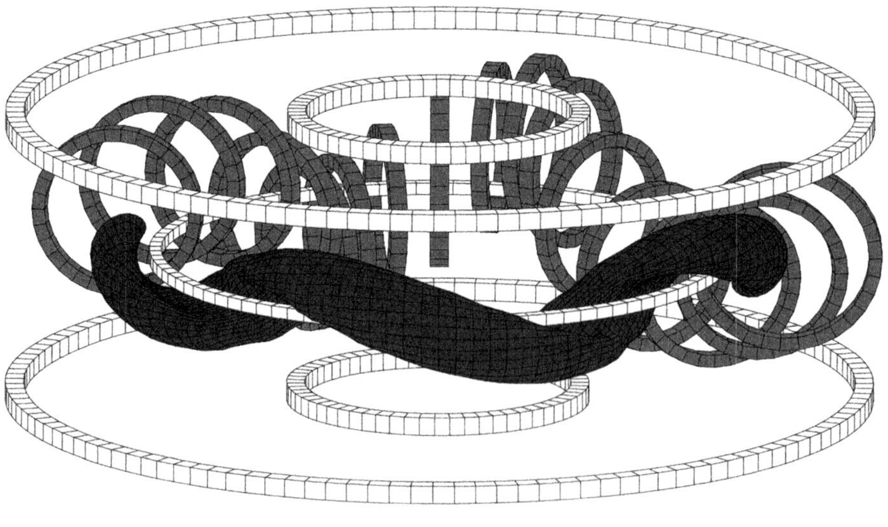

**Fig. 9.7** Coil schematic and plasma of TJ-II HELIAC. (From [77], Fig. 1)

**Table 9.2** Parameters of TJ-II, Uragan -2 M, H-1NF, WEGA/HIDRA, CTH and NCSX

| Parameter\machine | TJ-II | Uragan 2 M | H-1NF | HSX | WEGA/ HIDRA | Scyllac | CTH | NCXS |
|---|---|---|---|---|---|---|---|---|
| Major radius (m) | 1.50 | 1.70 | 1.0 | 1.2 | 0.72 | 4.0 | 0.75 | 1.4 |
| Minor radius (m) | 0.22 | 0.22 | 0...20 | 0.15 | 0.19 | 0.01 | 0.19 | 0.33 |
| Axial toroidal field (T) | 0.95 | 2.4 | 0.5 | 1.1 | 0.5 | 1.4 | 0.7 | 2.0 |
| Configuration type | Heliac | Torsatron | Heliac | QHS Helias | Stellarator | High β Stellarator | Compact toroidal hybrid | Helias QAS |
| ℓ poloidal periods | – | 2 | 1 | – | 2 | 0, 1, 2 | – | – |
| m toroidal periods | 4 | 4 | 3 | 4 | 5 | – | – | 3 |
| Rotational transform ι (2π) | 1.73 | – | 0.6–2.0 | 3 | – | – | 0.02–0.33 | – |
| ECH (MW) | 0.6 | – | – | 0.3 | 0.026 | – | Yes | – |
| ICR (MW) | – | Alfven | 0.06 | – | – | – | – | 6 |
| NBI (MW) H or D | 1.2 | – | – | – | – | – | – | 6 |

Stellarator optimization experiments [79] focused on impurity accumulation, turbulence, effects of the magnetic configuration on transport, plasma core fuelling, fast particle driven modes and power exhaust, coupled with modelling and validation of theory. Results were extended to LHD and W7-X.

### 9.7.2 Model Validation in Stellarators

Model validation in fusion plasmas in stellarators was carried out on TJ-II, especially in support of Wendelstein 7-X [80]. Using its flexibility and advanced diagnostics, experiments and models were compared in the following areas:

- Transport asymmetries;
- Turbulent & neoclassical entanglement & self-organization;
- Physics particle fuelling;
- Power exhaust;
- Edge-SOL coupling and SOL width;
- Fast particle physics.

### 9.7.3 Uragan-2 M and Uragan 3-M Torsatrons

The Ukraine Uragan-2M, summarized in Table 9.2, and Uragan-3M torsatrons/heliotrons [81] have carried out studies on Alfven wave heating and high-frequency wall conditioning.

Using Alfven heating at 9 MHZ in Uragan-3M, electron temperatures up to 0.5 keV at densities of up to $1.5 \times 10^{19} m^{-3}$ were achieved.

Since superconducting stellarators run at constant magnetic field, it was considered necessary to develop wall conditioning methods for steady-state fields [82]. The wave frequency was 140 MHz and produced a discharge with $10^{16}$ $m^{-3}$ density at 3–10 eV temperature, which was favourable for wall conditioning.

### 9.7.4 H-1NF Heliac

The H-1NF heliac had R = 1.0 m, a = 0.2 m and $B_T$ = 0.5 T [83]. Electron beam detection was used to map the magnetic surfaces which were accurate and closed.

Core magnetic islands and confinement were studied [84] in the vicinity of the core vacuum magnetic islands. Islands were observed to behave as improved confinement regions with a density peak near the island O-point.

### 9.7.5 HSX

HSX is a modular coil stellarator optimized for quasi-helical symmetry [17]. It was constructed with 48 modular coils which produced the quasi-helically symmetric (QHS) magnetic configuration. There are four straight sections connected into a square. There was almost no toroidal curvature in the QHS configuration and together with a high effective transform (~3), it produced small banana widths, low plasma currents and low neoclassical transport. There were also 48 planar coils which could alter the main configuration to degrade the QHS symmetry by increasing magnetic field ripple, viscous damping and neoclassical transport to a higher level. The main goals included investigation of transport, turbulence and confinement using ECRH heating. Results included:

- Electron temperatures of 2 keV were obtained with 100 kW of ECRH.
- Extensive equilibrium and plasma modelling was performed with V3FIT, VMEC, PENTA, GNET and SIESTA
- Understanding was improved of parallel flows, zonal flow, neoclassical transport, equilibrium reconstruction, bootstrap currents, density fluctuations, islands and impurity transport.

The mechanisms of toroidal flow during ECH were examined [85]. They evaluated the $J \times B$ and collisional forces and toroidal flows using the GNET code in order to clarify the mechanism of the toroidal flow changes in HSX and LHD. They observed that the obtained force by ECH was of the same order as the NBI force. Its direction agreed with experimental observations, indicating that the ECH force could drive the toroidal flow.

The HSX team in conjunction with Type One Energy have taken the HSX concept and extended it to a stellarator reactor design called STARBLAZER [86].

### 9.7.6 WEGA/HIDRA

The Hybrid Illinois Device for Research and Applications (HIDRA) was the former WEGA classical stellarator. HIDRA operated as both a stellarator and a tokamak. The machine is a five-period, $\ell = 2$, $m = 5$ stellarator, with major and minor radius 0.72 and 0.19 m [87]. Auxiliary heating was achieved with ECRH. Research involved dedicated studies on plasma-material interactions and magnetic fields measurements.

### 9.7.7 Scyllac

Fast linear theta pinches such as Scylla, which were invented by Fred Ribe, used a rapid compression of a linear field to heat a plasma. The confinement time was limited by plasma and heat loss via electron thermal conductivity and streaming out the ends. To combat this problem, the successor to the linear pinch was the toroidal high beta stellarator Scyllac [88]. Unlike other stellarators discussed in this chapter, Scyllac operated on the extremely short microsecond time scale.

The main result was the achievement of a beta of 70% held in equilibrium by a combination of $\ell = 0$, 1 and 2 stellarator fields plus rapid external feedback. Plasma densities up to $2 \times 10^{22}\ \mathrm{m}^{-3}$ were achieved at a temperature of 100 eV. Confinement was limited by end losses during experiments in a toroidal sector only. These experiments were not extended to a full torus.

### 9.7.8 CTH

A combination of plasma current and stellarator fields was explored in the Compact Toroidal Hybrid (CTH) designed to address 3D shaping effects on instabilities [89]. CTH had 7 independent control coils, including a single torsatron helical field coil wrapped around the vacuum vessel, plus a toroidal field coil, an ohmic transformer for plasma current, shaping field and vertical field pack. This coil set allowed the rotational transform to be changed over a very wide range from 0.02 up to 0.33.

The plasma current could provide up to 95% of the rotational transform in extreme cases, leading to observations that vertically unstable plasmas could be stabilized by the addition of a vacuum transform. Low q disruptions could be avoided with an additional vacuum transform [90].

### 9.7.9 NCSX

Princeton Plasma Physic Lab designed and began construction of the National Compact Stellarator Experiment (NCSX) [91, 92]. NCSX was intended to be a hybrid of stellarator and tokamak designs. NCSX would have used helias modular coils and a bootstrap current to augment the poloidal magnetic field, allowing it to be more compact than traditional stellarators and have similarities to tokamaks.

To avoid the major fragility of error fields in the plasma region, the highly twisted coils in their particular design required accuracy of 1/20,000th of an inch (1.27 μm, about 2 wavelengths of visible light) accuracy. These tight tolerances had to be maintained when cooling the copper conductor coils to liquid nitrogen temperatures. There were large cost and time overruns because of the extreme difficulties of meeting these tolerances.

Even though the project construction was 80% complete [92] and had spent over $70 million [93], it was cancelled in 2008 on the grounds that it could cost $170 million instead of its baseline cost of $102 million and would take at least until 2013 to complete. As recommended [93], work continued in the USA on smaller experiments so stellarator research opportunities [94] could be pursued. International collaborations involved participation at W-7X with substantial financial involvements.

### 9.7.10 Stellarator Robustness and Fragilities of Alternative Architectures

Several very different types of the stellarator family have been constructed and they have all been able to produce plasmas and do interesting experiments, with the very notable exception of NCSX which had construction tolerances much tighter than in other experiments. They all contributed to the stellarator knowledge base.

The overall fragility of these systems is difficulty of construction which is hard to quantify. Each design had its own challenges and some machines were built quickly and correctly.

The history of stellarator development has involved the identification of fragilities and the identification of mitigations to solve them, as follows:

- Error fields were mitigated by careful design, specification of tolerances and strict quality control.
- Particle losses were mitigated by an understanding of particle drifts and the design of appropriate quasi-symmetric field structures.
- Instabilities were mitigated by careful field design to control induced bootstrap currents.
- Power handling was controlled by controlling fields outside the separatrix to provide divertor structures.
- Access was provided to replace ohmic heating with external auxiliary heating.
- Experience from tokamaks was incorporated to establish internal transport barriers and H-mode confinement.
- The problem of interlocking stellarator coils was solved by the Helias configuration of combining toroidal and helical coils.

In summary, the stellarator family has lagged behind tokamaks due to the much higher complexity of field design which have produced many design weaknesses during development. However, advances in theoretical understanding and increased computational power have allowed stellarators to advance to a similar level of development as tokamaks.

### 9.7.11 Comparison of Stellarators and Tokamaks

A general comparison of the properties of tokamaks and stellarators had been made [95]. The advantages of tokamaks are:

- more technical simplicity;
- lower neoclassical transport (especially at high temperature);

- stronger toroidal rotation and associated flow-shear;
- weaker damping on zonal flows.

For stellarators, the advantages are:

- intrinsically steady-state operation;
- less MHD activity;
- nearly disruption-free;
- the stochastic magnetic boundary is beneficial for impurity retention in the divertor.

There are also some designs such as CTH which share the aspects of both systems. Characteristics which have advantages and disadvantages:

- Turbulence and turbulent transport are comparable in these two systems.
- Some drift-wave modes are more stable in stellarators.
- In the energy confinement scaling, an isotope effect appears in tokamaks but not in stellarators [96]. Hydrogen plasmas are similar to deuterium plasmas.
- As the number of degrees of the freedom is more for non-axisymmetric systems than axisymmetric ones [97], there are far more possible configurations in stellarators than in tokamaks giving flexibility but more difficulty in design.

## 9.8 Helias Fusion Reactor Concept HELIAS 5-B

### *9.8.1 Basic Design Considerations*

Over the years, there have been a wide range of stellarator fusion reactor designs [98] including ARIES-CS [99], a high field microsecond-scale pulsed approach [100] and a d-d catalyzed reactor [101].

Concentrating on extrapolation from the two most successful and large superconducting machines W-7X and the LHD, this section and the next examine two fusion concepts, the helias and heliotron.

The theoretical basis of a drift-optimized helias reactor has been developed [102]. It concluded that the helias concept developed at the IPP Garching and Griefswald, offered the chance of stable plasma confinement up to $\beta = 5\%$. Neo-classical transport losses would be below the critical level for ignition. This can be achieved by reducing the radial drift of trapped and circulating particles as far as possible. The theory of an optimized stellarator reactor mainly concentrates on the issues of plasma equilibrium, MHD-stability, neoclassical effects and fast particle behaviour. The main design feature was the optimization of the coil system as the main technical component of the reactor. The initial dimensions of a hypothetical helias reactor were R = 20 m, a = 1.6 m, B = 5 Tesla and rotational transform $5/6 < \iota < 1$.

The conditions for an optimized reactor configuration were specified as:

- Good magnetic surfaces of the vacuum field without major resonances, islands and stochastic regions. Islands or stochasticity in the boundary region may be utilized for divertor action.
- Low Shafranov shift and a stiff equilibrium configuration is needed up to $\beta = 5\%$. The rotational transform should depend only weakly on the plasma pressure.
- MHD stability should be up to $\beta = 5\%$.
- Sufficiently small neoclassical plasma losses are required so that the ignition condition is satisfied.
- Sufficiently small bootstrap currents should be obtained so that the magnetic field is not modified by this effect.
- Good confinement of $\alpha$-particles is essential.
- Good technical feasibility of the coil system is fundamental.
- Sufficiently large space is needed between coils and plasma for blanket and shield.

The geometrical form of the Helias configuration, the number of periods and rotational transform are consequences of these optimization criteria. In the method used to find an optimized helias configuration, a fixed plasma boundary is the starting point for computing the plasma equilibrium inside this boundary and the coil system outside. The plasma equilibrium in the domain is calculated using fixed boundary codes. The shape of the boundary and the pressure profile determine all properties of the equilibrium. By continuous variation of the boundary, the required optimum can be found.

To produce these fields, modular coils offer the possibility of building a reactor with coils of feasible size. The idea of modular coils [103] results from the fact that any toroidal vacuum field tangential to a surface can be produced by a current sheath on this surface. In designing the discrete coils, particular attention must be given to resonant perturbations which introduce islands on rational magnetic surfaces. In a helias configuration for a reactor, the equilibrium should be very stiff so that the profile of the rotational transform does not change with rising plasma pressure. The coil system should avoid those field errors which can resonate with the "lowest" rational magnetic surfaces in the plasma. Magnetic islands on high order rational surfaces are usually very small and therefore have little effect on plasma confinement. Only if many of these islands overlap, is enhanced plasma loss possible. Low shear in the Helias configuration is a possible remedy against this effect. The accuracy of a single coil seems to be of minor importance since deviations from the exact shape introduce high harmonics which either are statistically distributed and therefore are averaged out or they deviate systematically from the prescribed shape. If this deviation preserves the basic symmetry of the configuration this effect is of no concern. Thus the choice of field can have a major effect on the manufacturing tolerances required for the coils.

A key requirement is that fusion alpha particles should be well confined by the magnetic field. Orbits of the alpha particles in helias configurations were extensively studied using the Monte-Carlo technique. It was has shown that prompt losses were greatly reduced by the poloidal drift increasing with plasma pressure so that only a few percent of the fast particles are lost after one slowing-down time.

Another requirement is that the induced currents in the plasma should be as small as possible. As discussed above, the helias fields can be made quasi-symmetric to avoid these currents. These criteria formed the basis for proceeding with detailed reactor design.

### 9.8.2 *Updated Reactor Design*

The updated HELIAS 5-B fusion reactor concept [104], with major radius R = 22 m and minor radius a = 1.8 m, is an extension of the W-7 coil philosophy. It would produce 1.25 GW of electricity from 3 GW of thermal power, and is shown in Fig. 9.8 and its parameters are listed in Table 9.1.

The coil shapes were slightly changed from W7-X to a better configuration providing improved plasma confinement and more space for the blanket. A building block structure was adapted and optimized for better load distribution and reduced number of joints. The construction and maintenance strategy was to separate the whole torus for good access to the in-vessel components. A blanket maintenance concept was added, based on exchange of 400 large segments through five vertical ports.

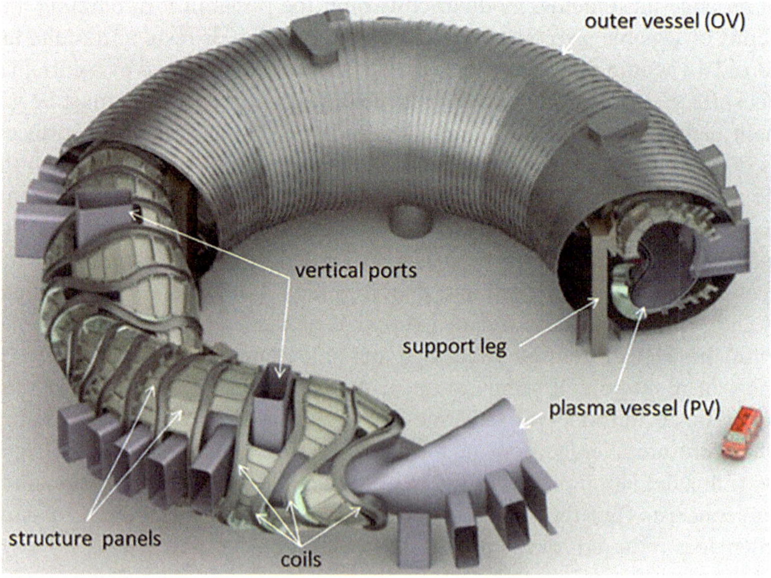


**Fig. 9.8** HELIAS-B5 overview. (From [104], Fig. 1)

The transition from the W7-X helias to a fusion reactor called HELIAS-5B was investigated along with the relevant engineering considerations, including the possibility of an intermediate machine [105, 106]. The main purpose of such a device would be to investigate burning plasma physics and to a limited extent the associated technologies, while the risk related to the extrapolation from W7-X results would have been kept at an appropriate level.

### *9.8.3 A Systems Approach to Optimization*

A systems approach to the design optimization of the five period HELIAS 5-B used the PROCESS code with modifications [107] to define the accessible reactor design window. In the multi-dimensional physics and engineering parameter space, sensitivity studies were carried out for optimizing different parameters and costs. Promising design points were analysed in more detail using the plasma operation contour approach to determine an optimum start-up path to ignition. Scans were performed to determine such quantities as optimum major radius and blanket and shield thickness. The optimum pathway to ignition required 55 MW of external heating to reach ignition.

The PROCESS framework allowed a direct comparison of tokamak and stellarator reactors, leading to similar construction costs.

### *9.8.4 Detailed Breeding Blanket Design*

A detailed study of the breeding blanket structure and options was made [108]. The Helium-Cooled Pebble Bed (HCPB) and the Water-Cooled Lithium Lead (WCLL) concepts which would be tested in ITER were considered for suitability of structural features to the stellarator geometry. Access and repair and replacement were examined to better orient the blanket segmentation. An assessment of the modules structural behaviour was performed. Numerical models were used to investigate the impact of vacuum vessel temperature and Young's modulus. It was found that vacuum vessel temperature and thermal expansion in the stellarator configuration could have major impacts on the design and the blanket attachment system.

A further neutronic design and assessment of the Dual Coolant Lithium-Lead DCLL breeding blanket, which had been studied in the EU DEMO tokamak design, was carried out [109]. Nuclear models and analyses of a test helias reactor were developed. Encouraging tritium breeding ratio results would allow important improvements on the lacking shielding performance of initial designs. The next steps will be to adapt DCLL for helias geometry.

### 9.8.5 Balance of Plant

A study was made of the balance of plant requirements for a helias power plant with a HCPB breeding blanket [110]. Cooling loops extracting heat from the blanket and heat exchangers were used to transfer the thermal power from the helium to solar salt, which transported the heat to an industrial thermal energy storage system. For electricity generation, the heat from the solar salt was transferred to feedwater which was boiled and superheated in the steam generator and then used in the turbine. A sensitivity study for the pressure drop of the cooling loops showed that small changes have significant impact on the pumping power.

A scenario was developed to demonstrate the capability of a stellarator fusion power plant operating flexibly in an envisaged dynamic future energy system. Considering demand forecasts and load following mechanisms, a fusion power plant seems well equipped to cope with fluctuating grid demands and could potentially replace fossil fuel power plants as a sustainable energy source.

### 9.8.6 Robustness and Fragilities

The main robustness of the system comes from the helias concept itself, which evolved from traditional stellarator design with continuous helical coils constructed and wound in place, in order to allow for modular superconducting coils that could be assembled, maintained or removed for repair. This concept has been extended from the successful W7-X experiments to a fusion reactor.

The main fragility remains the size and cost of stellarators and the difficulty in construction, which led to delays with W7-X, although many of the problems were due to moving to a new site and insolvency of the coil manufacturer. These fragilities can and have been overcome, since W7-X was successfully built with problems that might have occurred with any large project finally being resolved. There are also new methods for using correction coils and better coil design to minimize error fields and to relax tolerances in construction, which can have a huge influence on cost. The success of W7-X and other advances means that the helias is a real contender to become a successful fusion reactor.

## 9.9 Heliotron Fusion Reactor Concept FFHR-d1

### 9.9.1 Basic Design

The conceptual design of the heliotron reactor is an extension of the LHC coil design [73, 111] with the basic parameters shown in Table 9.1. Two cross sections half a toroidal period apart are shown in Fig. 9.9. A pair of $\ell = 2$ continuously-wound

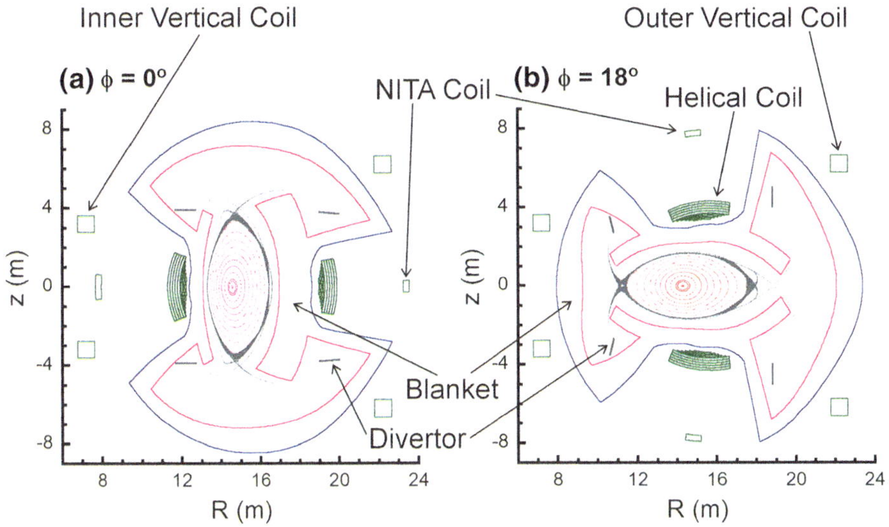


**Fig. 9.9** Heliotron FFHR-d1 fusion reactor vacuum magnetic field showing helical coils, vertical field coils and NITA coils plus blanket and divertor. (From [111], Fig. 2)

helical coils were designed with a toroidal pitch number of m = 10, along with two pairs of Outer Vertical and Inner Vertical coils, which were located above and below the helical coils to adjust the vertical field. The major radius of the helical coils was 15.6 m and the toroidal magnetic field was 4.7 T. In the standard operation scenario of FFHR-d1, the magnetic axis was shifted inward at R = 14.4 m to secure good energy confinement characteristics. A self-ignited fusion power of 3 GW (thermal) was calculated. The large cross section available for the helical divertor and the blanket around and behind it is also shown in Fig. 9.9, plus the "Newly Installed" Twist Adjustment (NITA) coils.

### *9.9.2 A Systems Approach to Reactor Design and Optimization*

The above design was taken as a base option for a full systems engineering optimization of the reactor called FFHR-d1A [111]. For FFHR-d1B (higher magnetic field version of FFHR-d1A), the operation point was explored using a design-integration systems code, HELIOSCOPE. This work followed on from earlier systems research [112] on FFHR-d1A which used the Direct Profile Extrapolation (DPE) method for predicting the plasma parameters based on the experimentally obtained data in LHD plasmas.

An integrated physics analysis using HELIOSCOPE determined a self-consistent operation point that assured the energy multiplication factor of Q = 10 which was also limited by MHD stability. The achievable beta was determined by the pressure driven instability using the Mercier stability criterion.

The variant FFHR-d1C was considered as a flexible design with configuration optimization. Slight modifications of the helical coils and vertical field coils provided many engineering benefits by using the shift-optimized (SO) configuration by adjusting the essential Fourier modes that define the outermost magnetic surface. This FFHR-d1C-SO configuration had favourable features of both inward and outward shifted configurations. The Shafranov outwards shift caused by an increase of the plasma beta became smaller in FFHR-d1A. Helical ripples were also smaller.

In the FFHR-d1A base case configuration, the distance between the innermost layer of the helical coil windings and the ergodic layer outside the last closed flux surface was found to be very marginal for a breeding blanket at 950 mm on the inboard side of the torus. This gap could be flexibly adjusted by employing a pair of sub-helical NITA coils located outside the main helical coils.

The latest design of the FFHR-d1C-SO configuration also employed a pair of NITA coils. The vacuum magnetic configuration of FFHR-d1CSO is similar to that shown in Fig. 9.9. The rotational transform was lower than 0.5 at the plasma core, although its impact on the emergence of a magnetic island was limited.

### 9.9.3 Use of Joints and High Temperature Superconductors

Very extensive research work and industrial testing was performed on the use of high temperature superconductor REBCO and the use of joints for connecting sections of superconducting coils [111]. REBCO provides a means of obtaining very high magnetic field at the coils and of solving the problem of assembling and separating helical coils. The entire process of coil winding, insulation, coil forms, stabilizer, joints, fabrication and assembly were considered in careful detail and feasible solutions were proposed. This technology was also adapted for the high field tokamak reactor proposal ARC [113] by MIT and Commonwealth Fusion.

### 9.9.4 Helical Divertor

The helical divertor inherent in the heliotron geometry had a number of natural advantages, but also important problems that remain to be fully resolved. The very large volume afforded by the field lines gave a low average heat flux. Also, the strike points were behind the helical coils and could be shielded from direct neutron damage. However, the heat flux is non-uniform in the divertor region. With a strike zone only 80 mm wide, the peak heat flux could reach 100 MW/m$^2$, instead of the acceptable averaged power of 8 MW/m$^2$. There are several proposed methods to reduce the heat flux.

Sweeping the vertical field has been successfully tested in LHD, but it also alters the plasma equilibrium which may have deleterious effects on confinement of plasma and of alpha particles. There are also problems with rapid sweeping of superconducting coils. Another method of sweeping was to introduce helical divertor coils to sweep the location of the strike zone, reducing peak loading.

Yet another method was to use Resonance Magnetic Perturbation (RMP) coils to spread the heat load. This approach which could work in steady state used 20 RMP coils similar to the correction coils installed in LHD called the Local Island Divertor coils. The calculated peaking was reduced to about 1/3rd. A further reduction could be obtained by radiation dispersion.

The possibility of a liquid divertor consisting of 10 units producing shower jets of liquid tin was considered, although problems remain.

Detailed calculations were performed to explore the very detailed field structure and motion of particles in these fields. It was concluded that the initial estimates of high concentrations of heat flux were too large, and that the problem is much more tractable than was thought.

### *9.9.5 A Liquid salt Breeding Blanket*

A liquid salt breeding blanket (LSBB) using FLiNaBe with melting point of 580 K was designed for tritium breeding. To increase hydrogen isotope solubility, it was proposed to add titanium powders. The "Oroshhi-2" facility was constructed to test these ideas in the presence of a strong magnetic field. Both liquid metal and cartridge blankets were under consideration. Access and construction solutions needed to be found for blanket and divertor.

### *9.9.6 Robustness and Fragilities*

The heliotron reactor designs FFHR-d1 and variants are highly robust due to the successful operation of the LHD which has confirmed the viability and reliability of such a reactor. Similarly to high-field tokamak reactors, high tech solutions such as REBCO with joints and liquid salt filled breeding blanket tanks have been proposed to solve the problems associated with continuous helical windings.

Fragilities are still associated with all the complexities of assembling and maintaining a complicated 3-D geometry, but there are reasonable solutions to all these problems, making the heliotron an attractive concept for a fusion reactor.

### 9.9.7 A Helical Volumetric Neutron Source FFHR-b2

An alternative device to FFHR-d1 was proposed as the smaller FFHR-b2 helical volumetric neutron source [114]. The FFHR-b2 would be relatively small and provide component testing for a helical fusion reactor. New technologies would be HTS magnets, ceramic pebble divertor, and cartridge-type liquid metal blankets operated in reactor conditions. They concluded that $Q \geq 1$ could be achieved in such a machine, producing a neutron flux sufficient to test components.

## 9.10 Prospects for Stellarators

Thanks to the success of the superconducting W7-X and LHD, the Helias and the Heliotron concepts have been established as viable prototypes for successful fusion reactors. The designs of HELIAS-5B and FFHR-d1 demonstrate that viable reactor concepts exist. Systems approaches are optimizing these designs.

The real strength of the stellarator family concept now is based on the high level of understanding of the magnetic field structure and how to optimize it. The concept of quasi-symmetry and the computer codes able to optimize fields to minimize particle trapping and losses, and to reduce energy transport to a minimum, makes the concept highly attractive. The absence of plasma currents and the avoidance of plasma disruptions is a major factor in favour of these intrinsically steady state devices. In addition, the ability to avoid the power losses associated with current drive in the tokamak may yet prove to be a decisive advantage.

## References

1. E. Morse, *Nuclear Fusion* (Springer, Cham, 2018)
2. G. McCracken, P. Stott, *Fusion: The Energy of the Universe*, 2nd edn. (Academic/Elsevier, Waltham, 2013)
3. A.A. Harms et al., *Principles of Fusion Energy* (World Scientific, Singapore, 2000)
4. J.L. Johnson, The stellarator approach to toroidal plasma confinement. Nucl. Technol. Fusion **2**(3), 340 (1982)
5. K. Miyamoto, *Plasma Physics and Controlled Nuclear Fusion* (Springer, Berlin/Heidelberg, 2005)
6. Shohet, Overview of the torsatron/stellarator laboratory program at the University of Wisconsin – Madison, Stellarator Physics, report IAEA-TECDOC-558. Proceedings of the 7th International Workshop on Stellarators, Oak Ridge, p. 56 (1989)
7. A.H. Boozer, What is a stellarator? Phys. Plasmas **5**, 1647 (1998)
8. D.V. Bartlett et al., Plasma confinement in stellarator systems. Philos. Trans. R. Soc. London **A300**, 579 (1981)
9. Magnetic field stellarator, Wikimedia commons, the free media repository, from principle of the stellartor, European Nuclear Society. http://www.euronuclear.org/info/encyclopedia/s/stellarator.htm. 5 Dec 2011

10. C.M. Braams, P.E. Stott, *Nuclear Fusion: Half a Century of Magnetic Confinement Fusion Research* (Institute of Physics Publishing, Bristol, 2002)
11. L. Spitzer, The stellarator concept. 2nd United Nations Conference on the Peaceful Uses of Atomic Energy, Geneva, Switzerland (1958)
12. G. Grieger, H. Renner, H. Wobig, Wendelstein stellarators. Nucl. Fusion. **25**(9), 1231 (1985)
13. D.J. Lees, Culham stellarator programme, 1965–1980. Nucl. Fusion **25**(9), 1259 (1985)
14. G. Grieger et al., Physics optimization of stellarators. Phys. Fluids B Plasma Phys. **4**, 2081 (1992)
15. A.H. Boozer, Establishment of magnetic coordinates for a given magnetic field. Phys. Fluids **25**, 520 (1982)
16. P. Helander, J.E. Geiger, H. Maaßberg, On the bootstrap current in stellarators and tokamaks. Phys. Plasmas **18**(9), 092505 (2011)
17. K.M. Likin et al., Overview of HSX stellarator, 52ndAPS-DPP Annual Meeting, November, 8–12, 2010, Chicago, IL (2010)
18. S.A. Henneberg et al., Properties of a new quasi-axisymmetric configuration. Nucl. Fusion **59**(2), 026014 (2019)
19. D. Palumbo, Some considerations on closed configurations of magnetohydrostatic equilibrium. Nuovo Cimento B. **53**, 507 (1968)
20. J. Nührenberg, R. Zille, Stable stellarators with medium β and aspect ratio. Phys. Lett. A **114**(3), 129 (1986)
21. F. Bauer et al., *The Beta Equilibrium, Stability and Transport Codes – Applications to the Design of Stellarators, Perspectives in Physics* (Academic, Boston, 1987)
22. S. Hirshman, W. van Rij, P. Merkel, Three-dimensional free boundary calculations using a spectral Green's function method. Comput. Phys. Commun. **43**(1), 143 (1986)
23. M. Drevlak, D. Monticello, A. Reiman, PIES free boundary stellarator equilibria with improved initial conditions. Nucl. Fusion **45**(7), 731 (2005)
24. J. Loizu et al., Equilibrium β-limits in classical stellarators. J. Plasma Phys. **83**(6), 715830601 (2017)
25. C. Nührenberg, Computational ideal MHD: Alfvén, sound and fast global modes in W7-AS. Plasma Phys. Control. Fusion **41**(9), 1055 (1999)
26. H. Yamada et al., Characterization of energy confinement in net-current free plasmas using the extended international stellarator database. Nucl. Fusion **45**(12), 1684 (2005)
27. S. Sudo et al., Scalings of energy confinement and density limit in stellarator/heliotron devices. Nucl. Fusion **30**(1), 11 (1990)
28. L. Giannone et al., Physics of the density limit in the W7-AS stellarator. Plasma Phys. Control. Fusion **42**(6), 603 (2000)
29. C.G. Albert, S.V. Kasilov, W. Kernbichler, Accelerated methods for direct computation of fusion alpha particle losses within, stellarator optimization. J. Plasma Phys. **86**, 85860201 (2020)
30. C.D. Beidler et al., Results from the international collaboration on neoclassical transport in stellarators (ICNTS). Paper TH/P8-10, Proceedings of the 22nd IAEA Fusion Energy Conference, Geneva, 13–18 October 2008
31. S.A. Henneberg, Designing the optimal fusion experiment. Greifswald. https://indico.mpp.mpg.de/event/6293/attachments/14745/16649/final_Henneberg.pdf. 29 Oct: IPP-MPG (2019)
32. M. Landreman, P.J. Catto, Geometry and neoclassical theory in a quasi-isodynamic stellarator. arXiv:1011.5184v1 [physics.plasm-ph], 23 Nov (2010)
33. P. Helander, Theory of plasma confinement in non-axisymmetric magnetic fields (review article). Rep. Prog. Phys. **77**, 087001 (2014)
34. C. Zhu et al., Identification of important error fields in stellarators using the Hessian matrix method. Nucl. Fusion **59**(12), 126007 (2019)
35. C. Zhu et al., New method to design stellarator coils without the winding surface. Nucl. Fusion **58**(1), 016008 (2018)

36. F. Wagner et al., Major results from Wendelstein 7-AS stellarator. IAEA Fusion Energy Conference, Paper OV/2–4, Lyon, France (2002)
37. M. Hirsch et al., Major results from the stellarator Wendelstein 7-AS. Plasma Phys. Control. Fusion **50**(5), 053001 (2008)
38. K. McCormick et al., Island divertor experiments on the Wendelstein 7-AS stellarator. J. Nucl. Mater. **313–316**, 1131 (2003)
39. A. Weller et al., Investigation of the beta-limit in the W7-AS stellarator. 19th IAEA Fusion Energy Conference, Paper EX/S3-1, Lyon, France (2002)
40. M.C. Zarnstorff et al., Equilibrium and stability of high-beta plasmas in Wendelstein 7-AS. IAEA Fusion Energy Conference, Vilamoura, Portugal, Paper EX/3-4 (2004)
41. T. T. Klinger on behalf of the enterprise Wendelstein 7-X, Stellarators difficult to build? The construction of Wendelstein 7-X, ITER Seminar, 14 April Cadarache, France https://www.iter.org/doc/www/content/com/Lists/Stories/Attachments/680/ITER_W7X.pdf (2011)
42. D. Clery, The bizarre reactor that might save nuclear fusion, Science Magazine. https://www.sciencemag.org/news/2015/10/bizarre-reactor-might-save-nuclear-fusion. 21 October (2015)
43. T. Klinger et al., Overview of first Wendelstein 7-X high performance operation. Nucl. Fusion **59**, 112004 (2019)
44. G. Grieger et al., Physics studies for helical-axis advanced stellarators. IAEA Fusion Energy Conference, vol. 2, p. 369, Nice, France (1988)
45. G. Beidler et al., Physics and engineering design for Wendelstein VII-X. IAEA 7th International Workshop on Stellarators, Oak Ridge, Tennessee, p. 479, 10–14 Apr (1989)
46. F. Wagner, Stellarators, Private Communication (2021)
47. M. Wanner et al., Status of Wendelstein 7-X construction. Nucl. Fusion **43**(6), 416 (2003)
48. H.-S. Bosch et al., Final integration, commissioning and start of the Wendelstein 7-X stellarator operation. Nucl. Fusion **57**, 116015 (2017)
49. K. Risse et al., First operational phase of the superconducting magnet system of Wendelstein 7-X. Fusion Eng. Des. **124**, 10 (2017)
50. F. Füllenbach et al., The Wendelstein 7-X trim coil system commissioning and first operational results. Fusion Eng. Des. **124**, 94 (2017)
51. R.C. Wolf et al., Performance of Wendelstein 7-X stellarator plasmas during the first divertor operation phase. Phys. Plasmas **26**, 082504 (2019)
52. C.D. Beidler et al., Demonstration of reduced neoclassical energy transport in Wendelstein 7-X. Nature **596**, 221 (2021)
53. T.S. Pedersen et al., Experimental confirmation of efficient island divertor operation and successful neoclassical transport optimization in Wendelstein 7-X. 28th IAEA Fusion Energy Conference, Nice, France (virtual) 10–15 May 2021, presentation OV/3-2, paper IAEA-CN-286-1153, submitted to Nuclear Fusion (2021)
54. M. Jakubowski et al., Overview of the results from the divertor experiments at Wendelstein 7-X and their implications for steady state operation. 28th IAEA Fusion Energy Conference, Nice, France (Virtual) 10–15 May 2021, Presentation EX/7-4, Paper IAEA-CN-286-0998, Submitted to Nuclear Fusion (2021)
55. K. Uo et al., Recent results on heliotron E. Nucl. Instrum. Methods Phys. Res. **207**(1–2), 151 (1983)
56. A. Iiyoshi, M. Wakatani, Heliotron E results and large helical project. J. Nucl. Sci. Technol. **26**(1), 204 (1989)
57. M. Yokoyama, Y. Nakamura, M. Wakatani, An optimized helical axis stellarator with modulated $\ell$=1 helical coil. Plasma Phys. Fusion Res. **73**, 723 (1997)
58. M. Yokoyama et al., Roles of bumpy field for collisionless particle confinement in helical axis heliotrons. J. Plasma Fusion Res. Ser. **1**, 445 (1998)
59. S. Fumimichi et al., Experimental program of Heliotron J. J. Plasma Fusion Res. Ser. **3**, 26 (2000)
60. K. Nagasaki, Recent results from the Heliotron J Experiment. Hangzhou international stellarator workshop, March 26th–28th, 2018, Zhejiang University, Hangzhou, China (2018)

61. S. Ohshima et al., Turbulent properties against hydrogen isotope ratio and zonal flow activities in Heliotron J. 28th IAEA Fusion Energy Conference, Nice, France (Virtual) 10–15 May 2021, Paper IAEA-CN-286-0790, Submitted Nuclear Fusion (2021)
62. K. Nagasaki et al., Effect of magnetic configuration on energy confinement and energeticparticle-driven MHD modes in Heliotron J. 28th IAEA Fusion Energy Conference, Nice, France (Virtual) 10–15 May 2021, Presentation EX/P6–34, Paper IAEA-CN-286-0740, Submitted to Nuclear Fusion (2021)
63. Large Helical Device, Wikipedia, https://en.wikipedia.org/wiki/Large_Helical_Device, Creative Commons 2.0 licence. Last accessed 17 June 2022
64. G. Motojima, Recent design considerations for the large helical device. IAEA 7th International Workshop on Stellarators, 10–14 April 1989, Oak Ridge, Tennessee, Report IAEA-TECDOC-558, 471 (1989)
65. A. Iiyoshi, Recent status of the large helical device project. IAEA 7th international workshop on stellarators, 10–14 April 1989, Oak Ridge, Tennessee, Report IAEA-TECDOC-558, 467 (1989)
66. M. Wakatani, Y. Nakamura, K. Hanatani, Theoretical study of the large helical device. IAEA 7th International Workshop on Stellarators, 10–14 April 1989, Oak Ridge, Tennessee, Report IAEA-TECDOC-558, 475 (1989)
67. Large Helical Device Project, National Institute for Fusion Science (NIFS) Annual Report, Toki, Japan, April 2017–March 2018 (2018)
68. S. Kido et al., Hitachi's activities in fusion device and particle accelerator development. Hitachi Rev. **58**(2), 113 (2009)
69. M. Fujiwara et al., Overview of LHD experiments, 18th Fusion Energy Conference, Sorrento, Italy, Paper OV1/4, p. 17 (1990)
70. H. Yamada et al., Impact of heat deposition profile on global confinement of NBI heated plasmas in the LHD. Nucl. Fusion **43**(8), 749 (2003)
71. O. Kaneko et al., Extension of improved particle and energy confinement regime in the core of LHD plasma. Plasma Fusion Res. Regul. Articles **4**(027), 1 (2009)
72. Y. Suzuki, K.Y. Watanabe, S. Sakakibara, Theoretical studies of equilibrium beta limit in LHD plasmas. Phys. Plasmas **27**, 102502 (2020)
73. Y. Takeiri, The large helical device: entering deuterium experiment phase toward steady-state helical fusion reactor based on achievements in hydrogen experiment phase. IEEE Trans Plasma Sci. **46**(7), 2348 (2018)
74. M. Isobe et al., Neutron diagnostics in the large helical device. IEEE Trans. Plasma Sci. **46**(6), 2050 (2018)
75. K. Ogawa, M. Isobe, M. Osakabe, Progress on integrated neutron diagnostics for deuterium plasma experiments and energetic particle confinement studies in the Large Helical Device during the campaigns from FY2017 to FY2019. Plasma Fusion Res. Rev. Articles **16**, 1102023 (2021)
76. M. Kobayashi et al., RMP induced H-mode transition during divertor detachment with enhanced edge radiation in deuterium plasmas in LHD. 28th IAEA Fusion Energy Conference, Nice, France (Virtual) 10–15 May 2021, Presentation EX/2-4, paper IAEA-CN-286-0837, Submitted to Nuclear Fusion (2021)
77. C. Alejaldre et al., First plasmas in the TJ-II flexible Heliac. Plasma Phys. Control. Fusion **41**(3A), A539 (1999)
78. T. Estrada et al., Sheared flows and transition to improved confinement regime in the TJ-II stellarator. Plasma Phys. Control. Fusion **51**, 124015 (2009)
79. E. Ascasibar et al., Overview of recent TJ-II stellarator results. Nucl. Fusion. **59**, 112019 (2019)
80. C. Hidalgo, Overview of the TJ-II stellarator research programme towards model validation in fusion plasmas. 28th IAEA Fusion Energy Conference, Nice, France (Virtual) 10–15 May 2021, Paper IAEA-CN-286-0969, Submitted to Nuclear Fusion (2021)

81. V.E. Moiseenko et al., RF plasma production and heating below ion-cyclotron frequencies in Uragan torsatrons. Nucl. Fusion **51**, 083036 (2011)
82. V.E. Moiseenko et al., VHF discharges for wall conditioning at the Uragan-2M torsatron. Nucl. Fusion **54**, 033009 (2014)
83. M.G. Shats et al., Experimental investigation of the magnetic structure in the H-1 Heliac. Nucl. Fusion **34**(12), 1653 (1994)
84. T.A. Santhosh et al., Core magnetic islands and plasma confinement in the H-1NF heliac. Phys. Plasmas **17**, 082503 (2010)
85. Y. Yamamoto et al., Mechanism of toroidal flow generation by electron cyclotron heating in HSX and LHD plasmas. 28th IAEA Fusion Energy Conference, Nice, France (Virtual) 10–15 May 2021, Presentation TH/P5-19, Paper IAEA-CN-286-0824, Submitted to Nuclear Fusion (2021)
86. Type One Energy, Website https://www.typeoneenergy.com/. Accessed 17 June 2022
87. R. Rizkallah et al., Latest results from the Hybrid Illinois Device for Research and Applications (HIDRA). IEEE Trans. Plasma Sci. **46**(7), 2685 (2018)
88. R.E. Siemon, A summary of Scyllac results, Los Alamos Scientific Laboratory report LA-7125-MS, Los Alamos New Mexico (1978)
89. G. Hartwell, The compact toroidal hybrid, a university scale fusion experiment, plasma physics workshop, SMF-PPD, Universidad National Autónoma México, October 12–14 (2016)
90. D. Maurer, Plasma disruption avoidance using non-axi-symmetric shaping with stellarator fields. ISHW, Greifswald, Germany, 8 October (2015)
91. B.E. Nelson et al., Design of the national compact stellarator experiment (NCSX). Fusion Eng. Des. **66-68**, 169 (2003)
92. NCXS Factsheet, National Compact Stellarator Experiment, Princeton Plasma Physics Laboratory. https://www.pppl.gov/sites/pppl/files/publication/file/NCSX.pdf. Last accessed 12 Mar 2021
93. R.L. Orbach, Future of the Princeton Plasma Physics Laboratory, US DOE Announcement, May 22, 2008. https://ncsx.pppl.gov//DOE_NCSX_052208.pdf. Last accessed 17 June 2022
94. D.A. Gates et al., Stellarator research opportunities: a report of the National Stellarator Coordinating Committee. J. Fusion Energy **37**, 51 (2018)
95. Y. Xu, A general comparison between tokamak and stellarator plasmas. Matter Radiat. Extremes. **1**, 192 (2016)
96. H. Yamada et al., Isotope effect on energy confinement time and thermal transport in neutral-beam heated stellarator-heliotron plasmas. Phys. Rev. Lett. **123**, 185001 (2019)
97. A.H. Boozer, Physics of magnetically confined plasmas. Rev. Mod. Phys. **76**, 1071 (2005)
98. C.D. Beidler et al., Stellarator fusion reactors – an overview. J. Plasma Fusion Res. Ser. **5**, 149 (2002)
99. J.F. Lyon et al., Systems studies and optimization of the ARIES-CS power plant. J. Vac. Sci. Technol. B. **54**, 694 (2008)
100. V. Queral et al., Initial exploration of high-field pulsed stellarator approach to ignition experiments. J. Fusion Energy **37**, 275 (2018)
101. J. Sheffield, D.A. Spong, Catalyzed D-D stellarator reactor. Fusion Sci. Technol. **70**, 36 (2016)
102. H. Wobig, The theoretical basis of a drift-optimized stellarator reactor. Plasma Phys. Control. Fusion **35**, 903 (1993)
103. S. Rehker, H. Wobig, Stellarator coil system without helical windings. 345A Proceedings of the symposium on fusion technology, Grenoble (1972)
104. F. Schauer, K. Egorov, V. Bykov, HELIAS 5-B magnet system structure and maintenance concept. Fusion Eng. Des. **88**(9–10), 1619 (2013)
105. F. Warmer et al., From W7-X to a HELIAS fusion power plant: motivation and options for an intermediate-step burning-plasma stellarator. Plasma Phys. Control. Fusion **58**, 074006 (2016)

106. F. Warmer et al., From W7-X to a HELIAS fusion power plant: on engineering considerations for next-step stellarator devices. Fusion Eng. Des. **123**, 47–53 (2017)
107. F. Warmer et al., System code analysis of HELIAS-type fusion reactor and economic comparison with tokamaks. IEEE Trans. Plasma Sci. **44**(9), 1576 (2016)
108. G. Bongioví, A. Häußler, et al., Advancements in the HELIAS-5B breeding blanket structural analysis. Fusion Eng. Des. **161**, 111928 (2020)
109. I. Palermo et al., Neutronic design and assessments of a DCLL BB: adaptation from DEMO tokamak to HELIAS stellarator. 28th IAEA Fusion Energy Conference, Nice, France. (Virtual) 10–15 May 2021, Presentation TH/P8-22, Paper IAEA-CN-286-0986, Submitted to Nuclear Fusion (2021)
110. F. Warmer, E. Bubelis, First considerations on the balance of plant for a HELIAS fusion power plant. Fusion Eng. Des. **146 B**, 2259 (2019)
111. N. Yanagi et al., Progress in the conceptual design of the helical fusion reactor FFHR-d1. J. Fusion Energy **38**, 147 (2019)
112. T. Goto et al., Design window analysis for the helical DEMO reactor FFHR-d1. Plasma Fusion Res. **7**, 2405084 (2012)
113. B.N. Sorbom et al., The engineering design of ARC: a compact, high field, fusion nuclear science facility and demonstration power plant. IEEE Symposium of Fusion Energy, Austin, Texas (2015)
114. J. Miyazawa et al., Conceptual design of the helical volumetric neutron source FFHR-b2. 28th IAEA Fusion Energy Conference, Nice, France. (Virtual) 10–15 May 2021, Presentation TECH/P8-20, Paper IAEA-CN-286-0717, Submitted to Nuclear Fusion (2021)

# Chapter 10
# Linear Magnetic Traps, Field Reversal and Taylor-State Configurations

**Abstract** This chapter examines linear magnetic traps as potential fusion reactor candidates. A case study of systems forward engineering was made of the proto-reactor scale MFTF-B tandem mirror experiment. It went through many design and construction changes, was built and never operated. A robustness and fragility analysis is performed. A design space is developed for the systems architecture of linear machines. Recent advances in tandem and multiple mirrors with gas boxes and field expanders have renewed reactor prospects. Toroidal mirror concepts are briefly examined. Configurations produced in a linear geometry such as Taylor state conserved helicity machines are analysed. The reversed field pinch provides another possibility for a fusion reactor, for which a design space is discussed.

## 10.1 Linear Mirror Systems

### *10.1.1 Basic Magnetic Mirror Machine*

Mirror machines were invented by Post and Budker and their physics was developed over the years [1, 2]. Mirror experiments were among the first fusion concepts and offered many economic and construction advantages compared to toroidal devices. The basic concept is to use coils with a common linear magnetic axis to create a magnetic field which is stronger at the ends and weaker between, as shown in Fig. 10.1.

Fast particles which are injected at right angles to the field between the mirrors are ionized and are trapped between high field ends. Confinement is due to conservation of total particle energy and the magnetic moment, which is proportional to the perpendicular energy divided by the magnetic field. As the particle moves into the high field mirror region, all of its energy becomes perpendicular and it is reflected. The ions are confined until scattering with other particles causes their parallel energy to increase which allows them to escape through the mirrors.

Early mirror machines were also found to be subject to instabilities which prevented density build-up. One was the DCX-2 [3], which had an overall length of 8.2 m with 2.65 m between mirrors, which produced 3.9 T and a ratio of mirror to



F. B. Marcus, *Systems Approaches to Nuclear Fusion Reactors*, Springer Series in Plasma Science and Technology, https://doi.org/10.1007/978-3-031-17711-8_10

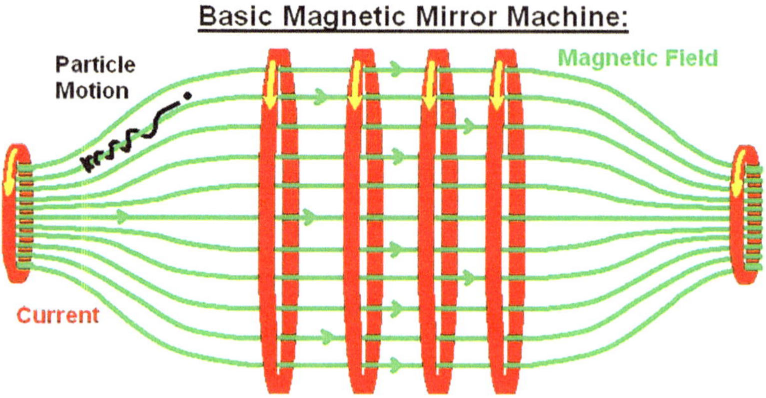


**Fig. 10.1** Basic magnetic mirror machine. (From [4])

central field of 3.3. A 600 keV molecular ion beam was injected, with beam trapping by ionization on a central arc. Plasma densities of up to $10^{16}$ $m^{-3}$ were obtained.

The electron density and temperature play an important role. Ions are confined by the mirror effect, but electrons are injected with the ions at much lower energy and scatter rapidly before much heating can occur. Hence, they would tend to escape along field lines and carry out heat transferred from the ions. To maintain plasma electrical neutrality, a large electric potential builds up along the field lines which retains electrons, usually at a few times the electron temperature in electron-volts. The potential then draws ions further out along the field lines by increasing their parallel energy. A balance is reached where the electron temperature is a significant fraction of the ion temperature. Any plasma streaming out the ends therefore represents a loss of both electron and ion energy. Equal numbers of ions and electrons stream out together, their energies altered by the electric potential.

## *10.1.2 Minimum-B Magnetic Wells*

### 10.1.2.1 Phoenix II

To suppress instabilities, mirrors increased in complexity by adding multipole coils to create a minimum magnetic field well. The Phoenix II minimum-B mirror machine used a superposition of a linear quadrupole cusp field with a simple mirror. Plasma was produced by magnetic field Lorentz trapping of injected neutral hydrogen atoms in an excited state [5]. The central magnetic field was 1.25 T. The magnetic well allowed higher densities than in a simple mirror, and there was evidence of exponential build-up of the plasma density as plasma trapping augmented the Lorentz trapping. However, the maximum density achieved was only

$4 \times 10^{15}$ m$^{-3}$, limited by the onset of ion cyclotron instabilities driven by the non-uniform trapped fast ion velocity distribution, which drove fast ion losses.

#### 10.1.2.2 Baseball II

The quadrupole magnetic well with a depth of 2:1 was generated in Baseball II by a superconducting baseball-seam shaped winding, which combined the types of coils used in Phoenix II into a single coil. The central field was 2 T and the plasma was created by neutral beam injection [6].

#### 10.1.2.3 2XIIB

The magnetic well 2XIIB mirror experiment had a central field of 0.67 T, a 2/1 mirror ratio, distance between mirrors 1.5 m and neutral beam injection with up to 370 A of deuterium at 19 keV, i.e. 7 MW [7]. Other features were slow compression of the target plasma and the use of a streaming plasma for stabilization, which was produced by a washer gun. The following results were achieved:

- Ion temperatures up to 13 keV by neutral beam injection;
- Plasma beta up to 40%;
- Electron heating from neutral beam injection;
- Electron density up to $2 \times 10^{19}$ m$^{-3;}$
- Reduction in ion cyclotron instability and noise with streaming plasma;
- Lawson criteria $n\tau$ increased with beam energy.

#### 10.1.2.4 Elmo

ELMO had a vacuum field that was a simple 2/1 magnetic mirror, but used electron cyclotron heating to create electron rings with beta of several tens of percents which were able to create a minimum in the magnetic field [8]. Full plasma stability required a certain amount of cold plasma which produced 'line-tying' stabilization to conducting end plates which could short out radial electric fields. The high beta rings provided a negative electrostatic potential compared to the positive one from hot ion rings.

### *10.1.3 Direct Conversion and Reactor Systems Analysis*

Even with ideal confinement, classical particle scattering out the ends of a mirror machine was too rapid, as calculated with the Fokker-Planck equation. The resulting energy gain Q was below one. Concepts were explored to recycle this lost power out of the mirrors, such as direct conversion of the energy of escaping charged particles to electricity.

One method [9] involved expanding the escaping fast ions over a fan with a 100 m radius to first magnetically separate the electrons and then decelerate the ions to collect each batch of different energies at maximum efficiency.

This method was investigated theoretically and experimentally [10, 11]. It was found that the space charge of the ions would rapidly increase the beam width as the escaping ion current increased. This could be countered by an alternating potential electrostatic field that also caused deceleration. This field was represented by a strong-focusing quasi-potential which allowed the calculation of the space-charge limited conversion efficiency. This efficiency increased with beam energy and decreased with exhaust current. Experiments were also carried out at Livermore [12, 13]. However, at very high ion temperatures in a mirror machine, more power transfer to the electrons would result in increased cyclotron/synchrotron radiation loss [14].

Detailed systems studies were made of incorporating direct conversion into a mirror fusion reactor system [15]. A model of a mirror reactor system was developed for minimizing systems costs with respect to the mirror ratio, magnetic field and injection energy. A d-$^3$He reactor was found to be considerably more expensive than d-t systems at that time, partly because of the poor neutralization efficiency of very high energy positive ion beams. There was a large cost penalty in using very high energy beams, several hundred keV, when direct conversion schemes become practicable.

### 10.1.4 Tandem Mirror TMX and TMX-Upgrade

The concept of the tandem mirror was that good confinement of a hot plasma in a central simple straight solenoid could be achieved by plugs at both ends with positive electrostatic potential. A high-beta hot-ion plasma in the end plugs both stabilized the entire tandem mirror and provided electrostatic confinement.

The Tandem Mirror Experiment (TMX) was proposed by Fred Coensgen [16] 1 year after the very large and superconducting MFTF was also proposed by him [17]. TMX [18] produced a plasma with NBI giving 40% beta in the central solenoid, with density of $3 \times 10^{19}\ m^{-3}$ and ion temperature of 0.25 keV in a magnetic field of 0.1 T. A certain amount of warm plasma was necessary to stabilize the plugs which came from the high density central cell. The plasma potential was positive in the central solenoid, but even more positive in the end plugs, thus confining ions inside the plugs and confining electrons outside the plugs. A detailed theoretical study of tandem mirrors was developed [19].

As a result of this success, TMX was upgraded to TMX-U, with larger magnets and higher fields. Auxiliary heating included ECRH and ICRH as well as high power NBI, with the purpose of creating and studying transport barriers [20]. The linear machine was inside a vacuum vessel 21.3 m long and consisted of an 8 m long central cell, two end cells 3 m long each, and rectangular shaped end fan tanks, using water cooled copper coils. The magnetic fields were 0.3 T in the central solenoid, 0.5 T at the minimum field in the end cells, and 2 T in the two mirror fields of each end cell. The field expander reduced the fields outside the confinement region to 0.01 T before hitting the conducting end walls, which could be electrically floated or grounded. The end plugs each had 6 neutral beam injectors to produce sloshing ion distributions. Another 6 injectors were used for barrier pumping. Ions were heated by ICRH in the central solenoid and ECRH was applied in the end plugs.

Thermal barrier experiments in TMX-U [21] showed that transport processes and end losses can be controlled at modest densities by using central cell heating by NBI and ICRF. Plasma potential measurement showed that deep thermal barriers were established. However, there were implied problems at high density.

Plasma was maintained in an axisymmetric region of TMX-U only by gas fuelling and neutral beam heating and was stabilized [22] by line tying through the endplates and finite Larmor radius edge stabilization.

### 10.1.5 Tandem Mirror PHAEDRUS-B

Application of ICRF in the Phaedrus-B tandem mirror end cells resulted in increases in central cell plasma density by a factor of three [23]. Radial potential trapping of ions in the plasma caused the increase in radial confinement.

### 10.1.6 Tandem Mirror GAMMA 10

The symmetric tandem mirror GAMMA 10 was constructed with a central cell, anchor cells, and mirror cells. It was already producing impressive results [24] in 1985 at the same time as TMX-U. The anchor cells used minimum-B mirrors to suppress MHD instabilities. Plugs and thermal barriers were produced in the axisymmetric mirrors. End plugging experiments were carried out by a combination of neutral beams and gyrotrons. The configuration of Gamma-10 is shown in Fig. 10.2 [25]. Thermal barriers were formed, giving electron temperatures of 0.8 keV.

Direct conversion to electricity of thermal ion energy from escaping particles was designed for Gamma-10 [26]. The direct convertor consisted of a cusp magnetic field and two stage decelerating electrodes. Electrons followed magnetic field lines and were separated from keV energy ions, whereas the ions were decelerated and captured by positive biased electrodes.

With high power plug electron cyclotron heating (ECH) up to 400 kW, the ion confining potential of more than 2 kV was confirmed [27]. The drift type low frequency fluctuations were suppressed by the positive radial electric field produced by the plug ECRH.

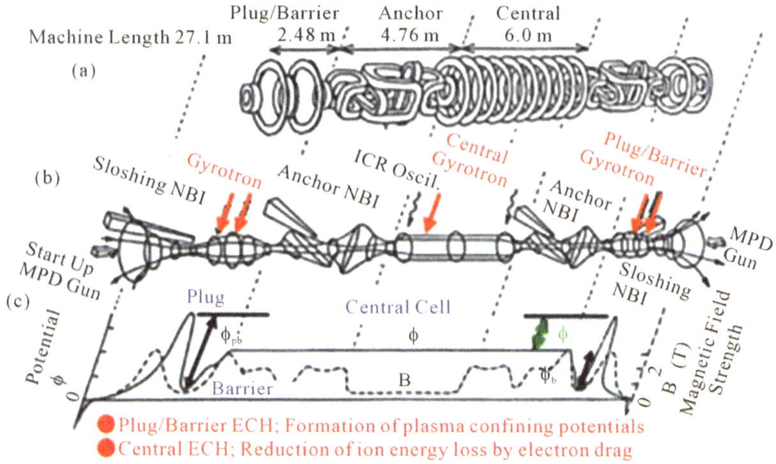


**Fig. 10.2** Schematic of Gamma-10 magnet and power systems, with typical axial magnetic field B and potential profile φ. (From [25], Fig. 2)

A divertor simulation experimental module was installed [28] in GAMMA 10/PDX and used to study the mechanisms of detachment which reduced heat and particle fluxes to the divertor plate under conditions equivalent to ITER.

### *10.1.7 Tandem Mirror KMAX*

The Keda Mirror with AXisymmetry (KMAX) was built as an axisymmetric tandem mirror consisting of a straight solenoidal central cell with mirrors at each end plus and two ring-cusp cells [29]. To facilitate the plasma gun operation, two small expanders were added to the two ends to achieve favourable field-line curvature. The axial length of the KMAX device was 11.2 m and the central cell length was 5.2 m. The inner diameters of the central cell and the mirror throats were 1.2 m and 0.3 m, respectively. The magnetic field at the midplane and the throats were 0.048 T and 0.320 T, respectively.

In the central cell, the magnetic beach configuration was adopted for slow wave launching and Ion Cyclotron Resonance Heating. The ring-cusp cells were referred to as divertor stabilizers and used for MHD stabilization. Plasmas were produced by two washer guns as ICRH targets. With a total ICRH power of 100 kW, the plasma diamagnetism increased by 15-fold, with a corresponding peak beta of 2%, density of $1.5 \times 10^{18}$ m$^{-3}$ and temperature 60 eV.

The KMAX machine philosophy was encouraged by a reactor relevant study on an axisymmetric tandem mirror configuration [30].

## 10.2 Mirror Fusion Test Facility (MFTF) and MFTF-B – A Case Study of Systems Forward Engineering

This section analyses the Mirror Fusion Test Facility (MFTF-B) as a case study by using Checklists 1.3A and 1.3B from Chap. 1. The MFTF was the largest and most expensive superconducting mirror proto-reactor ever built, and it was shut down on the day that construction was finished! It is therefore an essential subject for systems approach analysis and for studying lessons learned that may be relevant to the future of mirrors as possible fusion reactors.

### *10.2.1 Systems Forward Engineering and Architecture in MFTF*

The starting point for forward system engineering from Checklist 1.3B is to establish goals and review other machines, then choose options.

The MFTF [17] was designed at Lawrence Livermore National Laboratory (LLNL). It received funding and construction was begun in 1976, with completion expected in 1981, with goals to:

- extend the theoretical scaling laws for the microinstabilities characteristic of the mirror loss-cone ion distribution;
- demonstrate a high ion-energy, reactor-grade plasma

The machine was to act as a mirror fusion reactor prototype, with goals to:

- demonstrate the use of superconducting magnets;
- approach fully steady-state conditions;
- use neutral beam heating and fuelling.

One year after the proposal and start of construction, there were already upgrades being considered to reach reactor temperatures and steady state operation with higher energy and longer pulse neutral beam injection [31]. Lessons learned came largely from 2XII-B. An implicit goal of MFTF was to extend 2XIIB results to higher energy and to see if stabilization techniques would scale to higher energies. Superconducting technology had been tested in Baseball II.

The building blocks chosen from the mirror design space were initially far simpler than in the final version. The magnets initially were a single minimum B coil set, with neutral beams and plasma guns for stability and a starting plasma. The entire experiment was to be enclosed in a 9 m diameter vacuum vessel and cryostat, to be mostly filled by the giant superconducting magnet shown in Fig. 10.3. It became one of two end plugs in the redesigned tandem mirror MFTF-B.

The team was based on the successful 2XII-B staff, and the plasma operation scenarios were identical, but on a much larger scale. The initial ambition was high, to extend the 2XII-B results to near reactor conditions. However, even if fully

**Fig. 10.3** One of the two yin-yang mirrors arrives at LLNL. (From [32])

successful, the mirror reactor would still suffer from the Fokker-Planck calculated diffusion rate out of the end mirror which would result in a fusion Q energy gain too low for a reactor. The initial machine parameters were:

- Central and mirror magnetic fields – 2.0 T and 4.0 T (mirror ratio = 2) in a minimum-B configuration producing a double fan shaped plasma
- Length between mirrors = 3.4 m
- Start-up with stream from plasma guns plus NBI with 1000 A at 20 keV.
- Sustaining neutral beams with 750 A at 80 keV (60 MW) deuterium lasting 0.5 s

The plasma goals were:

- Lawson product $n\tau = 10^{18}\ m^{-3}\ s$
- Ion temperature 50 keV
- Electron temperature 1 keV
- Beta 50%

## *10.2.2 First Design Improvement of MFTF, One Year Later*

The first design improvement [31] kept all the same machine building blocks, but involved an upgrade of the neutral beam duration to 30 s with the idea of further demonstrating steady state operation. There was no time limitation from superconducting magnets or the capacity of the neutral beam cryopumps. The longer pulse would cause the vacuum vessel surfaces to come into gas equilibrium and the effects of recycling could be investigated on the titanium gettered surfaces. There would potentially have been some limitation on operation and pulse rate due to extensive d-d neutron production.

The original MFTF experiment [33] was built in existing building 431 at LLNL, which inside conveniently had a concrete vault with 2 m thick walls and dimensions of 20 m × 20 m by 60 m long. A substation was located outside with power supplies taking 250 MW of power from the electrical grid to supply the neutral beam injectors. A 1 kW liquid helium refrigerator was available for the superconducting magnets and cryopanels for pumping inside the vacuum vessel containing the magnets and plasma.

The main building was already being extended to take account of the possibility of a tandem mirror conversion of the type in the recently proposed TMX experiment. Hence a major modification of the entire machine philosophy was being considered just a year into construction.

To reach the high densities needed for a fusion quality plasma, scenarios for density build-up were investigated [34]. Hot plasma build-up by neutral beam injection into an initially cold solid or gaseous target was possible in large mirror machines, as was demonstrated in 2XIIB.

It was thought that with high neutral beam injection currents, a field-reversed plasma might occur with improved energy confinement [35].

### *10.2.3 Systems Review Process in Early Stages*

The unusual aspect of such a large project as MFTF was that the goals of the project and the design to accomplish them were continuously changing during the construction stage. Much of this change was driven by several factors.

- Traditional mirrors had limitations on fusion gain Q that meant that reactor construction for net power production was very difficult or even impossible.
- Tandem mirror experiments such as TMX were having considerable success and promised higher fusion gain.
- Experimental results from other machines were evolving very rapidly during construction.
- There was intense competition for funding and support from other concepts such as the TFTR tokamak which was aiming for high Q.

### *10.2.4 Alteration of MFTF to Tandem Mirror MFTF-B*

The tandem mirror experiments on TMX were successfully completed in 1980, and the MFTF team decided to completely alter the MFTF experiment to become the tandem mirror MFTF-B [36, 37]. The total project budget total for plant and equipment plus operation was \$350.4 million (1980 dollars), not including the available building and power from the grid. At the same time, an upgrade of the TMX experiment was being constructed.

#### 10.2.4.1 Overall System

The state of design and construction of MFTF-B in August 1983 and the design evolution was summarized [38] in four architectural stages:

- First – a single-cell, minimum-B mirror referred to as yin-yang coils producing a double fan-shaped magnetic field.
- Second – a simple tandem mirror similar to TMX and consisting of a straight solenoid with two minimum-B mirrors at each end to act as end-plug MHD anchors for stability and electrostatic plugs for confinement
- Third – a central solenoid with Auxiliary (A)-cells at each end to form an electrostatic thermal barrier plus end plugs of minimum B mirrors as before
- Fourth and final – a straight central solenoid with very high field axicells at its ends to throttle the flow of central cell plasma to the plugs, with transition expander coils outside those and ending with minimum B end plugs.

The final axicell magnet configuration for construction of MFTF-B is shown in Fig. 10.4, which depicts the coil configuration and profiles of magnetic field, electrostatic potential and plasma density. The length of MFTF-B approached 40 meters. The main aim of this complicated set of magnets and auxiliary heating systems was to maintain a high plasma density and temperature in the central solenoid, by means of magnetic mirror and electrostatic potential barriers.

#### 10.2.4.2 Axicell and Central Solenoid

Concentrating first on the axicells and the central cell, the density was expected to reach a peak in the very high magnetic field axicell and then drop rapidly in the transition and end regions to values well below the central cell density. Two factors contributed to this density drop: the decrease in magnetic field from the 12-T peak to 0.3 T in the transition region and the hump in electrostatic potential in the axicell that reflects some of the central-cell ions.

#### 10.2.4.3 Expander Region

The 15-fold drop in magnetic field strength allowed the magnetic field lines to expand as they emerged from the axicell. As a result, the plasma spread out and its density was reduced to one fifteenth. The hump in potential in the axicell was created by injection of neutral beams to increase the density locally. To sustain the

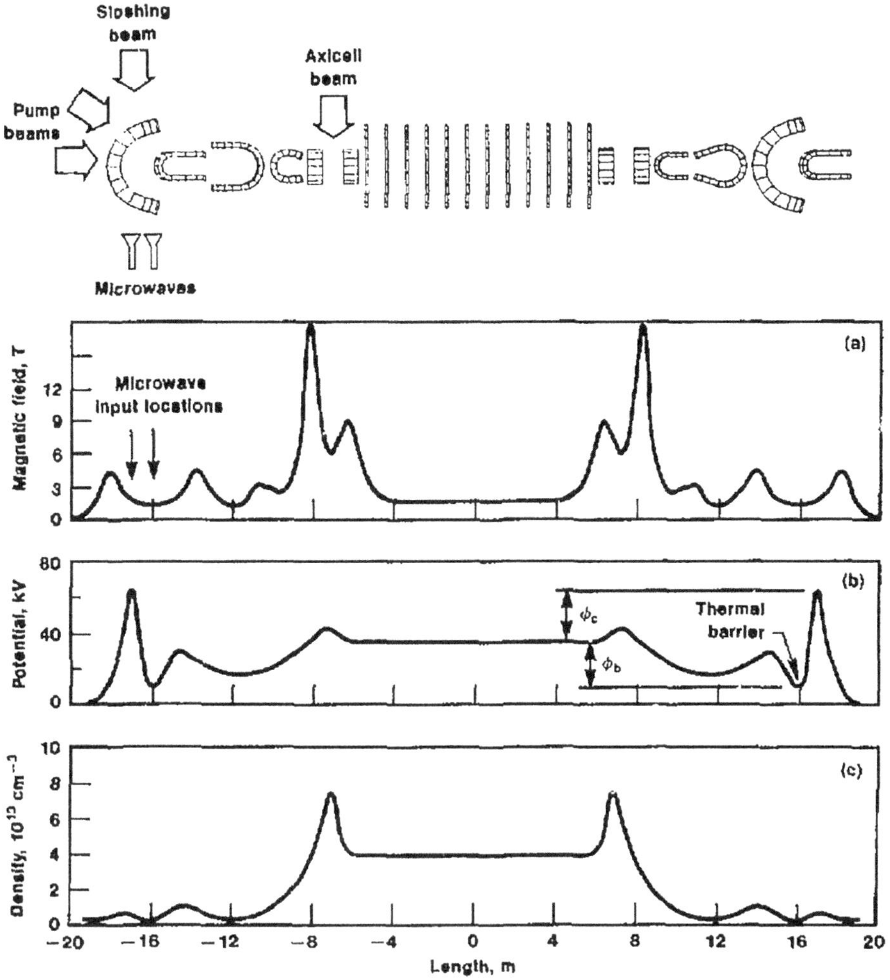


**Fig. 10.4** MFTF-B final design as an axicell thermal barrier machine with profiles (**a**) magnetic field, (**b**) potential, and (**c**) density. (From [38], Fig. 6)

density drop at the end region, the transition region had to be kept free of ions that would have clogged the potential "hole" between the central cell potential and the peak confining potential. Most ions would have passed over the hole as they bounced from end to end, but some scattered locally and would eventually have filled in this hole, raising the density if they were not removed. To remove these trapped ions, beams of neutral particles injected along the axis (or at a shallow angle) would provide the necessary pumping by charge exchange.

#### 10.2.4.4 End Plugs

In the yin-yang end plugs, a thermal barrier and confining potential would have been constructed with sloshing-ion beams and electron cyclotron resonance heating. The sloshing-ion distribution would have been set up by injecting ions into the mirror in such a way that they had velocity components that were both perpendicular and parallel (with respect to the magnetic field) at the midplane of the mirror cell. A double peaked density distribution (axially) with a minimum at the midplane would have been achieved. The distribution has a local minimum in potential at the midplane, which was the beginning of a thermal barrier. Even without enhancing this dip in potential, low-energy ions were locally trapped, providing at least partial stabilization against microinstabilities.

Completing the formation of the thermal barrier, ECRH would have been used to trap the electrons magnetically in the region around the mirror midplane, further lowering the potential and thereby creating a barrier. Since the potential formed in the first place because ions were magnetically trapped in the mirror and could not escape as fast as electrons are lost axially, any process that prevented electrons from escaping locally will lower the potential locally and contribute to the thermal barrier. ECRH would also have been applied to the plug electrons beyond the midplane to further build the confining potential peak.

#### 10.2.4.5 Vacuum and Cryostat Systems and In-vessel Components

The vacuum vessel went through many evolutions and eventually grew to 58 m long and filled the concrete vault surrounding the experiment [39]. The original single mirror led to a large bell-jar vacuum vessel in which it was possible to do a full field test of the original yin-yang minimum B mirror at superconducting temperatures and full forces and stresses.

The in-vessel components and the rest of the machine were described [40]. The vacuum vessel had the goals of providing a vacuum environment for the whole machine and for supporting all components inside the vessel, such as the superconducting magnets, and outside, such as the neutral beam injectors. The largest of the several vessel components were end vessels 10.6 m in diameter. The vessel segments were bolted together and a thin weld applied for maintaining vacuum integrity. The vessel had to support 42 superconducting magnets, plus all the internal components including a first wall warm water-cooled plasma liner with titanium gettering, liquid nitrogen thermal shields for the magnets and cryopanels for pumping the inside of the vessel.

### 10.2.5 Revised Goals of MFTF-B

The major changes in MTTF-B were the result of changes of goals in the program [40]. The central goal was to reach high Q with:

- axial ion confinement times in excess of 1 s with $n\tau = 5 \times 10^{19}\ m^{-3}$ s at Q = 0.5;
- central ion confinement potentials of 30 kV;
- plasma temperatures high enough to test thermal barriers and radial transport;
- high beta up to 50%;
- microstable sloshing ions and thermal barriers in the end magnets;
- central cell very high densities and temperatures to produce high fusion rates.

### 10.2.6 Final Construction Status Report and Project Cancellation

In a final status report on MFTF-B, construction would be finished in 1986 [41].

At the same time, yet another upgrade MFTF-$\alpha$ + T was proposed though not implemented, involving the use of many MFTF-B components but altering the facility to allow for full d-t operation. It would have produced a central-cell fusion power of 7 MW and a fusion gain Q = 2 with a wall loading of 2 $MW/m^2$ and an availability of 10% [42]. It was fully shielded and accessible to personnel 24 h after shutdown, with provisions for repair of all systems. The revised end plug design was based on the MARS tandem mirror fusion reactor design [43]. In MARS, the central cell fusion plasma was self-sustained by alpha heating while electron-cyclotron resonance heating and negative ion beans maintained the electrostatic confining potentials in the end plugs, giving a high fusion gain of Q = 26. Low-activation structures and convection cooling of the blanket enhanced the environmental and safety aspects.

The MFTF-B construction was finished in 1986 and the project was cancelled the day after, supposedly for budget considerations, and never operated. It was scavenged, dismantled and eventually the building was removed. At the same time, a summary of mirror experiments and interpretation was published [44].

### 10.2.7 Robustness, fragilities and Implications of Cancellation

The robustness of MFTF-B resided in its highly modular nature and the ability to construct and access a linear machine. However, the fragilities were inherent in the nature of a tandem mirror machine, which were to maintain high gradients and rapid variation in electrostatic potential along open magnetic field lines. Unfortunately, strong variations in density and temperature can be cancelled out by particle flows along field lines and electron thermal conductivity, hence the complexity in attempting to maintain these gradients.

The stated reason for cancellation of the MFTF-B project, and indeed the whole line of mirror and tandem mirror research in the USA except for university projects, was stated to be for reasons of balancing of the budget. From a systems point of view, there were several project fragilities that weakened the project and made it more subject to cancellation:

- Even after construction was approved, MFTF underwent major changes in fundamental machine physics, goals, machine size, cost and duration.
- The initial goal to test instability scaling was narrow for a large project.
- When very positive results were available from TMX, the MFTF-B did not scale up those results, but aimed at something even better based on a guess as to how TMX-U would perform.
- When TMX-U demonstrated success only at medium rather than high density, it weakened the physics basis.
- In an attempt to rival TFTR, MFTF-B kept raising its goals and changing the machine to compete with not yet existing results from a competitor.

In summary, the project was probably evolving too rapidly during construction, even though it had great promise and smaller machines to support the physics.

The implications of the cancellation were bad for the future of mirror and tandem mirror research. The machine cost nearly $400 million and had elements of a tandem mirror fusion reactor. To start again would be extremely expensive.

The great tragedy is that MFTF-B might have produced very interesting plasmas that could well have been upgraded to a d-t fusion experiment with Q over 1.

## 10.3 Gas Traps, Multiple Mirrors, Cusps and Field Reversal

### *10.3.1 Gas Dynamic Trap (GDT)*

An alternative but related approach to tandem mirror confinement is the Gas-Dynamic Trap (GDT) experiment [45]. GDT is shown in Fig. 10.5 [46].

GDT has a large mirror ratio, axial symmetry and a long distance between mirrors which exceeds the effective mean free path of ion scattering into a loss cone. Under these conditions, the plasma confined in GDT is isotropic and Maxwellian. The rate at which it is lost out of the ends is governed by a set of simple gas-dynamic equations. Stability is provided through a plasma outflow through the end mirrors into regions where the magnetic-field lines' curvature is favourable for stability.

The machine is over 10 m in length, with a mirror to mirror distance of 7 m, and mid-plane and mirror field strengths of 0.35 T and 12 T, giving a mirror ratio of 35.

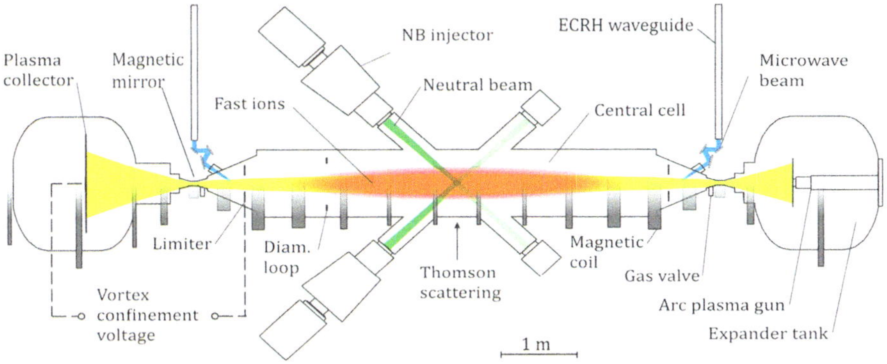


**Fig. 10.5** Schematic layout of the GDT device. (From [46], Fig. 1)

The neutral beam injectors provide 5 MW of power at 25 keV for up to 5 ms, and the gyrotrons provide 0.8 MW at 54.5 GHz.

The plasma results obtained were very encouraging. Bulk plasma densities up to $2 \times 10^{19}$ m$^{-3}$, electron temperatures of 0.9 keV, mean energy of trapped fast ions of 9–12 keV, and maximum plasma beta of 60% were obtained. The gas dynamics and favourable curvature in the expander helped to stabilize the plasma.

The best way of suppressing transverse losses was the vortex confinement method, which relied on producing a radial electric field which caused the plasma to rotate in a thin peripheral layer.

## 10.3.2 *Multiple Mirror GOL-3 and GOL-NB*

The GOL-3 [46] was a multiple mirror device with turbulent heating from a high power electron beam. It consisted of a 12 m long magnetic system with 110 coils producing a multiple mirror field with maximum and minimum magnetic field in each period of 4.8 T and 3.4 T. The solenoid terminated with single magnetic mirrors with a field of 8 T. In the output expander, the field decreased to 0.05 T to reduce the loading on the endplates.

Heating was produced by an electron beam generator producing 30 kA at 1 MeV for a power rating of 30 GW, lasting up to 12 μs. Turbulent heating of the plasma led to 1 keV electron temperatures and up to 4 keV ion temperatures. A 50-fold increase in energy confinement time over a simple mirror was found, demonstrating that the multiple mirrors do indeed enhance confinement.

GOL-NB was an upgrade to include neutral beam heating [47]. The machine included a low magnetic field region between two multiple mirrors to be heated with two neutral beam injectors of 0.75 MW each.

## *10.3.3 Cusps*

A cusp configuration is like a mirror machine, except the currents are in opposite directions, creating a field zero half way between the two coils. Cusps create a minimum-B stable configuration. Since particles passing through the weak field region will have their pitch angle randomized, there tends to be rapid losses through the ends, unless a beta of 100% can close the ends.

### 10.3.3.1 Cusp with High Energy Electrons

Experiments have been performed in cusp geometry with high energy electrons providing a highest beta (of about 100%) plasma [48], confirming the theoretical conjecture that field lines at the ends of the cusp close at high beta and give much improved confinement. The work was a follow-on to the concept of a "Polywell" fusion reactor [49] which relied on cusp geometry to confine high energy electrons while ions were electrostatically contained.

### 10.3.3.2 Lockheed Martin Compact Fusion Reactor (CFR), T4 and T5

Lockheed Martin conducted a series of multiple cusp experiments T4 and T5 heated by neutral beam injection, leading to a design for a Compact Fusion Reactor (CFR) [50, 51]. The work had elements in common with the Polywell.

## *10.3.4 Tri Alpha Energy C-2U and C-2W Field Reversed Configuration*

Field reversed configurations (FRC) in linear machines have a long history in fusion, beginning with the ASTRON concept and experiment invented by Nick Christofilous [52]. The concept was to use a high energy electron beam injected into a mirror geometry with the idea of creating a magnetic field reversal leading to closed toroidal confinement.

The FRC concept was pursued in experiments such as TCSU [53] which was the largest FRC at that time. The overall device was linear with a length of about 8 m and a central quartz tube of 0.8 m diameter to allow penetration of a rotating magnetic field. Electron temperatures of up to 200 eV were obtained, with current drive efficiencies approaching 100%.

A field-reversed configuration plasma sustained by 10 MW of neutral-beam injection was achieved on the C-2U device built by Tri Alpha Energy [54, 55]. The concept was based on the production of a hot stable FRC in the C-2 experiment by colliding and merging two high-$\beta$ plasmoids from a dynamic version of field-reversed θ-pinch technology [56]. Most of the kinetic energy was converted into thermal energy with temperatures exceeding 0.5 keV.

C-2U and C-2W (also Norman after Norman Rostoker) consisted of a central solenoid confinement chamber 3 m long with angled neutral beam injection, with formation chambers and divertors on each end. The central magnetic field value was 0.1 T. Over 10 MW of 15 keV hydrogen was available for beam injection.

Most recently [57], up to 21 MW of neutral beam power at 14–40 keV was injected into the target plasma, producing an FRC plasma sustained for up to 30 ms, limited only by NBI pulse duration. The electron temperature reached 0.3 keV with a total temperature of 3 keV at densities of $1–3 \times 10^{19}$ m$^{-3}$.

## 10.4 Linear Fusion Reactors and Systems Design Space

### *10.4.1 Physics of the Systems Architecture Design Space*

The systems architecture design space for linear fusion reactors has evolved considerably to include tandem mirrors, gas box, multiple mirrors, field reversed configurations and the axisymmetric tandem mirror. Basic concepts include:

- Magnetic mirrors – Magnetic mirrors provide confinement by trapping ions or electrons with large perpendicular energy. These particles can produce magnetic wells through the diamagnetic effect, which means currents flow to maintain force equilibrium which excludes magnetic field.
- Minimum-B mirror – Magnetic mirror fields plus stabilizing cusp fields can create a magnetic minimum which stabilize the plasma against instabilities.
- Electrostatic potentials – Ions or electrons trapped by mirrors can create electrostatic potentials which can attract or repel ions or electrons, depending on charge. The strength of the potential depends on relative densities of hot and cold ions and electrons in different regions.
- High beta wells – High beta wells can be produced either by high energy injected ions via neutral beams, or relativistic electrons produced by ECRH or by good confinement of plasma in low magnetic field regions.
- End losses – Plasma particles can stream out the ends of mirrors due to pitch angle scattering between mirrors or by instabilities, and energy can be carried out by particles, electron thermal conductivity along field lines, or radial transport when there is good axial confinement.
- Thermal conductivity – In high density linear theta pinches such as Scylla, plasma would be lost by flowing out the ends, but losses by electron thermal conductivity were also very significant. In tandem mirrors, high density plasma can be coupled throughout the system, resulting in important end losses
- Thermal plasma – In tandem mirrors, a high density, and high beta plasma can be built up in the central solenoid confined by plasma potential.

- Loss cones – With high energy nearly perpendicular to the magnetic field, hot ions or hot electrons can be confined between mirrors. This confinement can create an electrostatic potential which confines the other species, but a loss cone can lead to velocity space instabilities due to the non-uniformity in the distribution function.
- Stabilization – Stabilization can be produced in many ways for several instabilities, including minimum magnetic field configurations, cusps, line tying with conducting end plates, radial fields for plasma rotation, sloshing ions or feedback control, gas boxes.
- Radial diffusion – In the case of long confinement times, radial diffusion of plasma can become important. Rotating plasmas can slow these losses

## *10.4.2 Design-Space Building-Blocks for Systems Architecture of Linear Machines*

To design a successful experiment or reactor, a number of physical building-blocks are available to the systems architect of linear machines:

### 10.4.2.1 Magnetic and Electrostatic Configurations

- Simple mirrors – Simple mirrors are produced by a straight magnetic section with high field magnets at each end, DCX-2 [2, 3].
- Minimum-B mirrors – Minimum B mirrors can be produced by mirrors plus multipoles or by baseball shaped windings or double "yin-yang" type windings, typically with 2/1 mirror ratio and a relatively small volume plasma compared to the coils, Phoenix II, Baseball II or 2XII-B [5–7].
- Straight solenoid centre cells – A standard straight section solenoid can be created with circular magnetic coils in a straight line to produce a constant magnetic field in a cylinder. In a tandem mirror, the central straight section can be of any length and diameter, but it needs either transition or mirror coils at each end as in TMX [16].
- Magnetic compression – Compressing the plasma by dynamically increasing the magnetic field can cause heating, 2XIIB [7].
- Tandem Mirror – A tandem mirror utilizes a straight central section with auxiliary heating in the end plugs to create electrostatic potentials and thermal barriers, TMX [18] and TMX-U [21].
- Relativistic electrons producing minimum B – High beta relativistic electrons produced by ECRH can produce a minimum B configuration, ELMO [8].
- Anchors – An anchor is a minimum-B mirror at each end of a tandem mirror to act as an end-plug for MHD stability, MFTF-B [38] and Gamma-10 [25].

- Auxiliary (A)-cells – A-cells are located at each end of a straight solenoid in a tandem mirror to form an electrostatic thermal barrier, MFTF-B [38].
- Axicells – Very high field axicells at the end of a straight solenoid was used to throttle the flow of central cell plasma to the plugs, MFTF-B [38].
- Thermal barrier – Thermal barriers can be created in minimum B cells or A-cells using auxiliary heating by neutral beams or electron cyclotron heating to create a positive or negative potential, see above.
- Transition and adaptor sections – Adaptor coils, often C shaped and minimum B, were used to transition from a straight solenoid to end plugs or mirrors, MFTF-B [38].
- End sections and expanders – End sections expand the magnetic flux exiting the mirrors in either an uncontrolled or controlled way to reduce the magnetic flux to a certain field level with a certain shape before it encounters a conducting wall, beam dump, or direct convertor, MFTF-B [30, 38] and Gol-3 [46].
- Gas dynamic trap – GDT had a long distance between mirrors, a large mirror ratio and axial symmetry, where end loss was governed by simple gas-dynamic equations in an expander region providing stability, GDT [46] and GDMT [58].
- Multiple mirrors – A system with many tens of multiple mirrors in series terminated with strong mirrors gives enhanced confinement, GOL-NB [47].
- Filed reversed configurations – A field reversed configuration can be produced in a simple mirror with high current and high energy electrons, ASTRON [52] and fast ions from NBI, C-2W [57].
- Cusps – A vacuum configuration giving a null field in the centre of a minimum B field. When filled with high beta plasma, it produces a closed configuration, T5 [50].
- Normal, cryogenic, superconducting and high temperature superconducting magnets – Mirror machines operating at very high beta offer the possibility of using normal conducting coils, but very high field require superconducting or high temperature superconducting coils, GDMT [58].
- Electrostatic direct convertors – Electrostatic direct convertors decelerate escaping ions with electrostatic fields in one or more stages to change ion energy into electricity [11, 13].

#### 10.4.2.2 External Heating and Plasma Creation

- Plasma guns – Plasma guns create a burst of cold plasma that is injected from the ends of the mirror machine along the field lines so it reaches into the rest of the mirror machine to act as a plasma stabilizer or beam target, C-2W [57].
- Arcs – Arcs can act as target plasmas, DCX-2 [3].
- Electron beams – Electron beams can be used to produce field reversal in ASTRON or turbulent heating, GOL-3 [46].
- Neutral beam injection – Neutral beams inject atoms into the plasma which ionize into high energy ions and low energy electrons. These ions can be injected either into a minimum-B well field or an end plug of a tandem mirror to create a hot ion

distribution with a loss cone and a positive potential which retains electrons. The beams can be aimed at angles other than perpendicular to the field to produce a wider distribution in velocity space which helps to avoid instabilities, but which can lead to faster scattering into the loss cone than if injected perpendicularly. Fast neutrals are used to heat the thermal plasma in the central solenoid of a tandem mirror. The higher the beam energy, the higher the ion temperature and the longer the beam slowing down time, GDMT [58].

- Electron cyclotron heating – Electron cyclotron resonance heating can be used to heat electrons in either the central solenoid or in end plugs of a tandem mirror. In particular, as in the ELMO device, it can be used in end plugs to create a high beta electron ring that produces a magnetic well for stability, and also creates a negative potential to retain ions, MFTF-B [38].
- Ion cyclotron heating – Ion cyclotron resonance heating can be used to heat the ions in the central solenoid of a tandem mirror, KMAX [29].

## *10.4.3 Linear Fusion Reactors*

### 10.4.3.1 Advanced Tandem Mirror Reactor

An axisymmetric tandem mirror reactor design [30] took advantage of advances in superconducting magnet technology to minimize the size and cost of end plugs and in gyrotrons allowing intense electron cyclotron heating (ECH). The yin-yang magnets and thermal barriers were no longer needed. The concept is shown schematically in Fig. 10.6.

The version of the tandem mirror considered here is a variation on the axisymmetric kinetically stabilized tandem mirror (KSTM) [59] and confirmed in GDT [60]. MHD stability was provided by the favourable curvature of magnetic field lines in the expander where flux diverges as it emerges from the last mirror coil. End plugs with a positive potential from trapped, injected hot ions prevent leakage out of the ends of the central cell solenoid which acts as a fusion reactor. Charged fusion products such as 3.5 MeV alphas escaped out the ends along expanding magnetic field lines that spread the power over a large area in the expander. Ignition in the Centre Cell is achieved by mirror machines serving as end plugs. The potential profile along the length of the machine is shown in Fig. 10.6. Electron cyclotron heating (ECH) was added to the end plugs to increase electron temperature. The central cell beta was expected to approach 100%, allowing an economical power reactor to be developed.

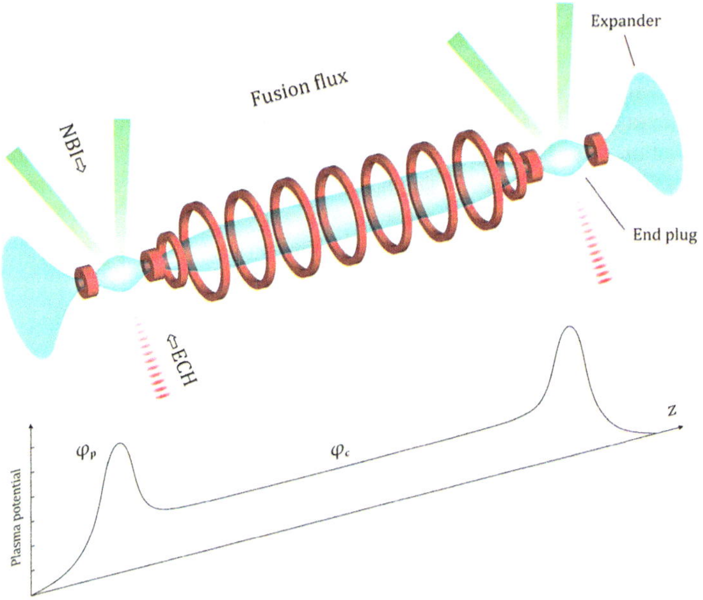


**Fig. 10.6** Schematic of axisymmetric tandem mirror coils and field with electrostatic confining potential produced by end plugs and expander. (From [30], Fig. 1)

#### 10.4.3.2 Gas Dynamic Multiple-Mirror Trap (GDMT) Reactor Prototype

Based on GOL-3, a project was being considered called the Gas Dynamic Multiple-Mirror Trap (GDMT) [61] using both multiple mirrors and a gas dynamic trap. The central solenoid would be like GDT with sloshing neutral beams and the end plugs would consist of multiple mirrors. Various methods were considered for enhancing confinement including helical windings.

The design includes high temperature superconductors producing a mirror field of 20 T with 5 s of 10 MW of NBI and 6 MW of ECRH [58].

### *10.4.4 Synthesis of Linear Mirror Design Space*

Mirror devices and reactor concepts have evolved strongly over the years. Starting from simple machines using magnetic mirrors with electrostatic potentials to confine the electrons, they have evolved to minimum B configurations and field reversed configurations for stability, tandem mirror systems using more complicated electrostatic potentials, to more sophisticated devices that provide pathways towards mirror reactors and/or neutron sources, including:

- Axisymmetric tandem mirror [30];
- C-2W with field reversal [57];
- Gas dynamic multiple mirror trap (GDMT) [58].

These concepts deal with and overcome the physics problems that led to the premature shutdown of MFTF-B. The mirror concepts should be regarded as real contenders for fusion reactors. Their weakness is that they have still not demonstrated the fusion triple product needed for a reactor.

## 10.5 Toroidal Versions of Linear Mirror Concepts

### *10.5.1 ELMO Bumpy Torus (EBT)*

The bumpy torus was an extension of ELMO by linking several mirror regions into a torus to reduce end losses. Stability would be provided by high beta electron rings produced by ECRH in each low field region between the mirrors.

The ELMO Bumpy Torus had a major radius of 1.5 m and minor radius 0.19 m, midplane and mirror fields of 0.7 T and 1.4 T, ECH heating power of 200 kW at 28 GHz and 60 kW at 18 GHz [62]. Each hot electron annulus reached toroidal field beta of 10–40% at temperatures of 100–500 keV. The bulk plasma reached beta values of 0.5% and $n\tau$ values of $10^{17}$ $m^{-3}s$ at electron and ion temperatures of up to 800 and 200 eV respectively. These promising results led to a superconducting upgrade with ECH and NBI which was designed, but never built.

### *10.5.2 Nagoya Bumpy Torus (NBT-1 M)*

The Nagoya Bumpy Torus (NBT-1 M) [63] had a major radius of 1.4 m and was a toroidal confinement device consisting of 24 canted mirror cavities with an axial magnetic field B of 0.34 to 0.41 T at the midplane. The plasma was produced and heated by microwaves of 30 kW at 18 GHz and 16 kW at 8.5 GHz. The results of the experiments were discouraging, because gradient-B reversal due to the hot electron rings was never observed. The plasma potential profile was the same as in a simple torus.

### *10.5.3 Auto-injection Mirror*

The auto-injection toroidal mirror reactor [64] was designed to solve the problem of pitch-angle scattering escape through the mirrors by using a re-entrant magnetic geometry with the same topology as that of the earth's magnetic dipole field. As particles emerged through the mirror, their energy was mostly in the perpendicular direction. As the ions followed the magnetic lines to the central region with a field strength equal to that of the mirror, their velocity became perpendicular again, therefore causing a long dwell time. This central region had neutral gas added so that the ions underwent charge exchange to produce fast neutrals that travelled radially outwards between the separated inner coils and were reinjected directly into the original reacting volume with optimum pitch angle. This reinjection was in principle capable of reducing external beam injection needs and hence increasing the fusion gain Q by an order of magnitude.

## 10.6 Taylor State $q < 1$ Torus: Spheromak and Reversed Field Pinch

### *10.6.1 Spheromak, Reversed Field Pinch and Taylor State*

As was the case with Field Reversed Configurations (FRC), it is possible to use a linear configuration to produce closed toroidal plasmas, in particular the spheromak [2], which is a torus characterized by an extremely low aspect ratio. Topologically, spheromaks have the plasma contained in a torus within a separatrix surface. Spheromaks were formed in various ways [65]:

- flux-core induction;
- combined theta-pinch and z-pinch;
- helicity injection from a coaxial, magnetized gun;
- a conical θ-pinch;
- kinked z-pinch.

Machines included:

- CTX
- SPHEX
- FACT
- HIT-SI

Spheromaks have properties in common with reversed field pinches, which are initiated in a torus with a weak toroidal magnetic field and a toroidal current induced by a central transformer as in a tokamak. The safety factor is much less than one and helicity plays an important role in determining plasma stability. Both type of plasma are contained within a conducting wall.

Stable plasma regimes described by the Taylor state [66, 67] were observed both in the ZETA reversed field pinch (RFP) and spheromaks. In an RFP, when moving from the toroidal axis outwards towards the edge, the toroidal magnetic field reverses its direction. It was shown that magnetic self-organization is important, and helicity provides a paradigm for describing the configuration in RFP and spheromaks. Force free configurations result from the minimization of the plasma magnetic energy subject to the constraint of constant total helicity, interpreted as a measure of linked magnetic fluxes. A fully relaxed plasma is referred to as being in the Taylor state. These observations encouraged further experiments.

### *10.6.2 SSPX Spheromak*

A schematic of the Sustained Spheromak Physics Experiment (SSPX) and a plasma equilibrium is shown in Fig. 10.7 [65].

The geometry was that of a linear coaxial gun firing into a chamber with a conducting shell and terminated by a high field magnetic mirror coil. The flux conserving boundary had a major radius of 0.5 m and a height of 0.5 m with a 1.2 cm thick conducting copper shell. The fields were powered by capacitor banks to produce a multi-millisecond experiment. The magnetic bias coils were first energized, then a strong current was made to flow between the inner and outer cylindrical

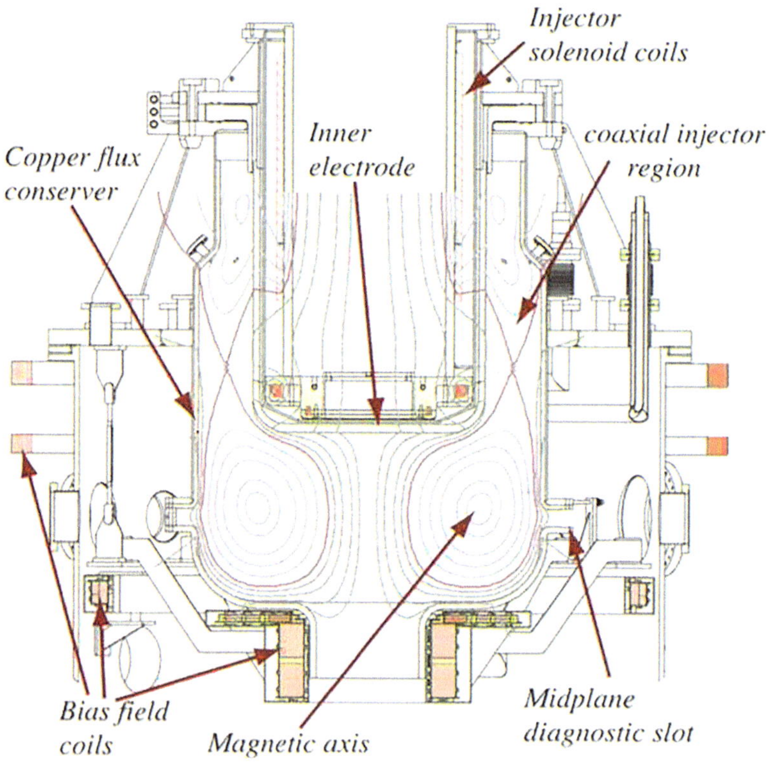


**Fig. 10.7** SSPX schematic with plasma equilibrium. (From [65], Fig. 1b)

electrodes using coaxial helicity injection (CHI), resulting in a spheromak plasma of about 1 m diameter trapped in the lower conducting shell chamber.

The unfortunate conclusion of a decade of experimentation on SSPX was that it was not possible to sustain the plasma against resistive decay, while at the same time to maintain good plasma energy confinement properties. Instabilities that allowed current transport to the inside of the plasma also enhanced energy loss [68], thus severely limiting achievable energy confinement times.

### *10.6.3 Reversed Field Experiment RFX-Mod and RFX-Mod2*

The reversed field pinch experiment RFX-mod had a major radius of 2.0 m and a minor radius of 0.459 m and produced plasma currents of up to 2 MA [69]. To sustain long pulse plasmas, the original conducting shell in RFX had been replaced by a matrix of $48 \times 4 = 192$ independent saddle coils so as to maintain the originally induced image currents.

Plasma discharges of up to ½ s in duration were achieved, with electron temperatures up to 1 keV. These discharges were characterized by repetitive reconnection events, detected by magnetic probes. These events were often associated with the partial or total transitions from a helical topology with conserved flux surfaces to a configuration characterized by a chaotic magnetic field. A large fraction of the input power was dissipated during reconnections and the electron thermal energy was reduced significantly (−30%), implying a significant loss of energy confinement during reconnection. This is reminiscent of the loss of confinement in spheromaks when attempts were made to inject helicity to maintain the plasma current. In RFX-mod, experimental measurements showed an increase of 250 eV in the central ion temperature during reconnection. It seemed that suprathermal ion heating and electron acceleration mechanisms were dominant.

A further upgrade called RFX-Mod2 is converting the conducting copper shell to be the vacuum vessel, thereby improving plasma stability [70].

### *10.6.4 MST and Reactor Studies*

Other large reversed field pinch experiments include MST [71, 72], which showed very promising results on energy confinement. More recent MST results revealed a wide range of instabilities during neutral beam heating.

An RFP with a relatively low toroidal field offers potential cost and technological advantages in a reactor. Based on the earlier good confinement results obtained on MST, a reversed field fusion reactor was explored [73]. The reactor was based on an extremely low aspect ratio $A = 2$, leading to very high plasma beta of 66% and a nearly (94%) self-sustaining plasma current profile. These conditions have yet to be demonstrated experimentally.

### *10.6.5 Review and Prospects for RFP*

A full review of RFP research [74] covered the machines RFX-mod, MST, EXTRAP T2-R, KTX, RELAX and TPE-RX A general summary of the plasma limits explored include plasma current up to 2 MA, electron density and temperature up to $10^{20}$ $m^{-3}$ and 1.5 keV, and energy confinement times up to about 10 ms, in discharges lasting up to 1 s. The machines are therefore a couple of orders of magnitude away from demonstrating fusion reactor conditions, but offer the prospect of reactors smaller than tokamaks. A number of conclusions still support the possibility of using the RFP as the basis for a fusion reactor:

- A high level of magnetic fluctuations is not intrinsic to configuration sustainment.
- Actively driven edge poloidal current was achieved with Pulsed Poloidal Current Drive (PPCD).
- Magnetic self-organization was achieved with Single Helicity States (SH).
- Substantially improved confinement was achieved.
- Improved confinement PPCD plasmas also corresponded to SH equilibria.
- The improved regimes were achieved due to reliable active control of magnetic instabilities in the RFP with active coils acting in a feedback process.
- RFP devices used active control of MHD stability by feedback controlled coils to cancel multiple Resistive Wall Modes (RWMs) and suppress tearing modes.
- RFP RWM control allows for reactor relevant machines, since a thick passive conductive would no longer be necessary.
- Tearing mode control is a crucial ingredient for the achievement of Single Helicity states, which allows for sustaining an RFP configuration with low magnetic stochasticity and an internal transport barrier.

The review concluded that scientific issues need to be resolved in order for the RFP to be a viable fusion reactor concept:

- The scaling of confinement with current and size needs to be demonstrated, since present experiments indicate favourable tendencies.
- Higher plasma density needs to be explored.
- Active plasma exhaust needs to be demonstrated in steady state conditions, since divertors have not yet been explored.

If an RFP reactor could be built, it would have several advantages:

- The toroidal field is relatively low.
- There are reduced aspect-ratio limitations.
- Ohmic heating makes a substantial contribution to overall power balance.

They concluded that the RFP was well worth pursuing as a potential reactor.

# References


1. D.J. Rose, M. Clark Jr., *Plasmas and Controlled Fusion* (MIT Press, Cambridge, 1961)
2. A.A. Harms et al., *Principles of Fusion Energy* (World Scientific, Singapore, 2000)
3. P.R. Bell et al., Oak Ridge multiple-pass injection experiment, DCX-2, Plasma physics and controlled nuclear fusion research, IAEA, Culham, vol II, paper IAEA-CN-21/112 (1965), p. 77
4. Magnetic Mirror, Wikipedia, basic magnetic mirror machine by M.J. Moynihan (2022). https://en.wikipedia.org/wiki/Magnetic_mirror. Accessed 17 June 2022
5. E. Thompson, J.G. Cordey, D.R. Sweetman, Experimental and theoretical investigation of plasma build-up and microinstabilities in a neutral-injection mirror machine. In: IAEA Fusion Energy Conference, vol. II, Madison, USA, paper IAEA-CN28/G-10 (1971), p. 689
6. K.H. Berkner et al., Plasma production by neutral beam injection in mirror systems, AEA Fusion Energy Conference, Madison, USA, Paper IAEA-CN28/G-11 (1971), p. 707
7. F.H. Coensgen et al., Recent experiments on the 2XIIB mirror machine. Seventh European Conference on Controlled Fusion and Plasma Physics. Lausanne, Switzerland (1975)
8. R.A. Dandl et al., High-beta relativistic electron plasmas in axisymmetric and non-axisymmetric mirrors, IAEA Fusion Energy Conference, vol. II. Madison, USA, paper IAEA-CN28/G-4 (1971), p. 607
9. R.R. Post, Mirror systems: fuel cycles, loss reduction and energy recovery. In: BNES nuclear fusion reactor conference. British Nuclear Energy Society. Culham, Oxfordshire (1969), p. 87
10. F.B. Marcus, Direct conversion of charged particle energy into electricity. D.Phil. Thesis, University of Oxford (1972)
11. F.B. Marcus, C.J.H. Watson, A computation study of space charge limitations in direct conversion systems. European Conference on Controlled Fusion and Plasma Physics, Moscow (1973), p. 239
12. R.W. Moir, W.L. Barr, R.P. Freis, R.F. Post, Experimental and computational investigations of the direct conversion of plasma energy to electricity, Fourth IAEA Fusion Energy Conference, Madison, USA, paper IAEA-CN-28/K-1 (1971), p. 315
13. W.L. Barr, R.W. Moir, Test results on plasma direct converters. Nucl. Technol. Fusion **3**, 98 (1983)
14. D.J. Rose, G.L. Flint, F.B. Marcus, Some topics on pulsed systems and mirrors. Nuclear Fusion Reactor Conference, Proceedings of a British Nuclear Energy Society Conference, UKAEA Culham Laboratory, Culham (1969)
15. J.G. Cordey, F.B. Marcus, D.R. Sweetman, C.H.H. Watson, Efficient Re-circulation of power in mirror reactors, Fourth IAEA Fusion Energy Conference, Madison, USA, Paper IAEA-CN28/K-3 (1971), p. 353
16. F.H. Coensgen, T.M.X. Major Project proposal, Lawrence Livermore National Laboratory, Livermore, CA, LLL-PROP-148 (1977)
17. F.H. Coensgen, Scientific Editor, MX major project proposal, MFTF. Lawrence Livermore Laboratory, Report LLL-Prop-142 (1976)
18. T.C. Simonen et al., Summary of results from the Tandem Mirror Experiment (TMX). Lawrence Livermore National Laboratory Report UCRL-53120 (1981)
19. J. Kesner et al., Introduction to tandem mirror physics, MIT Plasma Fusion Centre Report PFC/RR-83-35 (1983)
20. M.O. Calderon, H.H. Bell, Tandem Mirror Experiment Upgrade (TMX-U) Overview – Recent Events. Lawrence Livermore Laboratory Report UCRL-92549, Submitted to the 11th Symposium on Fusion Engineering Proceedings, Austin, Texas (1985)
21. T.C. Simonen et al., TMX-U tandem mirror thermal barrier experiments. IAEA Fusion Energy Conference, Kyoto, Japan, vol. 2, Paper IAEA-CN-47/C-I-1) (1986), p. 231
22. A.W. Molvik et al., Stable operation of an effectively axisymmetric neutral beam driven tandem mirror. Nucl. Fusion **30**(5), 815 (1990)
23. R.A. Breun et al., Radial transport in the PHAEDRUS-B tandem mirror, 12th IAEA Fusion Energy Conference, Nice, Paper IAEA-CN-50/C-III-1 (1988), p. 475

24. A. Mase et al., Plasma confinement in the axisymmetrized tandem mirror Gamma 10. 12th European conference on Controlled Fusion and Plasma Physics, Part I, Budapest (1985), p. 528
25. W. Horton et al., Tandem mirror experiment for basic fusion science. World J. Nucl. Sci. Technol. **4**, 53 (2014)
26. Y. Yasaka et al., Direct energy conversion experiment on the Gamma 10 tandem mirror. Fusion Sci. Technol. **51**(2T), 171 (2007)
27. T. Imai et al., Status and plan of Gamma 10 tandem mirror program. Fusion Sci. Technol. **59** (1T), 1 (2011)
28. M. Yoshikawa et al., Study of detached plasma profile in the divertor simulation experimental module of Gamma 10/PDX, 28th IAEA Fusion Energy Conference, Nice, France (Virtual) 10–15 May 2021, Presentation EX/P8–1, Paper IAEA–CN–286-0691, Submitted to Nuclear Fusion (2021)
29. M. Liu et al., Ion cyclotron resonant heating in the central cell of the Keda Mirror with AXisymmetricity (KMAX). Phys. Plasmas **25**, 082515 (2018)
30. T.K. Fowler, R.W. Moir, T.C. Simonen, A new simpler way to obtain high fusion power gain in tandem mirrors. Nucl. Fusion **57**(5), 056014 (2017)
31. C.C. Damm et al., Directions for possible upgrades of the Mirror Fusion Test Facility (MFTF), Lawrence Livermore Laboratory Report UCRL-80475 (1977)
32. Mirror fusion test facility, Wikipedia, One of the two yin-yang mirrors arrives at LLNL (2022) https://en.wikipedia.org/wiki/Mirror_Fusion_Test_Facility. Accessed 17 June 2022
33. K.I. Thomassen, The Mirror Fusion Test Facility (MFTF), Lawrence Livermore Laboratory Report UCRL-81154, IECE Conference, Sand Diego, California, paper CONF-780801-2 (1978)
34. J.W. Shearer, P.A. Willmann, Some estimates of mirror plasma startup by neutral beam heating of pellet and gas cloud targets, Lawrence Livermore Laboratory Report UCID-17845 (1978)
35. J.W. Shearer, Estimation of neutral-beam-induced field reversal in MFTF by an approximate scaling law, Lawrence Livermore Laboratory Report UCID-18615 (1980)
36. D.E. Baldwin, B.G. Logan, T.C. Simonen, Physics basis for MFTF-B, Lawrence Livermore Laboratory report UCID-18496-Part 2 (1980)
37. T.K. Fowler, K.I. Thomassen, Mirror fusion test facility status, Lawrence Livermore Report UCID-19241 (1981)
38. K.I. Thomassen, Evolution of the magnetic configuration for the MFTF-B, Energy and Technology Review, succeeded by Science and Technology Review, Lawrence Livermore Laboratory, Report CRRL-52000-83-8 (1983), p. 20
39. J.W. Gerich, Design and construction of the vacuum vessel for the tandem Mirror Fusion test Facility, Lawrence Livermore Laboratory Report UCRL-88964. 9th International Vacuum Conference, Madrid, Spain (1983)
40. V.N. Karpenko, The mirror fusion test facility: an intermediate device to a mirror fusion Reactor, Lawrence Livermore Laboratory Report UCRL-88950. 5th topical meeting on the technology of fusion energy, Knoxville, Tennessee, paper CONF-830406-113 (1983)
41. V.N. Karpenko, The Mirror Fusion Test Facility (MFTF-B) – Status, Lawrence Livermore Laboratory Report UCRL-92247. 6th Topical Meeting on the Technology of Fusion Energy, Knoxville, Tennessee, Paper CONF-850310 – 107 (1985)
42. W.D. Nelson, MFTF-α+T progress report, Oak Ridge National Laboratory Report ORNL-FESC-83/9 (1985)
43. C.D. Henning et al., Mirror Advanced Reactor Study (MARS) final report summary, Lawrence Livermore Laboratory report UCRL-89370. 10th Symposium on Fusion Engineering IEEE, Philadelphia, paper CONF-831203-38 (1983)
44. R.F. Post, Magnetic mirror fusion systems: characteristics and distinctive features, Lawrence Livermore Laboratory Report UCRL-97215. Course and Workshop on Physics of Mirrors, Reversed Field Pinches and Compact Tori, Varenna, Italy (1987)
45. A.A. Ivanov, V.V. Prikhodko, Gas-dynamic trap: an overview of the concept and experimental results. Plasma Phys. Control. Fusion **55**, 063001 (2013)

46. P.S. Bagryansky, A.D. Beklemishev, V.V. Postupaev, Encouraging results and new ideas for fusion in linear traps. J. Fusion Energy **38**, 162 (2019)
47. V. Postupaev et al., Status of activity on GOL-NB multiple-mirror experiment, 28th IAEA Fusion Energy Conference, Nice, France (Virtual) 10–15 May 2021, Presentation IAC/P4–4, Paper IAEA–CN–286-0664 (2021)
48. J. Park et al., High-energy electron confinement in a magnetic cusp configuration. Phys. Rev. X **5**, 021024 (2015)
49. R.W. Bussard, Some physics considerations of magnetic inertial-electrostatic confinement: a new concept for spherical converging-flow fusion. Fusion Sci. Technol. **19**, 273 (1991)
50. J. Trevithick, Lockheed Martin now has a patent for its potentially world changing fusion reactor. The War Zone (2018). https://www.thedrive.com/the-war-zone/29074/skunk-worksexotic-fusion-reactor-program-moves-forward-with-larger-more-powerful-design. Accessed 17 June 2022
51. T.J. McGuire, Encapsulating magnetic fields for plasma confinement, US Patent US2018/0047462A1 (2018)
52. E.R. Coleman, S.A. Cohen, M.S. Mahoney, Greek fire: Nicholas Christofilos and the Astron project in America's early fusion program. J. Fusion Energy **30**, 238 (2011)
53. H.Y. Guo et al., Improved confinement and current drive of high temperature field reversed configurations in the new translation, confinement, and sustainment upgrade device. Phys. Plasmas **15**, 056101 (2008)
54. H. Gota et al., Achievement of field-reversed configuration plasma sustainment via 10 MW neutral-beam injection on the C-2U device. Nucl. Fusion **57**(14), 116021 (2017)
55. H. Gota et al., Overview of C-2U/C-2W field-reversed configuration experiments, Tri Alpha energy, Inc. Exploratory Plasma and Fusion Research Workshop, EPR2017, Vancouver, Canada; (2017)
56. M.W. Binderbauer et al., Dynamic formation of a hot field reversed configuration with improved confinement by supersonic merging of two colliding high-beta compact toroids. Phys. Rev. Lett. **105**, 045003 (2010)
57. H. Gota et al., Overview of C-2W: High temperature, steady-state beam-driven field-reversed configuration plasmas. 28th IAEA Fusion Energy Conference, Nice, France (Virtual) 10–15 May 2021, Presentation IAC/1–1, Paper IAEA–CN–286-0749, Submitted to Nuclear Fusion (2021)
58. D. Yakovlev et al., The gas-dynamic multimirror trap project. 28th IAEA Fusion Energy Conference, Nice, France (Virtual) 10–15 May 2021, Presentation IAC/P4, Paper IAEA–CN–286-1280, Submitted to Nuclear Fusion (2021)
59. R.F. Post, Kinetically stabilized axisymmetric tandem mirrors: a faster route to economic fusion power? AIP Conf. Proc. **1154**, 75 (2009)
60. P.A. Bagryansky et al., Threefold increase of the bulk electron temperature of plasma discharges in a magnetic mirror device. Phys. Rev. Lett. **114**, 205001 (2015)
61. A. Beklermishev et al., Novosibirsk project of gas-dynamic multiple-mirror trap. Fusion Sci. Technol. **63**, 46 (2013)
62. L.A. Berry, C.L. Hedrick, N.A. Uckan, ELMO Bumpy Torus, Oak Ridge National Laboratory Report ORNL/TM-6743, Oak Ridge (1979)
63. A. Komori et al., Density fluctuations and confinement in the Nagoya Bumpy Torus (NBT-1 M). Phys. Fluids B. **3**(4), 1052 (1991)
64. F.B. Marcus, D.R. Sweetman, J.G. Cordey, V.A. Chuyanov, Direct re-injection as a solution to the mirror containment problem. Nucl. Fusion **11**(6), 661 (1971)
65. E.B. Hooper et al., Sustained Spheromak Physics Experiment (SSPX): design and physics results. Plasma Phys. Control. Fusion **54**, 113001 (2012)
66. J.B. Taylor, Relaxation or toroidal plasma and generation of reverse magnetic fields. Phys. Rev. Lett. **33**, 1130 (1974)
67. J.B. Taylor, Relaxation and magnetic reconnection in plasmas. Rev. Mod. Phys. **58**, 741 (1986)

68. Z. Gao, Compact magnetic confinement fusion: spherical torus and compact torus. Matter Radiat. Extremes. **1**, 153 (2016)
69. M. Gobbin et al., Ion heating and energy balance during magnetic reconnection events in the RFX-Mod experiment, 28th IAEA Fusion Energy Conference, Nice, France (Virtual) 10–15 May 2021, Presentation, Paper IAEA–CN–286-1052, Submitted to Nuclear Fusion (2021)
70. L. Marrelli et al., Status of the RFX-Mod2 reversed field pinch upgrade, 28th IAEA Fusion Energy Conference, Nice, France (Virtual) 10–15 May 2021, Presentation EX/P7-4, Paper IAEA–CN–286-1077, Submitted to Nuclear Fusion (2021)
71. J.S. Sarff et al., Tokamak-like confinement at a high beta and low toroidal field in the MST reversed field pinch. Nucl. Fusion **43**(12), 1684 (2003)
72. J.S. Sarff et al., Overview of results from the MST reversed field pinch experiment. Nucl. Fusion **55**(10), 104006 (2015)
73. S. Shoichi et al., Steady-state approach to low-aspect-ratio reversed-filed-pinch nuclear fusion reactor. J. Plasma Fusion Res. **81**(11), 932 (2005)
74. L. Marrelli et al., The reversed field pinch. Nucl. Fusion **61**(2), 023001 (2021)

# Chapter 11
# Inertial Fusion and Magnetic Fast Pulsed Systems

**Abstract** The NIF laser has obtained near breakeven with d-t fusion experiments using pellet compression by indirect-drive. These results and excellent historical documentation provide the basis for a case study with systems forward engineering. Other laser systems have been analysed which were investigating direct drive, fast ignition, shockwaves and non-thermal ion production. These experiments provided data for developing a design space for laser fusion reactors and the fundamental requirements for assembling the elements, leading to analysis of the HiPER fusion reactor design. Magnetic pulsed confinement systems including MagLIF and Z-pinches have produced results that are compatible with a path to a fusion reactor. These inertial confinement systems would have robustness advantages if various technological and physics challenges are met.

## 11.1 Systems Analysis Strategy for Laser Inertial Confinement

### *11.1.1 Introduction to Inertial Confinement*

Inertial confinement involves compressing a solid d-t fuel pellet (or high density plasma) so quickly that it converts the driving energy to densities and temperatures sufficient for self-heating by fusion alpha particles and burn-wave propagation. The fusion energy produced before the target blows itself apart can greatly exceed that of the input energy to the target. An inertial confinement reactor producing 100's of megawatts would probably require pulses of order $10^2$ MJ with a repetition rate of order $10^1$ per second.

The Lawson criteria for energy breakeven in a compressed spherical d-t pellet [1] with temperature about 10 keV, compressed density ρ and radius of the sphere R becomes the areal density (ρR):

$$\rho R \geq 1 \text{ g cm}^{-2} \tag{11.1}$$



F. B. Marcus, *Systems Approaches to Nuclear Fusion Reactors*, Springer Series in Plasma Science and Technology, https://doi.org/10.1007/978-3-031-17711-8_11

For a pellet mass $M_f$ and specific yield $Y^m = 3.39 \times 10^{11}$ J g$^{-1}$, then the total fusion yield Y is:

$$Y = M_f f_b Y^m \tag{11.2}$$

At the Lawson criteria, the fractional burnup $f_b$ is about 10%. Reactor level performance requires ρR in the range of 1–10 g cm$^{-2}$. Reaching fusion conditions requires laser heating powers of about a petawatt (PW), by delivering a few megajoules during a few nanoseconds (ns). The fuel is compressed in density by a factor of 10 in pellet radius, hence about 100–1000 in density. Pellet compression can be achieved by surrounding the pellet with an ablating material that compresses and heats the pellet when bombarded by a rapid energy source, such as a pulsed laser [1].

A high power laser beam incident on a solid creates a high density plasma. The interaction of the light with the plasma is highly non-linear, involving:

- Inverse Bremsstrahlung;
- Resonance absorption;
- Non-linear electron motion in high amplitude fields;
- Non-linear wave interaction and parametric instability;
- Stimulated Raman scattering;
- Stimulated Brillouin scattering;
- Two-plasmon decay;
- Intra-beam transfer.

The implosion is also subject to shock waves, quantum mechanical effects such as Fermi degeneracy, and to a range of hydrodynamic instabilities:

- Rayleigh-Taylor with heavier layer on top;
- Kelvin-Helmholtz;
- Richtmeyer-Meshkov;
- Hydrodynamic mix.

The laser beam itself is subject to a range of phenomena as it is amplified through its various stages which can distort the beam or damage the laser. Experiments and simulation codes need to take account of the full range of these phenomena.

### *11.1.2 Progress in Inertial Fusion Driver Concepts*

A summary of recent progress in the USA inertial fusion programme [2] indicated that there were three main paths to inertial fusion:

- laser indirect-drive (LID) of a solid d-t pellet by laser generated x-rays from a surrounding hohlraum (cavity) to compress the fuel at its centre;

- laser direct-drive (LDD) by direct laser irradiation of a shell around a solid d-t pellet which ablates and compresses the fuel;
- magnetic direct-drive (MDD) to rapidly compress a cylindrical plasma.

These approaches shared many common challenges, including how to:

- efficiently couple the kinetic energy of the implosion to internal energy of the fuel at stagnation (maximum compression);
- dynamically assemble the fuel at the necessary conditions, in pressure, temperature, and areal density ρR;
- initiate self-heating by d-t fusion produced alpha particles;
- obtain ignition and a burn wave.

A major challenge for laser fusion is achieving very high symmetry in the radiation imploding the pellet while producing a very rapid and very high energy pulse. Lack of symmetry leads to the pellet undergoing instabilities, with the result that the centre of the pellet does not sufficiently heat and compress to ignition.

Significant progress was made at three main facilities in the USA:

- The National Ignition Facility (NIF) at Lawrence Livermore National Laboratory with indirect drive;
- The OMEGA laser facility at University of Rochester Laboratory for Laser Energetics with direct drive;
- The Z Machine (linear current-induced pinch) pulsed power facility at Sandia National Laboratories with magnetic direct drive.

During the previous decade on NIF, improved performance was obtained as the target evolved to a high-density carbon-shell ablator. The gains resulted from reduced pellet instability, improved energy coupling and implosion symmetry.

Until the 2021 fusion energy conference, the best results were obtained using LID on NIF [2]. The hot spot temperature reached ~4.5 keV with areal density ~ 0.3 g/cm$^2$, pressure ~ 350 Gbar and fusion energy output >50 kJ. An August 2021 news release from NIF at LLNL claimed that 1.3 MJ of fusion energy were produced from a target implosion.

In the LDD approach on OMEGA, the best-performing spherical direct drive (SDD) implosion has a hydrodynamically-scaled predicted fusion yield in the 0.5 MJ range at NIF laser energy, approaching the burning plasma regime. Polar direct drive (PDD) implosions and laser-plasma coupling research were performed on both OMEGA and NIF for the LDD approach.

In the MDD approach, a hybrid concept, the Magnetized Liner Inertial Fusion (MagLIF) uses a combination of lasers for heating and Z-pinch for compression. MagLIF experiments on the Z machine have produced record yields and stagnation parameters through steady enhancements in the target initial magnetization, laser preheat energy delivered to the fuel, and electrical current delivered to the imploding shells. MagLIF implosions have also demonstrated significant potential for d-t alpha-particle confinement and stopping via the strong magnetic fields entrained in the fuel, a necessary requirement for self-heating and ignition.

### 11.1.3 Challenges for Inertial Fusion Reactor Development

Challenges of inertial fusion reactor development are to:

- achieve ignition and significant overall energy gain at manageable power and energy levels during each pulsed event;
- design a reactor that produces continuous electric power from repeated pulses.

Concepts for inertial fusion included HiPER and FIREX. These designs address the engineering challenges of turning a 1.85 MJ laser pulse of very low efficiency into a high frequency repetitive system producing ignition in a succession of pellets. They designed reactor chambers that contained pulsed fusion power from repeated explosions using circulating liquid lithium to absorb the neutrons and explosive power and breed tritium.

A potential advantage would be in the robustness of these systems, due to the separation of the driving laser from the pellet target and neutron absorbing fusion chamber. If the concepts could be made to work, repairs and maintenance are far simpler than in magnetic confinement systems intricately linked to the plasma.

Reactors based on Z-pinch and plasma focus physics have also been designed, and offer somewhat different challenges for pulsed reactors. A large range of commercial fusion companies have been established investigating new ideas which are analysed and compared from the point of view of robustness and fragility. These new concepts might provide successful solutions and could become important contenders as fusion reactors.

### 11.1.4 Systems Analysis Strategy

This chapter analyses inertial confinement design goals, functions and experimental results in order to examine robustness and fragilities and to learn lessons for the future. The method used for the largest experiment, NIF, is to carry out a case study that summarizes the design, assigns system elements and relations, uses forward engineering to determine the systems architecture and applies robustness analysis. The exercise is briefly repeated for other concepts. A systems approach is used to look at fusion reactor prospects and systems architecture of variants.

## 11.2 Case Study of Architecture and Forward Engineering of NIF

### 11.2.1 NIF Background

The National Ignition Facility (NIF) laser and target systems are comprehensively documented [3] and include a complete history of laser development at the LLNL. This provides the material for a case study of decades of detailed systems architecture and systems engineering. Reference [3] was part of an issue of Fusion Science and Technology in 2016 devoted to the NIF project, including articles especially relevant to systems analysis on control and information systems [4], target development [5], diagnostic systems [6] and operations [7].

The story of the design and construction of NIF is very much in line with the principles outlined for forward engineering of a fusion device in Checklist 1.3B. The history of laser development and systems approaches leading to NIF goes from 1960 when the laser was invented [8] to the completion of NIF in 2012. Extensive prototype development and systems approaches were used that led to NIF.

### 11.2.2 NIF Systems Approach

#### 11.2.2.1 Experimental Prototype Program

Approaches to design were taken that were consistent with a systems approach:

- Work began soon after the invention of the laser to develop powerful short pulse lasers to study the physics of laser operation and of pellet compression and heating.
- Inertial confinement fusion (ICF) was encouraged by fusion neutrons produced in 1974 by privately funded research [9].
- Further results indicated that MJ level lasers would need to be developed.
- Even in the 1970s, computer simulation codes were used to explore hundreds of thousands of potential systems designs of lasers and pellet targets to maximize fusion yield versus cost.
- Each major design cycle led to new lasers to test concepts (Cyclops, Janus, Argus, Shiva, Novette and Nova). Other laboratories also developed high energy lasers.
- These prototype experiments showed that target plasmas could absorb light better at high frequencies, leading to laser beam conversion to the third harmonic of $Nd^{+3}$ (Neodymium) at 0.351 μm to shine on the target.

#### 11.2.2.2 Systems Analysis Codes and Designs

- During the 1980s, systems design mathematical models and experimental results were combined to produce three designs for final consideration.
- Design models showed that cost of a laser scales with amplifier surface area and buildings cost scales with floor space, implying a 3D multi-floor structure.
- The US National Academy of Sciences was asked to review the research program and to make recommendations for studies leading to an ICF facility.
- Systems studies using detailed codes continued into the 1990s to optimize:
  - performance;
  - risk;
  - cost analysis.
- These modelling codes included:
  - laser medium pumping (flashlamp size, duration, reflectors);
  - beam extraction;
  - linear and non-linear beam propagation;
  - damage;
  - frequency conversion from 1ω to 3ω;
  - cost.
- The relatively fast code ChainOp [10] gave a survey of options:
  - laser apertures;
  - pumping intensity and duration;
  - disk counts and thicknesses;
  - disk doping.
- Prop92 was a more detailed physics model [11] and was used to develop the final NIF design [12] and to develop the NIF Level 1 primary design criteria:
  - Laser pulse energy 1.8 MJ;
  - Peak power ≥ 500 TW ($0.5 \times 10^{15}$ W);
  - Wavelength 0.35 μm;
  - Beamlet position ≤50 μm;
  - Recovery time ≤ 8 h between shots and ≥ 250 shots/y.
- and Level 2 functional requirements, including:
  - Laser pulse duration up to 20 ns;
  - Dynamic range ≥ 50:1;
  - Capsule irradiation symmetry of 2 cones with 8 beams each;
  - Laser pulse spot size ≤600 μm.

#### 11.2.2.3 Overall Plant Layout

The site of NIF is the size of three football fields, with the main areas being:

- Four capacitor bays;
- Two laser bays;
- Two beam switchyards including frequency triplers;
- Target bay;
- Optics assembly building.

The capacitor bays rapidly deliver over 300 MJ of energy to power the flashlamps that pump up the desired quantum states in neodymium-doped glass. The laser bays create a total beam energy at 1ω of 4 MJ with a peak power of over $10^{15}$ W, equivalent to over 100 kW for every person on earth. The switchyards convert the beams to 1.8 MJ of energy and focus it on an indirect drive hohlraum about 1 cm long, which converts the beam energy to x-rays that compress and heat a pellet of about 2 mm diameter, with a repetition rate of 1–3 shots per day.

#### 11.2.2.4 Laser and Target Layout

The next level of design is shown in Fig. 11.1, illustrating a schematic view of the laser systems and components along a beamline.

The high level terms in Fig. 11.1 include:

- PAM – preamplifier module;
- ISP – input sensor package;
- PABTS – preamplifier beam transport system;
- MOR – master oscillator room;
- Regen – regenerative preamplifier;
- RPO – relay plane zero;
- ILS – injector laser system;
- LM1, LM2, LM3 - main laser mirrors.

Each subsystem consists of further levels of highly complicated equipment and subsystems.

### *11.2.3 NIF Beamline Systems Architecture and Design Space*

The NIF laser has eightfold symmetry and 48 beams with four sources each giving 192 beamlets. A prototype laser called "beamlet" was constructed to fully prototype and test the concepts. A laser pulse from start to finish used the following components from ICF design space to develop the full system:

- Fibre oscillator – generates the pulse.
- Fibre-optic systems – provides pulse amplification, splitting, spectral conditioning and temporal pulse shaping.
- Diode-pumped regenerative oscillator – uses 58 trips to amplify to 10 mJ.
- Beam expanders and masks – provides rectangular cross section.

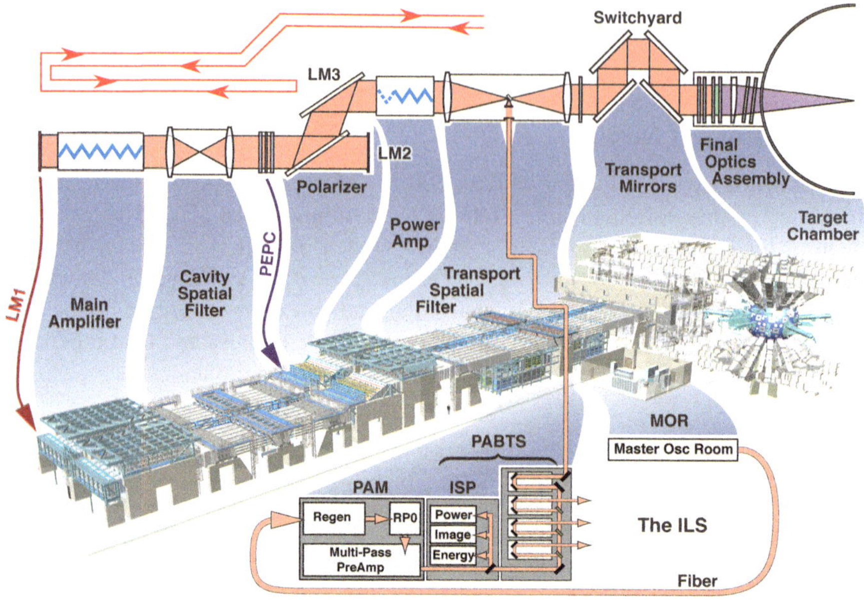


**Fig. 11.1** Schematic view of the laser systems and components along a beamline, indicating their locations on the corresponding CAD drawing. This CAD drawing illustrates one of the two laser bays of NIF. There are two clusters of lasers in each laser bay. Each cluster consists of six bundles; each bundle includes eight laser beams. (Reprinted by permission of Taylor & Francis Ltd, http://www.tandfonline.com on behalf of copyright © American Nuclear Society, http://www.ans.org, Fig. 6 from M. L. Spaeth et al., Description of the NIF Laser, Fusion Science and Technology, 69 (1) 25–145, DOI: 10.13182/FST15-144 (2016))

- Capacitor and flashlamp systems - transfer the energy from 192 capacitor modules, which charged during 60 s and stored 384 MJ, then delivered to the flashlamps during 400 μs.
- Flashlamp pumped Nd glass rods – creates multiple pass amplification.
- Preamplifier beam transport system – removes a small portion of light for diagnostics, splits beam in 4 and relays through the small-optics relay plane.
- Transport spatial filter telescope – establishes small angular separation of multiple laser passes.
- Multiple slab power amplifier – boosts beam to maximum 1ω power.
- Plasma electrode Pockels cell – switches beam linear polarization so beam can travel through previously reflecting polarized mirror.
- Optical switchyard and converter – focuses 4 MJ and 1 PW through converters of 1ω - 3ω and delivers 1.8 MJ and 0.5 PW to centre of target chamber.
- Fusion targets – with a hohlrahm containing a d-t target with coating [5].

### 11.2.4 Highly Modular and Robust Construction

A highly modular and robust construction philosophy of line replaceable units [3] was chosen for NIF. Each of several identical modules was carefully manufactured, assembled, transported and installed at the following stages:

- Innovative optics required many new procedures developed in common with private industry and other laboratories, including the Laser Megajoule project in France.
- The optical assembly building was used to receive optical parts in multiple level clean room environments and assemble them into high specification tolerances modules.
- The beampath infrastructure system (BIS) was used to provide a high specification environment for assembling and linking to modules.
- Robotic transport and assembly systems were used to install modules in BIS and were available for maintenance and replacement of damaged equipment.
- The entire system was carefully aligned, continually sampled and controlled.

### 11.2.5 Operation and Control Systems

- Laser performance operations module (LPOM) software calculated and defined pulse shape parameters, set laser diagnostics, provided analysis and archived performance data, using latest models of beam propagation codes.
- Several dozen diagnostics were used to measure beam and target performance.
- LPOM equipment protection optimized power and energy while keeping operation in a safe range.

### 11.2.6 NIF Physics Results

As NIF was approaching completion in 2009, predictions were made about its fusion energy performance [13]. The main goal was to achieve ignition of a d-t fuel capsule. The physics approach was to use central hot spot ignition of a highly compressed spherical fuel pellet. NIF hoped to achieve in the pellet:

- Temperature 10 keV;
- Density of 1000 g/cm$^{-3}$;
- Pressure of 100 billion times atmosphere.

NIF was designed to use indirect drive with laser beams entering each end of a nearly cylindrical 'hohlraum', have the laser energy absorbed in the walls surrounding the pellet and converted to x-rays which ablated the outer layer of the pellet-containing capsule, thereby providing an opposite force compressing the pellet inwards. The

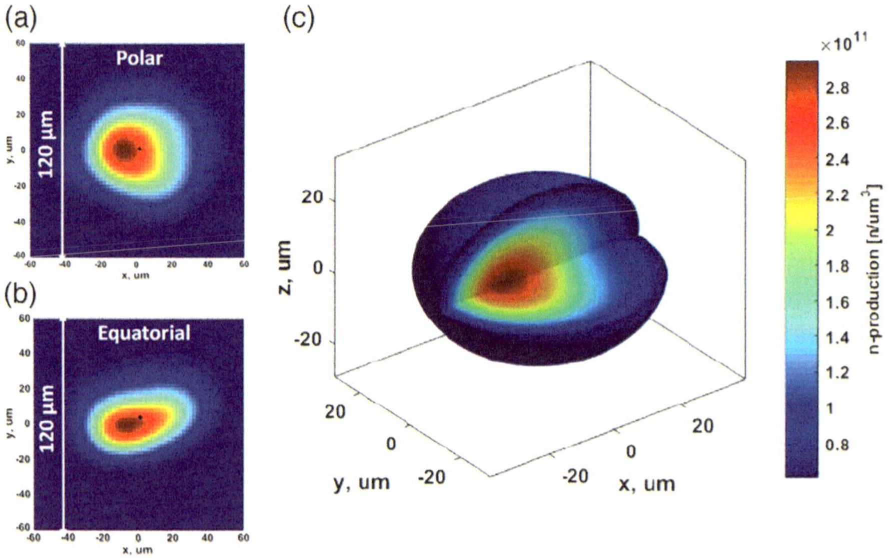


**Fig. 11.2** Neutron images from d-t fusion in compressed NIF pellet, (**a**) polar, (**b**) equatorial, (**c**) reconstructed 3-D volumetric emission. (From [14], Fig. 3)

central hot spot was to have ignited and caused a burn wave to propagate outwards through the remaining fuel.

After initially disappointing results, fusion energy output of twice the kinetic energy of an imploding shell was reported [14]. These experiments used high density carbon ablators with a three-shock laser pulse (1.5 MJ in 7.5 ns) to irradiate uranium hohlraums, resulting in a peak hohlraum radiative temperature ~290 eV. The hohlraum cylinder was 6.2 mm in diameter and 11.3 mm in length, and the target pellet had a radius of 0.91 mm.

The imploding shell composed of the nonablated high density carbon and the d-t cryogenic layer was compressed at 380 km/s resulting in a peak kinetic energy of ~21 kJ, producing a fusion yield of 54 kJ. Figure 11.2c shows a reconstructed 3-D d-t neutron emission profile from the compressed pellet.

Time-dependent low mode asymmetric plasma instabilities were suppressed, leading to an increased compression of the hot spot (which reached areal density of ~ 0.3 g/cm$^2$) and stagnation pressure ~360 Gbar. The maximum 14 MeV emissivity was $3 \times 10^{11}$ neutrons/µm$^3$.

The success of this experiment was the result of using a cryogenic liquid layer impregnated in foam [15] that allowed low to moderate convergence ratios. Output fusion energies were up to 50 kJ, where alpha heating provided a tripling of yield over PdV work [2, 16].

### 11.2.7 Computer Simulations of Inertial Confinement Laser and Target Interactions

The 2D/3D multiphysics inertial confinement fusion simulation code HYDRA [17, 18] developed at LLNL was continuously updated and has been used to model NIF experiments. It fully models the interaction of a time-dependent laser pulse on a hohlrahm inner surface, the generation of x-rays, and the interaction of the x-rays with the target ablator. It then follows the capsule compression, heating, fusion reactions, and ignition and outward burn waves if successful. HYDRA capabilities included:

- a Monte Carlo particle transport library;
- SN (discrete ordinates) multigroup radiation transport package (1D);
- constitutive models for elastic-plastic (strength) effects;
- a mix model;
- a Monte Carlo burn;
- models of diagnostic signatures of neutrons, gamma rays and charged particles;
- A 3D MHD package that treats resistive MHD;
- An implicit Monte Carlo photonics package, including
  - the addition of angular biasing;
  - integrated hohlraum simulations.
- A heavy ion beam deposition package;
- ion stopping power with improved accuracy below the Bragg peak;
- transport of neutrons, gamma rays and light ions;
- products they generated from nuclear and coulomb collisions;
- accurate simulations of nuclear diagnostic signatures from capsule implosions.

In NIF experiments, the full capsule solid angle was modelled in 3D, using a severely rough ablator perturbation and providing diagnostics signatures of capsule failure due to excessive instability growth.

### 11.2.8 Near Breakeven Production of 1.3 MJ of Fusion Energy

An August 2021 news release [19] and a 2022 IAEA webinar [20] reported that 1.3 MJ of fusion energy were produced in NIF from a target implosion, which exceeded the Lawson criteria. The fusion power production was 10 PW lasting 100 ps. This 1.3 MJ fusion yield was an order of magnitude higher than previous results. Continuous progress had been made in upgrading the hohlraum target for indirect drive and laser energy delivered, as well as diagnostics. A previous news release [21] had announced that the laser delivered energy had been increased from 1.8 to 2.1 MJ.

The clear implication is that core ignition had been achieved from fusion alpha particle heating and that an outward propagating burn wave had been initiated in the compressed pellet, which was the goal of NIF. In that case, further progress will certainly be possible by reoptimizing the laser pulse waveform and target to maximize the burn wave, using diagnostic measurements and fusion interpretation models.

### *11.2.9 Laser Inertial Fusion Energy (LIFE) Reactor*

Fusion reactor studies indicated that a pure fusion ICF reactor was possible with energy per event of hundreds of megajoules to several gigajoules, with repetition rates from several per second to once every 10 s in a fusion reactor chamber [22]. The Laser Inertial confinement Fusion-fission Energy (LIFE) reactor concept consists of a modest ICF neutron source surrounded by a subcritical fission fuel blanket. The idea was that the fission fuel would all be burnt and there would be no need for fuel reprocessing.

### *11.2.10 Overall Fusion Efficiency*

NIF has made excellent progress with pellet ignition and propagating burn towards the goal of high fuel burn up, but the overall power balance is still very far from a reactor. The energies involved at various stages of a single pellet shot every 8 h is as follows:

- Capacitor electrical stored energy for flashlamps: 384 MJ;
- Laser energy at 1ω: 4 MJ;
- Laser energy at 3ω: 1.8 MJ;
- Kinetic energy of imploding shell: 0.021 MJ;
- Thermal fusion energy produced: 0.050 MJ (late 2021 was 1.3 MJ).

Thus the current overall Q of the total system is under 1%, and the repetition rate is $3 \times 10^{-5}$ less than needed. If and when high pellet fuel burn up is achieved, new technologies will be required for a functioning fusion reactor.

The Mercury program was initiated [23] to develop a laser that could be used for ICF in a commercial fusion reactor. Hundred-joule, kilowatt-class lasers based on diode-pumped solid-state laser (DPSSL) technology were being developed. The project goals were to develop building blocks that could be scaled to beamlines at 10 Hz repetition rate, 10% wall-plug efficiency, and $10^9$ shot reliability, based on a transition from flashlamp pumping to diode pumping. The Mercury laser consisted of two amplifier assemblies pumped by four 100 kW diode arrays to produce a gain of $10^5$ with 100 J of potential output energy. This work established a pathway for workable ICF reactor concepts.

### *11.2.11 Laser Mega-Joule (LMJ) and ShenGuang III (SG-III) Large Scale Indirect Drive Lasers*

The Laser Mega-Joule (LMJ) project [24] was designed and built in France with similar technology and similar design parameters to NIF, 1.8 MJ delivered at 3ω, and for similar purposes. The two projects shared some optics development. LMJ was to be accompanied by the ultrafast PETAwatt Laser (PETAL). One of the LMJ goals was to obtain ignition and burn of DT-filled capsules via an indirect-drive scheme inside a rugby-shape hohlraum. The PETAL project would add one short-pulse (ps) ultra-high-power, high-energy beam (kJ) to the LMJ facility. PETAL offered a combination of a very high intensity multi-petawatt beam, synchronized with the nanosecond beams of the LMJ. This combination was planned to expand the LMJ experimental range to include fast ignition scenarios. Similarly to the "Beamlet" project at NIF, a prototype called Ligne d'Intégration Laser (LIL) was constructed and tested.

The high power laser in China is called ShenGuang III [25]. Its research also emphasizes inertial confinement by indirect drive. It has 48 beams with an aperture of 400 × 400 mm. Its maximum energy was designed to be about 180 kJ within 3 ns duration and 351 nm wavelength. Some of their experiments explored the laser-target coupling process including investigations of hohlraum radiation flux and laser energy coupling efficiency, as well as ablation and implosion physics, including shell asymmetry and implosion trajectory. Other work explored stagnation, hot-spot dynamics and the nuclear phase of the implosion.

Other high power lasers have been built, including ORION [26] in the UK.

## 11.3 Direct Drive, Fast and Shock Heating and Alternate Laser Technologies

As alternatives to indirect drive as in NIF, groups have been investigating other capsule drive mechanisms and other laser technologies to approach pellet ignition with lower energies which might lead to an optimum solution for a reactor.

### *11.3.1 PETAL*

The PETAL project [24] involved the addition of one short-pulse (500 fs to 10 ps) ultra-high-power, high-energy beam (a few kJ) to the LMJ facility. PETAL was also directly aimed at testing fast ignition concepts for the European HiPER laser fusion reactor project (see below). PETAL was based on chirped pulse amplification (CPA) plus optical parametric amplification (OPA).

The PETAL front end consisted of a Ti:sapphire oscillator delivering 3nJ and 100 fs pulses at 1.053 μm wavelength. The pulse was stretched to 9 ns in order to avoid optical damage at full energy. The PETAL main amplifier section had the same architecture as LIL and LMJ using a single 370 × 370 $mm^2$ beam able to deliver up to 6 kJ. A first compressor was used to reduce the pulse duration from 1.8 ns to 350 ps in an equivalent double pass configuration. The output mirror was segmented in order to divide the initial beam into 4 sub-apertures which were independently compressed and synchronized into the second compressor in a single pass configuration under vacuum. These sub-apertures were coherently added using a segmented mirror with three interferometric displacements. The pulse duration was adjustable from 0.5 to 10 ps.

## *11.3.2 Direct Drive Laser Fusion with OMEGA*

### 11.3.2.1 OMEGA Laser

The OMEGA laser [27] reached a performance level of 60 beams producing a laser pulse of 30 kJ at 0.351 μm [28]. OMEGA was built as a Nd glass laser, using rod amplifiers up to a beam diameter of 9 cm, followed by disk amplifiers up to an output aperture of 20 cm. Each amplifier was followed by a spatial filter. The 60 beams were expanded in the final spatial filters to a 28-cm aperture and converted to 0.351 μm using KDP crystals in the frequency-tripling configuration developed at University of Rochester Laboratory for Laser Energetics. The beams were uniformly distributed around a 3.3 m diameter target chamber and focused directly onto the target.

### 11.3.2.2 Direct Drive Experiments

In experiments on OMEGA [28], a record fuel hot-spot pressure of 56 Gbar was obtained using direct-drive onto a d-t pellet. Relative to symmetric, one-dimensional simulations, the inferred hot-spot pressure was approximately 40% lower. One-dimensional simulations by the code LILAC [29] and three-dimensional simulations by the code ASTER [30], suggested that hydrodynamic instabilities and low-mode drive instabilities were responsible for reduced target performance, as illustrated in Fig. 11.3. The mass density profile just before stagnation and just afterwards as modelled in 3D show the instability build-up. The different time dependencies of 1D and 3D show how these instabilities affect the neutron rate.

More recently, statistically informed physics understanding was used for the design optimization of direct drive experiments [31]. This analysis implied that beam-target radius degradation could be ameliorated by drive-phase zooming, leading to high performance. Higher yields could be obtained by minimizing mispointing of laser beams and by reducing $^3$He build-up in the target due to

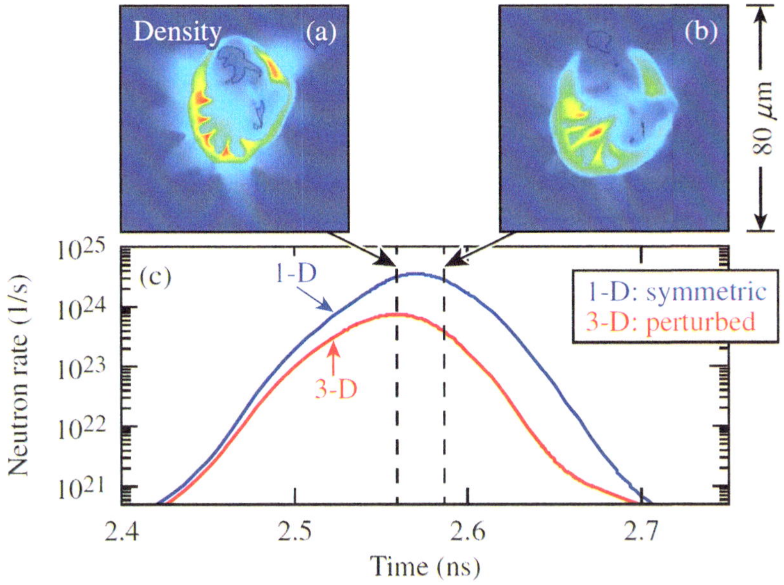


**Fig. 11.3** Simulations of hot spot (**a**) mass density near stagnation, (**b**) mass density at later time, (**c**) 1-D and 3-D calculated neutron rate. (From [28], Fig. 3)

radioactive decay of tritium. Simulations implied that direct drive used on NIF could lead to megajoule yields.

#### 11.3.2.3 Shock Wave and Blast Wave Heating

Direct-drive laser absorption, mass ablation rate, and shock heating were experimentally studied [28] on OMEGA to validate hydrodynamics simulations. Shock heating was diagnosed in planar-CH-foil targets. The shock wave heating [32] was examined where massive cryogenic shells were first imploded by direct laser light with a low implosion velocity. Next, a spherically convergent ignitor shock and the return shock created a hot-spot pressure greater than the surrounding dense fuel pressure. The thermonuclear gain could be significantly increased.

Shock wave experiments [33] on OMEGA also observed elevated heating in dynamically compressed CH foam.

Shock waves form an important part of programming pulses at NIF [34] where four timed shocks are part of the laser pulse delivered to the target.

The First Light Fusion Company [35] also has shock waves as a main part of their program.

## *11.3.3 Gekko and LFEX Lasers for Fast Ignition Realization EXperiment (FIREX)*

### 11.3.3.1 Firex Concept

At the Institute of Laser Engineering of Osaka University, the goal was to develop lasers and targets that could produce fast ignition in a pellet target, leading to the Fast Ignition Realization EXperiment (FIREX) project [36]. During fast ignition, the fuel is first compressed and then ignition is initiated by a very short, very high power laser pulse incident on the already compressed fuel core. Fast ignition might achieve ignition and burn with about one tenth of laser energy required otherwise. FIREX's goal was to demonstrate ignition temperatures of 5–10 keV.

### 11.3.3.2 Gekko and LFEX Lasers

The method used to study fast ignition was to construct two lasers, Gekko-XII and Laser for Fast Ignition Experiment (LFEX) [37, 38]. A block diagram of LFEX is shown in Fig. 11.4. The Gekko- XII laser, which was used for pellet compression, produced

- 0.527 μm light;
- energy of 1.5–4.5 kJ in total in a 1.5 ns pulse;
- both Gaussian and nearly flat-top pulse;
- nine beams among 12 available.

The LFEX laser used for fast heating, emitted:

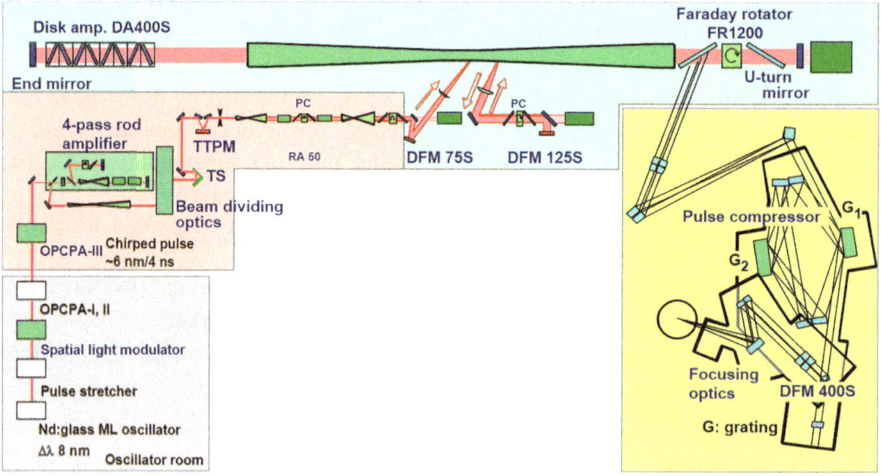


Fig. 11.4 Block diagram of LFEX. (From [37], Fig. 1)

- 1.053 μm light;
- energy up to 1–2 kJ in 1–5 ps;
- beam(s) that were focused and injected into a cone attached to a shell target to reach the target centre.

Shell targets were composed of:

- CD: deuterated polystyrene;
- 500 μm in diameter;
- 7 μm in thickness;
- A 10–20 μm wall-thickness gold cone with an opening angle of 30° to 45°;
- distance from the centre of the shell to the cone tip was 50 μm.

By 2016, LFEX had been upgraded to four beamlines and was able to deliver 2 kJ pulses in 1 ps for a 2 TW peak power, the world's highest [39]. It was designed to increase the central plasma temperature to 5 keV.

There are a number of special features of this laser system, consistent with achieving extremely high power that is concentrated in the tiny volume of a compressed pellet core. The LFEX was the first chirped-pulse amplifier (CPA) laser system with a kilojoule pulse energy. The front end generated a kilojoule nanosecond chirped pulse with a mode-locked fibre oscillator, pulse stretchers, optical parametric amplifiers, rod amplifiers, and a main amplifier. The rear end was a large pulse compressor using gratings and mirrors for the chirped pulse. About 70% of the pulse energy could be concentrated onto the target, with a spot size diameter of 5Fλ, where F is the F number of the lens, in this case F/10, and λ is the wavelength of the laser of 1.053 μm, giving a spot diameter of about 53 μm, of the order of the size of a compressed pellet hot spot.

#### 11.3.3.3 Fast Ignition Experiments

LFEX was used to create a source of intense ions for ion-driven fast ignition [40]. The ps pulses from LFEX were focused simultaneously with ns pulses from Gekko-XII onto the target chamber. The Gekko pulse compresses the pellet. The LFEX beam travels through a cone shielding the beam from plasma surrounding the target. Instead of being focused onto the target, the beam hits a thin foil which generates fast ions heading for the compressed core.

During experiments on fast ion generation, the laser pulse produced a peak intensity of $1.2 \times 10^{19}$ W/cm$^2$. Protons with energies exceeding 50 MeV were produced. The goal was to use a thin film to deliver the ions to the core.

Experimental results for direct fast heating of a compressed pellet core [41] investigated the heating efficiency, defined as the increment of internal core energy divided by heating laser energy on target. Six beams of the GEKKO XII (GXII) green laser at 1.6 kJ imploded a spherical CD (Deuterated polystyrene) shell target of 500 μm in diameter and formed a dense core of just under 3 g/cm$^3$, with a core temperature of about 700 eV. The core was then directly heated by LFEX, either on or perpendicular to the GXII bundle axis. The efficiencies reached 5% for the perpendicular mode, compared to 3% for a uniform implosion mode. The high power mode yielded temperature increases of the core of about 200 eV and energy of about 15 J. They have also proposed a counter-illumination fusion reactor, named CANDY.

## 11.3.4 The UK Central Laser Facility (CLF) and DiPOLE

### 11.3.4.1 CLF

The UK Central Laser Facility (CLF) [42] concentrates the UK's high power research lasers (apart from ORION):

- Artemis – ultrafast XUV science facility;
- OCTOPUS – imaging cluster;
- Gemini – high power, ultra short pulse (15 J, 30 fs, every 20 s);
- Vulcan – petawatt laser system;
- Ultra – laser, detector and sample manipulation technology.

### 11.3.4.2 Diode Pumped Optical Laser for Experiments (DiPOLE)

The DiPOLE 100 laser system [43] was delivered by the CLF team to the HiLASE Centre in Prague in 2016 and was recently operated for an hour at 10 pulses per second with a pulse energy in excess of 145 J. This is an increase of nearly 50% more than the system's original specification. In addition to demonstrating high efficiency laser pumping with cryogenic cooling relevant to laser fusion reactors, the system can help experiments achieve better data by detecting weak signals against background noise.

## 11.3.5 Petawatt and Exawatt Lasers

Laser fusion reactor designs that rely on developing fast ignition lasers are supported by the rapid progress in extremely high power lasers. A general review of such lasers [44] noted that with the 2018 Nobel Prize in Physics being awarded to Donna Strickland and Gerard Mourou for chirped pulse amplification (CPA), ultra-high-power lasers and their technology has rapidly developed, resulting in multi-petawatt facilities. Lasers up to 100 PW are under construction. Technology for much higher repetition rates and high-average powers with higher efficiencies were being developed. In addition to the CPA technique, a stretched pulse can be amplified by using three-wave mixing, called Optical Parametric Chirped Pulse Amplification (OPCPA) [45].

Whole new domains of physics and plasma heating are accessible with these lasers [46]. Ultra-short high-power lasers such as CoRelS in Korea can provide intensities up nearly $10^{23}$ W/cm$^2$, which can instantaneously turn matter into plasma. The electric field can drive an electron to relativistic speeds in a fraction of an optical period. These lasers offer the prospect of fast ignition.

### 11.3.6 Non-thermal p-$^{11}$B Fusion from Picosecond Lasers

The fusion reaction proton boron p-$^{11}$B generates three $^4$He (alpha) particles plus 8.7 MeV, thus making it a prospect for a completely non-radioactive fusion reactor. However, the fusion cross section peaks around 600 keV and achieving ignition of a thermal fusion plasma is much more difficult for p-$^{11}$B. Using fast laser pulses to create non-thermal ions [47] was investigated on the Prague Asterix Laser System (PALS). The energy delivered was 500 J in 0.3 ns [48] and produced $10^9$ fusion alphas per steradian from p-$^{11}$B. With a thick boron nitride target, a yield of $10^{11}$ $\alpha$-particles was obtained [49].

HB11 Energy [50] was founded to develop this concept using a chirped-pulse amplifier (CPA) laser [47, 51]. Marvel Fusion [52] also explored these ideas.

A proposal for a fusion reactor [53] used a first laser with a coil to produce an ultra-high magnetic field and a second laser ignited the reaction, based upon:

- Non-thermal fusion reactions were observed [48].
- The p-$^{11}$B fusion rate [54] indicated an avalanche multiplication.
- Kilo-Tesla fields could be created using a pulsed laser [55].

However, interpretation of the PALS experiments ruled out [56] any contribution coming from avalanche fusion. Further calculations [57] showed that avalanche reactions could only enhance the fusion rate by 1% at most.

## 11.4 Design Space and Systems Approaches for Laser Fusion

### 11.4.1 Design Space for a Laser Inertial Confinement Fusion Reactor

The fundamental problem in developing a design space is that current lasers such as NIF use technology suitable for a single shot. The technologies required for a reactor will need significant re-engineering of the laser and target systems and their interactions. The processes of laser and target development and their detailed operation are highly iterative, interactive and complex.

#### 11.4.1.1 Elements of a Design Space

In developing a design space, the current research paths are summarized and organized, even if some paths are more or less promising at the current time. Examples of a possible design space for laser fusion reactors are:

- type of target
  - indirect-drive with hohlraum

    cylindrical shape
    rugby ball shape
    uranium plus gold hohlrahm
    liquid d-t fuel foam imbedded
    solid d-t fuel

  - direct drive

    uniform ablator
    uniform ablator plus cone for fast ignition

  - fuel

    d-t
    p-$^{11}$B if ultrafast ignition plans are successful
    d-d catalysed t production
    tritium breeding in target or in surrounding lithium

- characteristics of target
  - total fuel content and fusion yield
  - thickness and spatial distribution of fuel
  - thickness, distribution and composition of ablator
  - shape, thickness and composition of hohlraum (if indirect drive)
  - fuel composition and distribution
- type of laser
  - single shot
  - high repetition rate

    solid state pumped diode
    KrF gas

- characteristics of laser
  - total beam pulse energy, pulse shape and instantaneous power waveform
  - number of beamlets and spatial distribution on target

- wavelength or mixture of wavelengths
- spatial profile and polarization of individual beamlets
- single or multiple types of lasers and optics for their combination (e.g. compression laser plus fast ignition laser)

• highly simplified characteristics of laser and target interaction

- with indirect drive, interaction of laser with hohlraum for plasma and x-ray production and spatial distribution and timing of x-rays hitting pellet ablator
- laser-plasma interactions and instabilities and reflections
- compression timing and shock waves
- adiabat profile with combines density compression and heating
- physics of core compression and heating and ignition
- burn wave outwards

• type of fusion chamber

- Dry-wall
- Wetted-wall
- Thick-liquid wall
- Solid/voids

• characteristics of fusion chamber

- single pellet and overall power and energy handling capability
- first wall design and maintenance
- tritium breeding blanket, shield, cooling
- equipment and personnel shielding
- remote handling and maintenance requirements
- activation
- gas pumping and recovery after shot for next shot
- target delivery and tracking
- diagnostic access and diagnostic systems

• balance of plant buildings and systems

- power supplies
- gas and tritium handling
- tritium breeding and extraction
- remote handling and maintenance
- laser diagnostics and maintenance
- fusion chamber diagnostics and maintenance
- coolant circulation
- heat extraction and electric power generation
- operations and data analysis computers
- staff operations

#### 11.4.1.2 Fundamental Requirements for a Laser Fusion Inertial Confinement Reactor

For any laser fusion reactor, a certain number of criteria need to be demonstrated:

- optimizing pellet and laser design together – To reach ignition and create a burn wave, all aspects of the laser and pellet target must be optimized consistently.
- target pellet ignition and burn wave – a target pellet, most probably of d-t composition, needs to be ignited and a burn wave produced that produces much more fusion energy than was input by laser.
- large plant energy gain – For each pellet explosion, on average, a total fusion power gain must be large enough to generate net electricity economically. For example, with diode flashlamps at 15%, 1 to 3ω efficiency conversion of 50% and thermal conversion efficiency of fusion products of 30%, a 2 MJ laser pulse to a target would need to produce a fusion energy of 89 MJ just to obtain energy breakeven.
- reactor relevance – Once a target pellet ignition and burn is demonstrated, it must be replicated with reactor-relevant components, especially laser and target. Repetitive pellet target positioning and orientation is a huge challenge.
- a rapid pulse repetition rate at full energy – With a repetition rate of 10 per second, to produce 1000 MW (e) would require 333 MJ per pulse over the breakeven energy, or about 422 MJ per pulse. Multiple chambers could reduce this.
- cost effectiveness – The typical cost of 1 kWh = 3.6 MJ is about $0.15 in 2021 and higher in Europe, with both much higher in 2022. The 100 MJ of electricity per pulse is worth about $4. Therefore, the cost of each pellet plus shot averaged costs for other systems must be much less than this.
- pellet target manufacture, delivery to target chamber, positioning, and tracking systems and laser aiming systems – The whole pellet manufacture and delivery and tracking and delivery chain must be highly reliable and economical at 10/second repetition rates

In addition to demonstrations of fusion reactor relevance of any concepts, the following are needed for any laser fusion reactor:

- laser protection and maintenance – all parts must be monitored, testable and easily replaceable.
- target chamber – The target chamber needs to be ready anew for each pulse, guide the target to where lasers can produce fusion, absorb the plasma and neutron blast to the walls, protect the lasers from backscatter and neutrons, pump out used fuel including tritium, breed tritium in the surrounding neutron blanket and coolant, protect outside personnel
- cooling systems – the heat must be carried away and transferred to electric power generators and heat removal systems.
- cryogenic systems – are needed for pellet manufacture

- tritium breeding and gas handling systems – tritium needs to be separated from both breeding blanket and other gas sources.
- remote handling and maintenance – are essential.

## 11.4.2 *HiPER Laser Inertial Confinement Fusion Reactor Project*

### 11.4.2.1 Baseline HiPER Design

The HiPER (European High Power laser Energy Research facility) project [58–60] was intended to demonstrate the feasibility of inertial fusion energy production by lasers. An IFE reactor would consist of a driver delivering pulses of a few MJ of energy at rate of 5–20 Hz, inducing compression, ignition and burn of targets containing a few mg of deuterium-tritium mixture. High gain could potentially be achieved by direct laser compression and advanced ignition schemes, such as fast ignition and shock ignition.

The project studied targets compressed by means of a multi-beam, multi-ns laser pulse of about 250–300 kJ and either:

- fast ignited by an ultra-intense, pulse of about 100 kJ delivered in 15 ps and focused onto a spot of diameter about 30 μm or
- shock ignited by 300–500 ps multi-beam pulses of similar total energy, but irradiating a large fraction of the target surface.

Compression beams were considered that delivered 250 kJ and a peak power of 50–60 TW in a 10 ns pulse at 3ω to be focused as 48 focal spots on the capsule, providing uniform irradiation. A main objective was to identify the least expensive and most useful driver but not to preclude any alternative laser. HiPER decided to use DPSSL technology [23] compared it to NIF type flashlamps [61]. The HiPER target chamber, final optics and shielding are shown in Fig. 11.5.

The target chamber had 48 apertures and included a first wall able to resist the shock of plasma and neutrons, a breeding blanket and coolant. The final optics assembly had an image relay and a pinhole. A 4-m focal length was used together with a 5 m radius target chamber. The pinhole location was 16 m from the chamber centre. There are four sets of shields to protect optics and workers:

- The first one was placed against the reaction chamber, consisting of a spherical shell of 0.8 m of thickness, made of borated concrete.
- The final optics assembly shield was 16 m from the origin and 2.5-m thick, made of standard concrete.
- The 48 pinhole shields are inside the beam tubes with square sections of $1 \times 1\ m^2$ and 2.5 m long.
- Outside there was a cylindrical shell shape bio-shield of 25 m inner radius and 2 m thick.

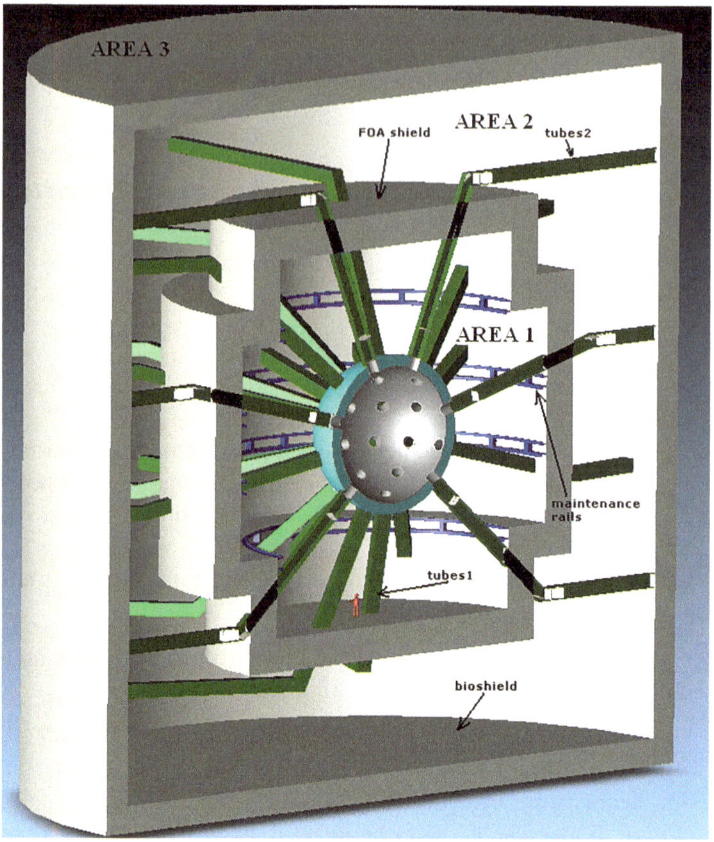


**Fig. 11.5** HiPER target chamber, shielding and beam access optics. (From [59], Fig. 5)

The protection of final optics was still an open question, since fast ions must be removed and the effect of neutrons, gamma and X-rays needed to be addressed.

#### 11.4.2.2 Systems Approaches to HiPER Reactor Aspects

The HiPER team considered all aspects of reactor design, and published detailed papers of their studies [60] in the following areas:

- Reactor Concepts and Materials
- Ignition Physics Reviews and Project Status
- HiPER Baseline Target: Preliminary Definition
- Shock Ignition: HiPER Baseline Target
- Shock Ignition: Modelling, Scaling, Target Design
- Shock Ignition on NIF or LMJ
- Individual Issues

- Target Irradiation Schemes
- Fast Ignition
- Alternative Concepts: Fast Ignition
- Other Topics in Ultra-Intense Laser Plasma Interaction
- Target fabrication and diagnostics
- Laser Physics and Architecture
- Laser Systems
- Target engagement and tracking

They also explored the physics and technology of reactor systems:

- Laser Driver
- Fusion Chamber Design
- Target Design and Physics Modelling
- High Volume Target Manufacture
- Target Injection and Tracking
- Control Systems
- Supporting Plant and Machinery
- Building

Some years previously, a detailed examination of target production and delivery of a target stream to the fusion chamber concluded that there were no major obstacles in that area [62]. As an example of subsystem examination, HiPER considered all aspects of target design and delivery. Details of pellet design and links to laser design were examined [58].

### *11.4.3 Detailed Systems Modelling of HiPER*

A detailed systems model described as a "Virtual Reactor Model" was developed to model the entire plant aspects of HiPER [63]. The modelling involved:

- Operational & Functional Analysis
  - Identification, organization and characterization of operations and functions;
  - Production of supporting models (trees, scenarios...);
  - Derivation of "system" requirements.
- Model Based System Engineering
  - Formal modelling of the operations and functions (scenarios, flow ...);
  - Formal modelling of the functional and physical architecture;
  - Derive "technical" requirements;
  - Modelling of specifications.

This general framework was used to develop models of:

- Materials supply
- Target fabrication & storage
- Tritium reprocessing
- Materials Handling & Treatment
- Target injection
- Chamber
- Laser
- Heat exchanger & steam turbine generator
- Energy Bank & Auxiliary devices
- Building
- Grid

The interactions between these systems were considered in terms of:

- Materials
- Synchronism
- Alignment
- Energy

In detailed model development for HiPER [60], the goals were to define the main functions that constitute the reactor assembly, simulate and optimize them, including the interfaces and the workflow of the functions with The Virtual Reactor Model. For a specific function, the model development would have consisted of developing a specific platform or Virtual Lab. Each of these platforms would have integrated 2 levels:

- The first level included complex detailed models that would be analysed through multiple simulation runs. Reduced models or meta-models were created in association with the results of these runs.
- The second level capitalized on the reduced models to speed up the run times to feed a knowledge data base as the key element for generating large scale numerical experiments.

## 11.4.4 The Future for Laser Fusion Reactors

In summary, extensive and continuing progress has been made on the path to pellet ignition and an outward burn wave and that further optimization is possible.

Many groups are developing technologies that could lead to more efficient ignition techniques at lower laser energy. Technological developments in improving laser efficiency hold out the prospect for net energy production in an eventual laser fusion reactor. Laser inertial confinement fusion therefore seems to be a viable option for future power producing reactors.

## 11.5 Magnetic Fast Pulsed Systems

Over many decades, rapidly pulsed magnetic confinement systems powered by capacitor banks, such as theta pinches and z-pinches, have been explored as potential neutron sources and fusion reactors. In some cases force equilibrium has been established for brief times, so whether they are inertial systems is debatable. They are included here because of their highly pulsed nature.

### *11.5.1 Dense Plasma Focus*

A dense plasma focus (DPF) [64] is a coaxial plasma gun which utilizes a pulsed high-current electrical discharge to heat and compress the plasma to very high densities and temperatures with energy densities in the range of $1–10 \times 10^{10}$ J/m$^3$ to act as an intense neutron source. Experiments conducted on several devices showed that neutron production scales much more slowly than hoped for. It is not currently expected that the dense plasma focus can scale to produce net energy gain in a fusion reactor. However, the DPF has important prospects for use as a neutron source or in nanotechnology applications.

### *11.5.2 Plasma-Jet-Driven Magneto-Inertial Fusion (PJMIF)*

A concept known as plasma-jet-driven magneto-inertial fusion (PJMIF) [65] involved a $\beta$ greater than 1 magnetized plasma as a target to be compressed to fusion conditions by a spherically imploding plasma liner. Compressional heating dominates over thermal transport in the target and magnetic amplification due to compression dominates over dissipation over the entire implosion. The feasibility of a proof-of-concept experiment using presently achievable plasma jets was evaluated. The Plasma Liner Experiment (PLX) [66] tested these concepts with two merging supersonic plasma jets formed and launched by pulsed-power-driven railguns. They concluded that much gun development was still needed for a fusion-scale demonstration of PJMIF.

### *11.5.3 MagLIF Z-pinch Experiments*

Z-pinches used a large current flowing through linear plasma to generate a poloidal magnetic field which compressed the plasma which was intensely heated by the large current density in the pinch. The name Z comes from the current flowing in the z-direction in cylindrical coordinates. Initially these pinches were highly unstable,

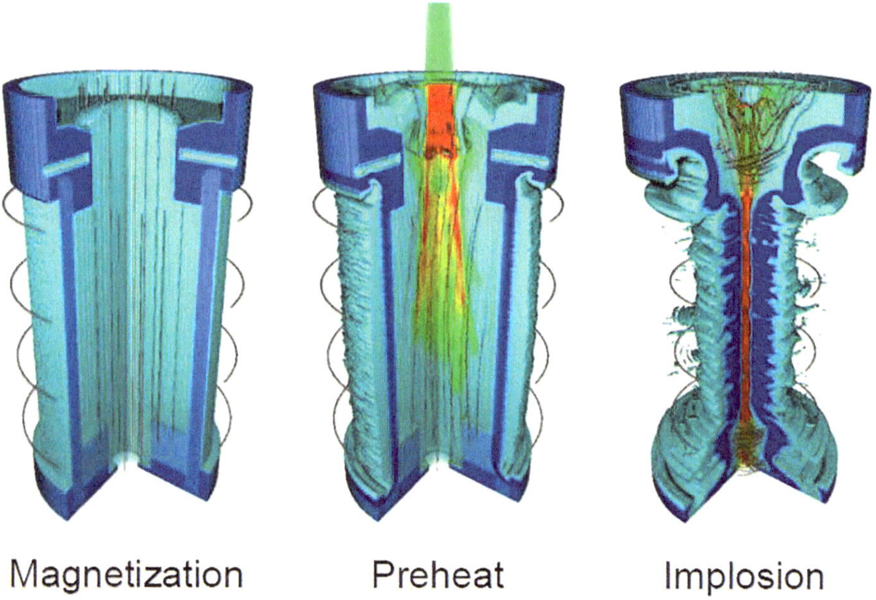


**Fig. 11.6** MagLIF stages of operation. (From [68], Fig. 1)

but various stabilization techniques were developed. Advances were sufficiently impressive that a couple of decades ago, Z-pinch based fusion reactors [67] were being designed.

Advances in plasma heating and confinement were achieved using the Magnetized Liner Inertial Fusion (MagLIF) concept [68]. Magneto-inertial fusion (MIF) utilizes magnetic fields to improve confinement and to limit thermal conduction. On the Z Machine at Sandia National Laboratories, MagLIF was the first MIF concept to demonstrate fusion-relevant temperatures, significant neutron production, and magnetic trapping of charged fusion products. They consider that it has the potential to generate multi-MJ yields and significant self-heating of the fuel.

In MagLIF experiments, a cm-scale metal liner was filled with deuterium and pre-magnetized using an axial magnetic field of 10–20 T. The plasma was then pre-heated to 100–200 eV via a kilojoule laser and finally radially imploded over ~100 ns via a current in the liner to velocities of ~70 km/s using 15–20 MA of current from the Z Machine. This process is schematically demonstrated in Fig. 11.6.

The axial magnetic field, which was flux-compressed to >1000 T near peak convergence, limited thermal conduction losses from the hot fusion fuel to the comparatively cold liner walls during the implosion and simultaneously increased trapping of charged fusion particles in the narrow radial direction during stagnation. Ion temperatures up to 3 keV and d-d neutron yields up to $10^{13}$ were achieved. The magneto Rayleigh-Taylor instability was relatively stabilized by dielectric coated liners. Experiments with simultaneous improvements in laser preheat, current delivery to the line and axial applied magnetic field would have increased performance by an order of magnitude to the equivalent of 2 kJ of energy if d-t fuel had been used. Simulations indicated that a next-generation Z machine with 60 MA could give a fusion gain of 70, making MagLIF a contender to be a viable source of fusion energy.

### *11.5.4 Z-pinch Fusion Reactor*

Another approach to Z-pinch fusion reactors is to use sheared-flow stabilization [69] as is being developed by ZAP Energy. This stabilization method produced an equilibrium Z pinch, which sustained the compressed plasma state for durations much longer than other plasma timescales. Recent experimental and simulation results provided encouragement for reactor development.

Earlier reactor studies [67] indicated that the Z-pinch could become an economically-attractive power plant using high-yield targets (~3 GJ) with low repetition rate per chamber (~0.1 Hz). The power plant would have used a repetitively pulsed power linear transformer driver, a recyclable transmission line (RTL), a dynamic hohlraum z-pinch-driven target, and a thick liquid-wall chamber. The RTL connected the pulsed power driver directly to the z-pinch-driven target and was made from frozen coolant or a material that was easily separable from the coolant. The RTL would have been destroyed by the fusion explosion, but its materials recycled and a new RTL inserted on each shot.

## 11.6 Prospects and Systems Robustness of Inertial Fusion Reactors

Recent experiments with lasers, especially at NIF with more than a megajoule of fusion power produced, have indicated that there are real prospects for d-t burn that could lead to a fusion reactor. Laser technology is being developed that gives much better overall energy efficiencies than are currently used in the largest lasers. New schemes are being developed to allow the possibility of laser fusion ignition with lower input energies. The relative separation of the laser and the fusion chamber offer important technological advantages over magnetic fusion devices. The robustness of an inertial confinement laser fusion system could be high, since the laser itself is very highly modular. The target chamber is relatively separate and easily accessible for repair and maintenance, as well as being relatively low cost compared to an integrated magnetic fusion system. However, there are still considerable interface challenges.

MagLIF and Z-pinch machines, although very closely coupled to the target and still operating on a single shot basis, do have some technological solutions to these problems and a fusion reactor is also conceivable. The need to replace part of the system after each pulse is a considerable challenge for the Z-pinch.

In summary, inertial confinement schemes seem to have viable paths towards fusion reactors.

## References


1. E. Morse, *Nuclear Fusion* (Springer, Cham, 2018)
2. P. Patel Progress in the U.S. inertial confinement fusion program. 28th IAEA Fusion Energy Conference, Nice, France (Virtual) 10–15 May 2021, Presentation OV/3-1, Paper IAEA-CN-286-1168, Submitted to Nuclear Fusion (2021)
3. M.L. Spaeth et al., Description of the NIF laser. Fusion Sci. Technol. **69**, 25 (2016)
4. G. Brunton et al., Control and information systems for the National Ignition Facility. Fusion Sci. Technol. **69**, 352 (2016)
5. A.V. Hamza et al., Target development of the National Ignition Campaign. Fusion Sci. Technol. **69**, 395 (2016)
6. J.D. Kilkenny et al., The National Ignition Facility diagnostic set at the completion of the national ignition campaign, September 2012. Fusion Sci. Technol. **69**, 420 (2016)
7. B.M. Van Wonterghem et al., Operations on the National Ignition Facility. Fusion Sci. Technol. **69**, 452 (2016)
8. T.H. Maiman et al., Optical and microwave – optical experiments in ruby. Nature **187**, 493 (1960)
9. KMS Fusion, Wikipedia. https://en.wikipedia.org/wiki/KMS_Fusion. Accessed 17 June 2022
10. R.A. Sacks et al., Propagation modelling in two transverse dimensions of the National Ignition Facility baseline performance. 2nd Annual International Conference on Solid State Lasers for Application to Inertial Confinement Fusion, 9610025-38, Paris (1996)
11. R.L. Gelinas et al., Optical propagation modelling for NOVA and beyond. Lawrence Livermore National Laboratory Inertial Confinement Fusion Annual Report, UCRL-LR-105820-92, 11 (1992)
12. J.T. Hunt et al., Laser design basis for the National Ignition Facility. Fusion Technol. **26**, 767 (1994)
13. E.I. Moses, Ignition on the National Ignition Facility: a path towards inertial fusion energy. Nucl. Fusion **49**, 104022 (2009)
14. S. Le Pape et al., Fusion energy output greater than the kinetic energy of an imploding shell at the National Ignition Facility. Phys. Rev. Lett. **120**, 245003 (2018)
15. R.E. Olson et al., Liquid DT layer approach to inertial confinement fusion. IAEA FEC Conference, Paper IFE/1-3, Ahmedabad (2018)
16. J.L. Kline et al., Progress of indirect drive inertial confinement fusion in the United States. Nucl. Fusion **59**(11), 112018 (2019)
17. M.M. Marinak et al., Advances in HYDRA and its application to simulations of inertial confinement fusion targets. Eur. Phys. J. Conf. **59**(January) (2008). https://doi.org/10.1051/epjconf/20135906001
18. M.M. Marinak et al., Advances in HYDRA and its applications to simulations of inertial confinement fusion targets. EPJ Web Conf. **59**, 06001 (2013)
19. LLL News, National Ignition Facility experiment puts researchers at threshold of fusion ignition, 8 August 2021. https://www.llnl.gov/news/national-ignition-facility-experiment-putsresearchers-threshold-fusion-ignition. Last accessed 17 June 2022
20. O. Hurricane, Lawson criteria for ignition exceeded in an inertial fusion energy experiment. IAEA Webinar on Building Stars: Breakthroughs in Fusion R&D, March 24, 2022. https://youtu.be/5KkVGASnPJ0. Last accessed 20 July 2022

21. LLL News, NIF sets new laser energy record, 10 July 2018. https://www.llnl.gov/news/nif-setsnew-laser-energy-record. Last accessed 17 June 2022
22. J.F. Latkowski et al., Chamber design for the laser inertial fusion energy (LIFE) engine. Fusion Sci. Technol. **60**, 54 (2011)
23. A. Bayramian et al., The Mercury project: a high average power, gas-cooled laser for inertial fusion energy development. Fusion Sci. Technol. **52**, 383 (2007)
24. J.-L. Miquel, C. Lion, P. Vivini, The laser mega-joule: LMJ & PETAL status and program overview. J. Phys. Conf. Ser. **688**, 012067 (2016)
25. S. Jiang et al., Experimental progress of inertial confinement fusion based on Shen Guang III Laser Facility in China. Nucl. Fusion **59**, 032006 (2019)
26. ORION laser, Wikipedia article ORION (laser). https://en.wikipedia.org/wiki/Orion_(laser). Last accessed 17 June 2022
27. T.R. Boehly et al., Initial performance results of the OMEGA laser system. Opt. Commun. **133**(1–6), 495 (1997)
28. S.P. Regan et al., Laser absorption, mass ablation rate, and shock heating in direct-drive inertial confinement fusion. Phys. Plasmas **14**(5), 056305 (2007)
29. J. Delettrez et al., Effect of laser illumination nonuniformity on the analysis of time-resolved x-ray measurements in UV spherical transport experiments. Phys. Rev. A **36**, 3926 (1987)
30. I.V. Igumenshchev et al., Three-dimensional modelling of direct-drive cryogenic implosions on OMEGA. Phys. Plasmas **23**, 052702 (2016)
31. V. Gopalaswamy et al., Statistically informed physics understanding and design optimization of direct drive inertial confinement fusion experiments. 28th IAEA Fusion Energy Conference, Nice, France. (Virtual) 10–15 May 2021, Presentation IFE/P4-6, Paper IAEA-CN-286-0942 (2021)
32. R. Betti et al., Shock ignition of thermonuclear fuel with high areal density. Phys. Rev. Lett. **98**, 155001 (2007)
33. K. Falk et al., Experimental observation of elevated heating in dynamically compressed CH foam. Plasma Phys. Control. Fusion **62**, 074001 (2020)
34. R. Betti, O.A. Hurricane, Inertial-confinement fusion with lasers. Nat. Phys. **12**, 435 (2016)
35. First Light Fusion Company. https://firstlightfusion.com/. Last accessed 17 June 2022
36. H. Azechi et al., The FIREX program on the way to inertial fusion energy. J. Phys. Conf. Ser. **112**, 012002 (2008)
37. J. Kawanaka et al., 3.1-kJ chirped-pulse power amplification in the LFEX laser. J. Phys. Conf. Ser. **112**, 032006 (2008)
38. H. Shiraga et al., Implosion and heating experiments of fast ignition targets by Gekko-XII and LFEX lasers. EPJ Web Conf. **59**, 01008 (2013)
39. J. Kawanaka et al., LFEX-laser: a multi-kilojoule, multi-petawatt heating laser for fast ignition. 26th IAEA Fusion Energy Conference, Kyoto, Japan, 17–22 October 2016, Presentation IFE/1-6 (2016)
40. A. Yogo et al., Laser-driven ion acceleration on LFEX for fast ignition: state of the art and applications. 26th IAEA Fusion Energy Conference, Kyoto, Japan, 17–22 October 2016, Presentation IFE/1-1 (2016)
41. Y. Kitagawa et al., Demonstration of direct fast heating of counter-imploded core plasma by LFEX laser. 28th IAEA Fusion Energy Conference, Nice, France. (virtual) 10–15 May 2021, Presentation IFE/P4-15, Paper IAEA-CN-286-0784 (2021)
42. CLF, Central Laser Facility. https://www.clf.stfc.ac.uk/. Last accessed 17 June 2022
43. DiPOLE. A new world record for UK-built laser technology. https://www.ukri.org/news/new-world-record-for-uk-built-laser-technology/. Last accessed 17 June 2022
44. C.N. Danson et al., Petawatt and exawatt class lasers worldwide. High Power Laser Sci. Eng. **7**(e54), 1 (2019)
45. A. Dubietis, G. Jonušauskas, A. Piskarskas, Powerful femtosecond pulse generation by chirped and stretched pulse parametric amplification in BBO crystal. Opt. Commun. **88**(4–6), 437 (1992)

46. V.B. Pathak et al., Strong field physics pursued with petawatt lasers. AAPPS Bull. **31**(4), 1 (2021)
47. H. Hora, Extreme CPA laser pulses for igniting nuclear fusion of hydrogen with boron-11 by non-thermal pressures for avoiding ultrahigh temperatures. J Energy Power Eng. **14**, 156 (2020)
48. D. Margarone et al., Advanced scheme for high-yield laser driven nuclear reactions. Plasma Phys. Control. Fusion **57**(1), 014030 (2015)
49. L. Giuffrida et al., High-current stream of energetic α particles from laser-driven proton-boron fusion. Phys. Rev. E **101**, 013204 (2020)
50. HB11 Energy, HB11 Energy laser boron fusion. https://hb11.energy/. Last accessed 17 June 2022
51. H Hora et al., H-11B fusion reactor with extreme laser pulses for non-LTE ignition. IAEA FEC Conference, Paper IFE/4-4, Ahmedabad (2018)
52. Marvel Fusion, The ultimate clean energy solution: quantum-enhanced fusion. https://marvelfusion.com/. Last accessed 17 June 2022
53. H. Hora et al., Road map to clean energy using laser beam ignition of boron-hydrogen fusion. Laser Part. Beams **35**, 730 (2017)
54. S. Eliezer et al., Avalanche proton-boron fusion based on elastic nuclear collisions. Phys. Plasmas **23**(5), 050704 (2016)
55. S. Fujioka et al., Kilotesla magnetic field due to a capacitor-coil target driven by high power laser. Sci. Rep. **3**, 1170 (2013)
56. F. Belloni et al., On the enhancement of p-11B fusion reaction rate in laser-driven plasma by α→p collisional energy transfer. Phys. Plasmas **25**, 020701 (2018)
57. F. Belloni, On a fusion chain reaction via suprathermal ions in high-density H–11B plasma. Plasma Phys. Control. Fusion **63**(5), 055020 (2021)
58. B. Le Garrec, HiPER laser: from capsule design to the laser reference design. Proceedings of SPIE – The International Society for Optical Engineering 7916, February (2011)
59. B. Le Garrec et al., HiPER laser reference design. Proceedings of SPIE – The International Society for Optical Engineering, May (2011)
60. HiPER Project Team, HiPER preparatory phase final report. http://www.hiper.org/Resources/HiPER_Preparatory_Phase_Completion_Report.pdf (2013). Last accessed 29 July 2022
61. B. Le Garrec, Laser-diode and flash lamp pumped solid-state lasers. Am. Inst. Phys. Conf. Proc. **1228**, 111 (2010)
62. K.R. Schultz, D.T. Goodin, A. Nobile Jr., IFE target fabrication and injection – achieving "believability". Nucl. Instrum. Methods Phys. Res. Sect. A **464**(1), 109 (2001)
63. M. Decroisette et al., A virtual reactor model for inertial fusion energy. Forum Teratec 2012, Complex Systems Design Workshop, Ecole Polytechnique, Palaiseau, France (2012)
64. R.S. Rawat, Dense plasma focus – from alternative fusion source to versatile high energy density plasma source for plasma nanotechnology. J. Phys. Conf. Ser. **591**, 012021 (2015)
65. S.C. Hsu, S.J. Langendorf, Magnetized plasma target for plasma-jet-driven magneto-inertial fusion. J. Fusion Energy **38**, 182 (2019)
66. S.C. Hsu et al., Laboratory plasma physics experiments using merging supersonic plasma jets. J. Plasma Phys. **81**, 345810201 (2015)
67. C. Olson et al., Development path for Z-pinch IFE. Fusion Sci. Technol. **47**(3), 633 (2005)
68. D.A. Yager-Elorriaga et al., An overview of magneto-inertial fusion on the Z Machine at Sandia National Laboratories. Nucl. Fusion in press. https://doi.org/10.1088/1741-4326/ac2db (2021)
69. U. Shumlak, Z-pinch fusion. J. Appl. Phys. **127**, 200901 (2020)

# Part V
# Synthesis and Conclusions

# Chapter 12
# Synthesis and Conclusions on the Applications of Systems Approaches to Fusion Reactors

**Abstract** This synthesis of the monograph begins with a summary of the systems methods developed and the lessons learned from analyzing a full range of fusion concepts, machines and reactor designs. An overall systems analysis strategy is implemented for studying different system architectures to compare the different concepts. A synthesis is made of the progress towards building one or more types of fusion reactor. The robustness and metrics of the various concepts are compared, and methods are suggested of mitigating fragilities. Finally, some general comments are made about the utility of systems approaches. The overall conclusion is that there are now several viable pathways to a fusion reactor, and systems methods are essential for success.

## 12.1 Methods Developed and Lessons Learned

This monograph has provided a comprehensive past, present and future survey of work towards commercial fusion energy and has demonstrated that the systems approaches bring new and useful insights and methods. Checklists are provided that embody general principles of systems engineering, architecture and emergent properties. These approaches contribute to improving good practice in experimental physics and engineering design. Common systems-wide challenges have a strong influence on device robustness and fragility in areas such neutron management, power loading from neutrons and plasma exhaust, tritium breeding and recycling, plasma characteristics and instabilities and commercial feasibility. Creating robustness can lead to fragilities which must be mitigated.

The main points of the monograph are summarized as follows:

- In Part I on introductory systems and plasma fundamentals:

Chapter 1 develops the principles of systems engineering, architecture and emergent properties. The fundamentals of tokamak plasma physics are summarized to provide a basis for future chapters and as a starting point for examples to illustrate systems principles. Precise methodologies for forward and reverse engineering and analysis of systems are developed and summarized in a series of checklists which are applied

 
F. B. Marcus, *Systems Approaches to Nuclear Fusion Reactors*, Springer Series in Plasma Science and Technology, https://doi.org/10.1007/978-3-031-17711-8_12

throughout the monograph. Case studies are chosen to illustrate the characteristics of important machines and experiments.

Chapter 2 assembles the elements of a design space and describes how their form, function and inter-relationships can be combined in the systems architecture of machines. A series of tokamaks with plasma current under 0.7 MA are examined for their design characteristics and principal operational scenarios which contribute to a catalogue compiled for all tokamaks. Plasma and machine simulations codes are essential components of systems design. Another catalogue is compiled of plasma analysis, integrated workflow and systems engineering codes, leading to a full integration of the systems design space.

- In Part II on high current tokamaks:

Chapter 3 carries out a full scale reverse engineering case study on the large 2 MA tokamak Doublet III and its upgrade DIII-D to understand their components, relationships and operation. Detailed analysis shows how the systems were designed to be both highly robust and agile, so that they could examine doublets, d-shaped plasmas, divertors, high beta, high confinement, internal transport barriers, current drive and advanced tokamak scenarios. Other 1–2 MA class tokamaks are examined and compared. A detailed analysis is also presented of very high field and low aspect-ratio spherical tokamaks, which have some intrinsic fragilities. In many areas, these have been mitigated to allow high performance and to provide a basis for compact fusion reactors.

Chapter 4 provides a case study in systems forward engineering of the MA class tokamak TCV to illustrate goal specification, multi-level design, optimized procurement, construction and operation of a new fusion machine that is highly flexible for plasma shaping and control. A crucial innovation for robustness was a thick, continuous vacuum vessel. Although it presented a potential fragility concerning plasma initiation, this was fully mitigated. The machine allowed plasma elongations up to 3 to 1 because of 16 fully independent shaping coils and 2 innovative stabilization coils. The resulting design was highly robust, inexpensive, flexible and even more agile than DIII-D in its shaping and control capability, which has allowed an extremely wide range of successful experiments. Along with high field and low aspect ratio tokamaks, it was an example of how one unusual design feature, if its fragilities are fully mitigated, can transform the overall machine design and operation.

Chapter 5 presents reverse engineering of JET, which is the largest operating tokamak in the world. It reached a plasma current of 7 MA before its internal divertor installation. Its systems architecture led to a highly robust machine. When there was a major failure of three short-circuited toroidal field coils, the highly modular design and spare parts allowed a rapid repair as part of a scheduled maintenance. A case study was made on the forward engineering and the optimization of planning of the first preliminary and then full power deuterium-tritium fusion experiments. The experimental results showed how a decade of careful planning and experiments led to a highly successful demonstration of near power breakeven of 16 MW of fusion power. Integrated neutron diagnostics played a crucial role in these experiments. After those experiments, JET has played a major role in prototyping plasma physics and technology in support of ITER, producing a 5 s fusion pulse producing 59 MJ. The high power d-t fusion experiments with circular plasmas in TFTR are discussed and compared.

Chapter 6 presents the systems architecture of superconducting tokamaks that are prototyping ITER, quasi-steady-state operation and future fusion reactors. Superconducting coils were a major design innovation necessary for fusion reactors. Several of these tokamaks are aimed at direct technology prototyping of ITER. A case study of reverse engineering is made of the Tore Supra tokamak. It was upgraded to WEST to be a more relevant prototype for ITER. The largest of these devices, JT-60SA which is a follow-on to the highly successful JT-60U, is of a comparable size and potential performance to JET. Through careful testing and quality control, they discovered a fragility in a coil feed and were able to mitigate the problem before major damage occurred.

- In Part III on prototype tokamak fusion reactors:

Chapter 7 on ITER presents the world's largest future fusion experiment as an example of full scale integration, where systems codes are an essential part of the design and construction process. A case study of reverse engineering on ITER explores its systems and its expected performance of 15 MA plasma current, $Q = 10$ in d-t, and its plans to study both inductive and advanced tokamak current drive scenarios. It is considered by many that ITER could not be built without a full systems engineering approach. ITER will also serve as a prototype and technology test bed for demonstration fusion reactors.

In Chap. 8, this integrated approach is extended to tokamak demonstration reactor designs. Many DEMOs are being designed as an extension of ITER. The EU DEMO concept is used as a case study in how systems codes and methods are being effectively used in engineering design. Large scale demonstration reactors planned by ITER partners and based on its baseline or advanced tokamak scenarios are analysed in detail and compared. Compact fusion reactor designs have been proposed using high temperature superconductors and unique neutron blanket and shielding concepts, which are scrutinized for robustness and fragilities and the mitigations chosen by the designers. The process of choice of goals and systems integration of tokamak fusion reactors is explored and summarized.

- In Part IV on helical, linear and inertial fusion reactor concepts:

Chapter 9 examines the family of stellarators, which were designed as steady-state toroidal confinement devices by providing a rotating field from external helical coils instead of internally by plasma current. A design space is established and a case study in systems forward engineering is made of the superconducting helias W7-X as a successor to the pioneering Wendelstein series. Reverse systems engineering is used to examine the goals, design choices and operational results for the superconducting LHD which evolved from the Heliotron series. A wide range of other stellarators are examined to catalogue the choices made in helical design space. The successful W7-X and LHD experiments have led to fusion reactor designs, the

HELIAS-5B and the FFHR-d1 which provide viable alternatives to tokamaks and are examined for their robustness and fragility properties.

Chapter 10 examines linear magnetic traps as potential fusion reactor candidates. A case study of systems forward engineering is made of the proto-reactor scale MFTF-B tandem mirror experiment which went through many design and construction changes, was built and ready to operate when its budget was removed. A robustness and fragility analysis is conducted. Recent advances in tandem and multiple mirrors with gas boxes and field expanders and field reversed configurations have renewed reactor prospects. A design space is developed for the systems architecture of linear machines. Toroidal versions of linear concepts are briefly discussed. Taylor state conserved helicity machines are presented, including those formed in a linear external magnet geometry. A design space is developed for the reversed field pinch which relies on the Taylor state.

Chapter 11 makes use of the extensive documentation and the excellent outcomes of the NIF laser indirect-drive d-t fusion experiments for a case study in systems forward engineering. Other laser systems have been analysed which used direct drive, fast ignition, shockwaves and non-thermal ion production. These experiments provided a basis for developing a design space for laser fusion reactors and the fundamental requirements for assembling the elements, leading to analysis of the HiPER reactor design. Magnetic pulsed confinement systems including MagLIF and Z-pinches also provide a path to a fusion reactor. These inertial confinement systems would have robustness advantages if various technological and physics challenges are met.

- In this Part V on synthesis and conclusions:

This Chap. 12 presents a synthesis of the various methods.

## 12.2 Overall Systems Analysis Strategy for Comparing Concepts by Systems Architecture

### *12.2.1 Methods*

General principles of systems approaches over a very wide range of examples have been developed, including planning, analysis, detailed design, implementation, robustness, fragility, performance, resource demands, systems control, fault-tolerance, modularity and decoupling. These general principles of systems and robustness optimization and protection against fragilities has been applied to tokamaks both standard and advanced, helical toroidal concepts, mirrors, linear magnetic traps and inertial fusion. These concepts were shown to be crucial in examining the interface between physics and engineering design concepts.

Even though the tokamak currently seems most promising, there are still many steps to be taken towards a commercial fusion reactor and alternatives are being actively pursued. The perspectives for commercial fusion reactors look promising, but the final form of fusion reactors is not yet determined. Systems approaches provide a methodology for optimizing the paths for design and optimization of fusion reactors.

### *12.2.2 Systems Strategy for Comparing Different Fusion Reactor Concepts*

The rest of this chapter examines the commonalities between various fusion reactor concepts and compares systems architecture characteristics of fusion reactors. Checklist 1.8 from Chap. 1 provides a framework for this approach and its conclusions.

## 12.3 Progress in Existing and Planned Machines

### *12.3.1 Systems Level Performance Metrics: Lawson Criterion and Triple Product*

The traditional and still central metric for evaluating fusion machines is the Lawson criterion [1] consisting of the product of plasma density and energy confinement time. When multiplied by ion temperature, it becomes the fusion triple product. A summary has been made of progress of machines (not including laser fusion) towards energy breakeven and ignition [2, 3], where a plot is presented of $n\tau$ versus $T_i$. The best results were obtained by the largest and highest current tokamaks, especially JET in d-t and JT-60U in d-d steady state. The highest energy gain Q in d-t fusion power is still held by the DTE1 experiments at JET, recently equalled by NIF laser inertial fusion, both on a transient basis. JET has recently concluded the DTE2 experiments which produced 59 MJ of fusion power during 5 s. The projected operating space for ITER is $Q = 10$ in quasi-steady state. Other tokamak concepts such as the spherical tokamak and very high magnetic field tokamaks may offer promise for more economical fusion reactors.

During the decade following 2010, the main emphasis in tokamaks and in stellarators was working towards steady state operation. The stellarator W7-X extended both fusion performance and pulse length [4]. In Fig. 12.1, a comparison was made of the triple-product performance versus duration of the first W7-X divertor plasmas with other helical devices and tokamaks. W7-X obtains the best result among long pulse devices, although still well short of the projected operating space for ITER and DEMO.

Other types of fusion experiment have been making important progress in the last decade. Results obtained during the last decade from linear traps [5] verified that

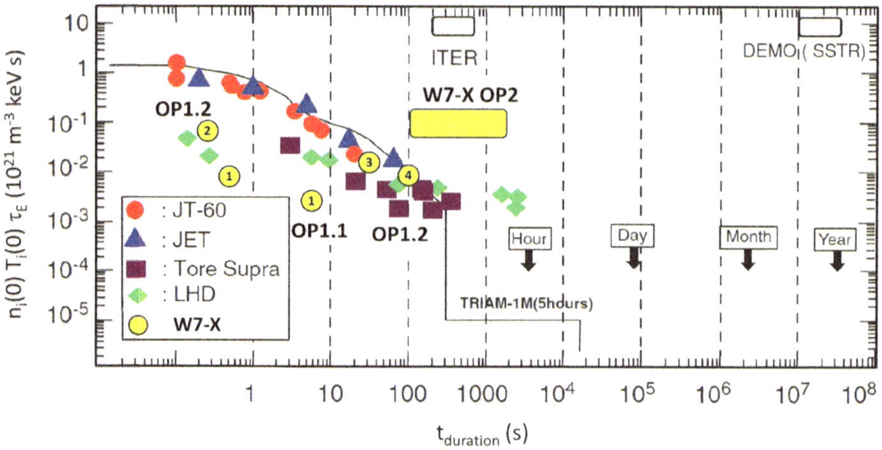


**Fig. 12.1** Triple product of stellarators and tokamaks versus time (s). (From [4], Fig. 14)

multiple-mirrors could work together with gas-dynamic traps to be developed into fusion reactors. The Tri-Alpha "Norman" field reversed concept also offers a new path towards fusion.

In the field of inertial confinement, NIF has demonstrated fusion energy output greater than the kinetic energy of an imploding shell [6], and a Lawson equivalent of ρR of about 0.3 at an ion temperature of 4.5 keV, which already involved significant alpha particle heating [7]. In 2021, NIF announced the production of 1.3 MJ of fusion energy, nearly comparable to the laser input energy to the target, with more progress expected.

In the domain of Z-pinches, simulations [8] indicated that a next-generation Z machine with 60 MA could give a fusion gain of 70. The MagLIF concept is also a contender to be a viable source of fusion energy.

## 12.3.2 *The Cost Metric of Fusion Experiments and Reactors*

A major concern about fusion reactors research and development is cost, especially in competition with renewable energy.

Fusion experiments showed a very wide range of costs, but an even wider set of circumstances. The largest operating tokamak JET [9] was built on time and on budget. Other machines had massive cost overruns and delays of decades. Some were ultimately very successful, others were cancelled. Still other machines were relatively inexpensive due to already existing buildings and high power grid connections or pre-existing large motor generators. Another strategy was used in Japan, where machine construction was outsourced to one or more private companies for a turn-key operation.

The cost of the largest fusion device under construction ITER was claimed to be unknowable [10] because many of the international partners made equipment contributions to the project paid by their internal budgets. However, the project is certainly late and far over the original budget. Nevertheless, the project is a major research undertaking essential to the whole field, and its costs are not indicative of a final reactor.

Fusion experiments can also be constructed very inexpensively. At the CRPP (now the Swiss Plasma Centre), we designed and built TCV to have the very high performance of a MA tokamak due to vertical elongation and shaping, extreme flexibility, ease of maintenance and upgrade. Its cost was perhaps 1/3 of comparable MA machines. TCV achieved this success due to an extensive systems approach that examined a wide range of variants at different levels to determine the optimum architecture.

The ingredients for high performance and low cost were:

- Key choices in the tokamak design space were to rely on low voltage start-up, which allowed a simple, thick walled stainless steel vacuum vessel
- Stability was maintained of highly vertically-elongated plasmas by wall stabilization and simple in-vessel coils, which were driven by a low power, high frequency power supply.
- There were 16 individually powered shaping coils, giving maximum plasma shaping and elongation control
- Demountable, low field toroidal field copper coils were used for ease of construction and maintenance
- This design led to the use of inexpensive, commercially available components, including:
  - a standard motor generator with a frequency doubled winding to operate at 120 Hz;
  - standard thyristor rectifier power supplies in 12 phases giving a 1 ms response time;
  - a continuous and thick stainless steel vacuum vessel with an 8 ms time-scale, without double walls or bellows, to stabilize the plasma and allow baking and high vacuum with large numbers of access ports for heating and diagnostics.
- The mechanical simplicity of the machine was compensated by the sophisticated control system, which was easily upgradeable as digital technology advanced.
- An international call for tender was launched for all major components, sometimes resulting in high quality bids three times lower in cost than the highest.
- Auxiliary heating by high-power gyrotrons was developed by an in-house research programme.
- New buildings for experiment, motor generator and staff were provided by the EPFL (Swiss Federal Institute of Technology at Lausanne) as part of moving the entire campus to a new site

A modern fusion reactor will be far more complicated, but TCV is an example of successful procedures leading to a machine still making major contributions.

The economics of a pulsed DEMO tokamak reactor have been considered in detail by using a systems model [11] to optimize the design parameters and then arrive at costs. The design considered had a net output power of 500 MW (e) and plasma pulse duration of about 2 h. Systems codes are commonly used in the design process as numerical tools for optimization studies to obtain a quantitative understanding of the most suitable optimization criteria. Scans in parameter space show the cost of electricity had a minimum for low applied heating power at an aspect ratio of $A \sim 3.5$. Cost minima of around €0.25/kWh seems to be accessible. In Spring 2022 in Belgium, the price was about €0.33.

Another detailed study of the economy of fusion energy [12] used economic analysis of the fusion power plant model DEMO2 to determine the cost of electricity. The model investment and operating costs were determined and the limit sales price of electricity was found using the net present value method. The levelized cost of electricity method with the inclusion of external costs was used for a comparison of selected power plant types based on the OECD statistical data and the EU ExternE project results. The comparison showed the levelized cost of electricity of fusion power plants competitive to the present renewable resources. After internalisation of external costs, fusion power plants might become even the second cheapest power source.

### 12.3.3 Metrics, Scaling Laws, Systems and Predictive Codes, Innovation and Breakthroughs

This section discusses the limitations of metrics derived from current scaling laws in systems and predictive codes, and to emphasize the roles of innovation and breakthroughs. In the analysis of the physics behaviour of fusion plasmas, a wide range of physics laws and scaling laws have been devised and implemented for machine design in analysis, systems and predictive codes, for example TSC for tokamaks and HYDRA for laser fusion. These codes are essential for design optimization and for planning future machines, as has been discussed throughout this monograph. However, most or all of the concepts considered have had important advances in innovation or even major breakthroughs that have rapidly advanced the concept towards success. Detailed analysis with codes can sometimes indicate areas of special fragility that need innovation, but sometimes the breakthroughs just happen, occasionally rapidly, often slowly, but always as a result of long and dedicated work and analysis by scientists and engineers. For each concept, some of the main breakthroughs and innovations included:

- Tokamak H-mode and internal barriers – The construction of the divertor tokamak ASDEX led to the discovery of the H-mode improved confinement regime and of transport barriers.

- Spherical tokamak high beta operation – The use of low aspect ratio and high vertical elongation tokamaks to achieve high plasma pressure beta with good confinement opened a new route for fusion reactors.
- Stellarator possibility of high confinement – Single particle confinement was demonstrated on the high precision stellarator CLASP. Plasma confinement much better than Bohm diffusion was observed on the early WENDELSTEIN stellarators.
- Reversed Field Pinch quiescent state – On the original RFP ZETA experiment, a quiet high confinement state was observed, which was explained by J.B. Taylor as representing helical flux conservation.
- Tandem mirror experiments – On the TMX tandem mirror, it was shown that the confinement of mirror machines could be improved by using electrostatic potentials in combinations with mirrors to contain plasmas.
- Indirect drive laser fusion – By steady innovation over the years in laser power, uniformity of illumination and careful target design, near breakeven fusion conditions are being approached.
- Liner Z-pinches – The concept of liner implosion of a Z-pinch called MagLIF and its experimental results has revived the fortunes of the historic Z-pinch, showing a path to plasma ignition.

A wide range of technological innovations are also bringing the possibility of more compact and less expensive fusion reactors, including:

- High-temperature superconductors – The commercial availability of REBCO and other high temperature superconductors may greatly improve the economics of fusion reactors.
- Demountable joints for high temperature superconductors – Fusion reactors built with high temperature superconductors with demountable coil joints could lead to greatly reduced assembly problems and cost and improve the possibility of repair and maintenance. This flexibility is especially important for linked tokamak coils and heliotron continuous windings.
- Diode-pumped solid-state laser (DPSSL) – This technology offers the prospect of high efficiency pumping of lasers delivering repetitive pulses.

### *12.3.4 The Central Role of ITER*

Even though there is a great deal of research on fusion devices other than ITER or ITER prototypes and support machines, ITER itself remains a central focus and a major contributor to the entire fusion programme, both magnetic and non-magnetic. It is assembling all of the innovation, research, development, theory and simulation models relevant to a functioning fusion power plant:

- large-scale superconducting magnets;
- cryogenic systems;

- high-power auxiliary heating systems;
- plasma and neutron and fast particle diagnostics;
- tritium breeding, blanket cooling and tritium reprocessing systems;
- plasma facing components and heat exhaust systems;
- material testing and development for radiation damage;
- plasma equilibrium and machine control;
- plasma instability control;
- remote handling, maintenance and radioactive material management;
- nuclear licensing of a fusion plant;
- systems codes for design, performance predictions, quality control, machine assembly and operations.

Even for inertial confinement systems, many of the above systems are relevant, and the developments for ITER will benefit their development, especially in the realms of tritium breeding and processing plants.

## 12.4 Pathways Towards an Operating Fusion Reactor

### *12.4.1 Pathways Towards a Working Fusion Reactor for Each Concept*

Although both tokamaks and indirect drive laser fusion have achieved significant plasma self-heating by fusion alpha particles, no operating fusion machine has yet achieved high fusion energy gain compared to input energy, nor ignition nor continuous burn or burn waves consuming much of the fuel. Tokamaks have demonstrated long-pulse operation under high power plasma conditions, but none have demonstrated true steady state operation under full plasma power output and fusion neutron conditions. Inertial fusion devices are still at the single shot stage, and have yet to demonstrate rapid repetitive pulsing.

A full systems approach to a fusion reactor requires a number of steps to be designed for and demonstrated in working machines. The tokamak field is most advanced in this respect with the current generation of long-pulse superconducting machines in operation and ITER under construction. The use of systems codes such as PROCESS and BLUEPRINT is essential to fully scope and design fusion reactors. The stages towards a fusion reactor for both inertial and magnetic confinement include:

- self-heating of plasma by fusion alpha particles;
- net production of fusion energy compared to heating energy;
- highly-efficient auxiliary heating or laser pumping systems;
- steady state production of net electric power;
- tritium breeding ratio greater than one and self-sufficiency in tritium with tritium extraction and recycling plant;

- shielding of external machine components from neutrons;
- power handling of plasma exhaust and neutrons to components;
- long term material resistance to damage;
- cooling of plasma facing components;
- cryogenic plant for repetitive frozen fuel pellets;
- superconducting magnet feed systems and cryogenic plant for magnetic confinement;
- nuclear site licensing;
- remote handling and radioactive material processing and disposal.

This list is not exhaustive, but indicates the number of issues that need to be considered in reactor development.

### 12.4.2 Robustness and Fragilities of Different Concepts

In summarizing robustness and fragilities of each fusion reactor concept, it is important that such analysis becomes not a way of choosing winners and losers. It is rather an assessment of the nature of each system. Each fragility provided an opportunity to mitigate against that fragility. Faults tend to occur in systems that are thought to be either conventional technology or robust, since they will often receive less design attention.

An example occurred in superconducting coils, where the coils operate at design specifications, but the connectors experience faults, spectacularly at CERN [13]. A weld in a superconducting wire connecting two magnets heated above its operating temperature and the resistance and subsequent heating caused an arc through the liquid helium and puncture into the surrounding vacuum vessel. The arc vaporized a connector between two magnets. Such arcs can also cause stresses and failures directly inside the magnets as well. The liquid helium flowed through the hole, and the pressure wrenched magnets off their concrete supports.

As one way of comparing concepts among all fusion systems, their robustness and fragility properties are listed and compared.

### 12.4.3 Tokamak Reactors

The tokamak concept seems to be the most robust magnetic confinement concept due to the many operating experiments and the highest achieved plasma parameters. There are a wide range of design variations, each with their own special advantages. The concept is fundamentally simple, with plasma confined in a torus and MHD equilibrium maintained by a vertical field. However, with the need for toroidal plasma shaping, divertor and instability coils surrounding a plasma inside a toroidal vacuum vessel, tritium breeding blanket and shield, careful design is needed to avoid major fragilities in case of malfunction requiring repair.

- ITER-like reactor
  - robustness – The stability of machine operation is backed by highly prototyped and tested systems components and proven experimental scenarios.
  - fragilities – Two of the major fragilities involve the superconducting coils and plasma disruptions, and they are linked. The plasma disruption can induce large voltages in the coils. Other fragilities involve high cost and a high electrical load for auxiliary heating for current drive which makes net power gain difficult. The alternative of inductive pulsed operation has problems of rapid plasma restart.
  - mitigation – Disruption mitigation is provided by shattered pellet injection, impurity gas injection and disruption detection and plasma control systems. Power supply inefficiency is mitigated by negative ion beams and RF and ECH.
- Advanced tokamak
  - robustness – Advanced tokamaks are designed to operate at lower plasma current, higher poloidal beta, a large fraction of bootstrap current, and have continuous operation through residual current drive, allowing better reactor economics.
  - fragilities – Advanced tokamaks rely on detailed control of plasma profiles to achieve high performance and it may be difficult to maintain these in true steady state.
  - mitigation – A wide range of methods of adjustment of auxiliary heating profiles in the plasma are being developed.
- High-field
  - robustness – The use of very high field REBCO high-temperature superconductor and toroidal field coil joints allow a compact and more economical tokamak reactor to be constructed, especially when using a tank containing liquid metal which breeds tritium, simplifying shielding and maintenance.
  - fragilities – The REBCO technology in tokamaks is far less tested than the $Nb_3Sn$ and NbTi technology in ITER and the joints, high fields and stresses provide added risks of coil failure. The smaller dimensions risk high neutron penetration to coils producing heating and damage.
  - mitigation – The technology for a reactor is being prototyped in a smaller experiment.
- Spherical tokamak
  - robustness – By using REBCO high field high-temperature superconductor and very low aspect ratio, a very high current, high performance spherical tokamak with high beta can be constructed, resulting in a very economical machine.

- fragilities – Current designs use normal coils inside the vacuum vessel for start-up, and experiments have already had major failures of these coils. Neutron damage and heating of the central coil due to small spacing for shielding is also an issue.
- mitigation – A large prototype is being built to try to resolve these problems.

### 12.4.4 Stellarator, Helias and Heliotron

A toroidal plasma contained entirely by external coils without plasma current offers many potential advantages for fusion reactor operation.

- Stellarators and Helias
  - robustness – To solve the problem of steady state operation in tokamaks, the stellarator offers the prospect of true steady state plasma operation without external current drive and plasma disruptions by creating all fields with external coils. Difficulties of construction and repair of a classical stellarator were overcome by many separate helias coils.
  - fragilities – Such machines are highly susceptible to construction requirements for tight tolerances in coils and assembly. One large machine was cancelled due to difficulty satisfying tolerances.
  - mitigation – New techniques have been developed to ease and manage tolerance requirements, and successfully construct and assemble such magnet systems.
- Heliotron
  - robustness – Heliotrons and torsatrons have relatively simple, continuously-twisted coils. The development of REBCO and coil joints makes reactor construction much more possible.
  - fragilities – The joints are still a fragility, and need to be demonstrated in operating plasma conditions.
  - mitigation – Laboratory level tests indicate that the technology can be successful.

### 12.4.5 Mirrors, Linear Traps and Field Reversal

The classical mirror machine, especially when supplemented with minimum-B coils, offered the simplicity of an open ended systems with great ease of access for maintenance and repair, and an automatic method of exhausting plasma heat. Unfortunately, the end losses were too great to make net electricity production. Solutions have been explored:

- Tandem mirrors
  - robustness – The tandem mirror offered the prospect of a successful fusion reactor by a combination of mirror fields and electrostatic potentials in plug regions to greatly improve plasma confinement. Maintenance and plasma exhaust would continue to be easily managed.
  - fragilities – Experimentally, the simultaneous control of all the density and temperature profiles proved to be very difficult in the desired plasma regime.
  - mitigation – The evolution of various mirror plus plug concepts was explored in the MFTF-B design, but unfortunately they were never tested.
- Gas box
  - robustness – A dense plug of plasma and exhaust power handling and plasma stability is provided by boxes at the machine ends with a high gas pressure.
  - fragilities – The neutral gas can leak into the main chamber and cause plasma losses.
  - mitigation – The gas box concept can be combined with a long multiple mirror concept to improve confinement and isolate the gas box from the central plasma.
- Multiple mirrors
  - robustness – A very high number of multiple mirrors reduces the possibility of loss from scattering into a single mirror loss cone.
  - fragilities – There are still end losses in such systems.
  - mitigation – Combination with a gas box can alleviate the problems of both machines.
- Expanded loss flux
  - robustness – By having a simple mirror with a very large expanded flux region, the curvature of the loss region can provide stability to the whole system.
  - fragilities – There are still problems with losses through the loss cone.
  - mitigation – Prototype experiments need to be conducted.
- Field reversed experiment
  - robustness – A field reversal in a mirror machine can be produced by ECRH or by neutral beams. In the field reversal by ion beams concept, a stable plasma is both formed and heated by these beams.
  - fragilities – The results still need to be extended to larger scale and the fusion regime achieved.
  - mitigation – Larger machines are being planned.
- Bumpy torus
  - robustness – The problems of end loss can in principle be overcome by linking several simple mirrors back into a torus, and using ECRH to create field minima.

- fragilities – Experiments showed that stable configurations were not achieved.
- mitigation – These experiments were abandoned

- Reversed field pinch
  - robustness – Another method is the reversed field pinch using a flux conserved stable regime.
  - fragilities – A conducting shell needs to be right next to the plasma, and steady state operation needs to be demonstrated.
  - mitigation – A series of large scale experiments is continuing.

### 12.4.6 Inertial Fusion

- Laser indirect drive LIFE
  - robustness – The problems of highly interlinked systems and neutron damage to magnets is avoided by a laser fusion system which separates the laser system from the target system, allowing far easier construction of first wall and breeding blanket and shielding.
  - fragilities – The economics and technology of high repetition rate at high laser efficiency and high fusion gain remain to be demonstrated. Efficiencies are low due to energy needed to power laser flashlamps and to uniformly heat the hohlraum wall.
  - mitigation – Target improvement and laser refinements are showing a continuous path to pellet burn. Repeated pellet ignition and burn has yet to be even tested. High efficiency diode pumped solid state lasers are being developed.
- Laser direct drive HiPER
  - robustness – Direct drive plus fast ignition might lead to much lower laser energies for pellet burn and ignition.
  - fragilities – High level performance of direct drive remains to be demonstrated.
  - mitigation – Demonstration is required on a large laser system.
- Z-pinch MagLIF
  - robustness – High fusion performance has been demonstrated, and a larger machine could reach ignition. Technical solutions have been proposed for a large pulse every few seconds.
  - fragilities – The energy of a single event, about 1 GJ, is so large that major challenges are present for a sustainable reactor concept.
  - mitigation – Prove the concept in a larger scale experiment.

In summary, the application of robustness and fragility studies emphasize the ways that problems in individual concepts can be mitigated, and provide an overall view of the fragilities and mitigations over a full range of fusion devices.

## 12.5 Overall Conclusions: Optimization of Fusion Reactors by Systems Approaches

It is sometimes argued that fusion will be available too late to help in the climate crisis and cannot compete economically with renewables [14]. In fact, with the great diversity of approaches, fusion reactors are now closer to being built than ever before. Demonstration reactors are on the drawing board, and private enterprises are making considerable investments in alternative approaches.

Even on the mainline tokamak path represented by ITER, its success in design and innovation has been an inspiration and a central technical source for reactor designers. High field compact tokamaks such as SPARC [15] and ARC [16] have begun design and construction and they potentially offer a rapid path to conventional tokamak reactors if some major problems can be overcome. Low aspect ratio spherical tokamaks hold similar promise with similar major challenges.

Economic studies have shown that fusion reactors can be competitive. The cost of developing fusion is comparatively a drop in the ocean of oil being purchased in the world, currently for a cost in mid-2022 of about 10 billion US dollars or EU euros per day.

The general conclusion is that systems approaches are highly relevant and important to the fusion experiment and reactor design process. They provide methodologies vital for understanding and planning and are already considered essential in large scale projects.

The overall conclusion is that there are now several pathways to a fusion reactor, and systems methods are essential for success.

## References


1. E. Morse, *Nuclear Fusion* (Springer, Cham, 2018)
2. S.E. Wurzel, Measuring progress in fusion energy: the triple product. Fusion Energy Base, 19 December 2019
3. S.E. Wurzel, S.C. Hus, Progress toward fusion energy breakeven and gain as measured against the Lawson criterion, arXiv:2105.10954v4 (2022)
4. R.C. Wolf et al., Performance of Wendelstein 7-X stellarator plasmas during the first divertor operation phase. Phys. Plasmas **26**, 082504 (2019)
5. P.S. Bagryansky, A.D. Beklemishev, V.V. Postupaev, Encouraging results and new ideas for fusion in linear traps. J. Fusion Energy **38**, 162 (2019)
6. S. Le Pape et al., Fusion energy output greater than the kinetic energy of an imploding shell at the National Ignition Facility. Phys. Rev. Lett. **120**, 245003 (2018)
7. J.L. Kline et al., Progress of indirect drive inertial confinement fusion in the United States. Nucl. Fusion **59**(11), 112018 (2019)
8. D.A. Yager-Elorriaet et al., An overview of magneto-inertial fusion on the Z-machine at Sandia National Laboratories. 28th IAEA Fusion Energy Conference, Nice, France (Virtual) 10–15 May 2021, Presentation OV/P-7, Paper IAEA-CN-286-1217, Submitted to Nuclear Fusion (2021)

9. E.N. Shaw, *Europe's Experiment in Fusion: The JET Joint Undertaking* (North-Holland, Amsterdam, 1990)
10. M. Claessens, *ITER: The Giant Fusion Reactor: Bringing a Sun to Earth* (Springer, Cham, 2020)
11. W. Biel, Systems code studies on the optimization of design parameters for a pulsed DEMO tokamak reactor. Fusion Eng. Des. **123**, 206 (2017)
12. S. Entler et al., Approximation of the economy of fusion energy. Energy. **152**, 489 (2018)
13. M. Brice, Eight-month delay for LHC, Nature News **455**, 23 October (2008)
14. L.J. Reinders, *The Fairy Tale of Nuclear Fusion* (Springer, Cham, 2021)
15. A.J. Creely et al., Overview of the SPARC tokamak. J. Plasma Phys. **86**, 865860502 (2020)
16. B.N. Sorbom et al., The engineering design of ARC: a compact, high field, fusion nuclear science facility and demonstration power plant. IEEE Symposium of Fusion Energy, Austin, Texas (2015)

# Glossary

**Alfven time** plasma radius divided by Alfven velocity
**Alfven velocity** proportional to magnetic field divided by square-root density
**Alfven wave** low frequency plasma oscillation
**anchor cell** minimum-B mirror for suppressing instability
**appm** atomic parts per million
**auxiliary cells** volumes of field that form electrostatic thermal barrier
**AWH** Alfven wave heating
**axicells** very high field region to throttle flow out of mirror machine
**axisymmetry** unchanged by rotation in the toroidal direction
**B** magnetic field
**ballooning** pressure gradient induced instability
**bootstrap** pressure induced current
**BSSCO** bismuth strontium calcium copper oxide superconductor
**burn wave** propagating fusion burn region in compressed pellet
**CERN** European centre for nuclear research
**CHI** coaxial helical injection
**chopper** switching power supply
**CPA** chirped pulse amplifier
**CS** central solenoid
**cusp** field from parallel opposite-current coils
**Dalpha** line radiation from deuterium atoms
**DCLL** dual-coolant lithium lead tritium breeding blanket
**d-d** deuterium-deuterium fusion reaction
**DEMO** demonstration fusion reactor
**disruption** rapid loss of plasma
**divertor** removes plasma energy outside separatrix
**DMS** disruption mitigation system
**dpa** displacement per atom
**DPSSL** diode-pumped steady state laser
**d-t** deuterium-tritium fusion reaction



F. B. Marcus, *Systems Approaches to Nuclear Fusion Reactors*, Springer Series in Plasma Science and Technology, https://doi.org/10.1007/978-3-031-17711-8

**ECCD** electron cyclotron current drive
**ECRH** electron cyclotron resonant heating
**EDA** enhanced Dalpha mode
**electron** negatively charged fundamental particle surrounding nucleus
**element** a unit in a system, also atom with a specific atomic number
**ELM** edge localized mode
**EPED** peeling-limited pedestal
**expander** escaping field lines from mirror expansion volume
**finger** interlocking coil joint
**fishbone** rapidly repeating spike instability
**FRC** field reversed configuration
**fs** femtosecond ($10^{-15}$ s)
**HCCR** helium-cooled ceramic reflector tritium breeding blanket
**HCD** heating and current drive system
**HCPB** helium-cooled pebble bed tritium breeding blanket
**heliac** stellarator with central conductor on plasma axis
**helias** twisted coils around torus making toroidal and helical fields
**heliotron** toroidal device with mono-direction helical windings
**HHCL** helium-cooled lithium lead tritium breeding blanket
**hierarchy** multiple levels of complexity in a system
**H-Mode** high energy confinement mode
**hohlraum** target chamber where x-rays are produced
**IBL** ITER baseline scenario
**ICRH** ion cyclotron resonant heating
**I-Mode** improved confinement ELM-free H-mode
**ion** positively charged atom or nucleus
**isodynamicity** radial drift of all particles averages to zero
**ITB** internal transport or temperature barrier
**ITG** ion temperature gradient instability
**kA** kilo-ampere (current)
**kHz** kilohertz
**l** periodicity in poloidal plane
**Lawson** criterion for energy breakeven
**LBO** laser blow-off
**LDD** laser direct-drive induced fusion by rapid compression
**LH** lower hybrid
**L-H** low confinement transition to high confinement H-mode
**LHCD** lower hybrid current drive
**LHEP** lower hybrid enhanced performance
**LID** laser indirect-drive induced fusion by rapid compression
**limiter** structure to define plasma edge
**LLCB** lithium-lead ceramic tritium breeding blanket
**L-mode** low energy confinement regime

**locked mode** non-rotating magnetic island
**loss cone** empty space in velocity distribution function where particles escape
**LSBB** liquid salt tritium breeding blanket
**LSBB** liquid salt breeding blanket
**m** periodicity in toroidal plane, also particle mass, also azimuthal wave number
**MA** mega-ampere (current)
**major radius** distance from machine axis to plasma centre
**Maxwellian** velocity distribution of particles in thermal equilibrium
**MDD** magnetic direct-drive induced fusion by rapid compression
**MHD** magneto-hydrodynamics
**MHz** megahertz
**minimum-B** magnetic field increases in any direction a particle moves
**minor radius** distance from plasma centre to plasma boundary
**mirror** high magnetic field region that turns around particles
**ms** millisecond
**ms** millisecond
**NBCD** neutral beam current drive
**NBI** neutral beam injection
**n** density, also toroidal wave number$n_e$density of electrons
**$n_i$** density of ions
**ns** nanosecond ($10^{-9}$ s)
**NTM** nonlinear tearing mode
**Ohmic** heating by current flow
**omnigenous** see isodynamicity
**P** heating power
**pedestal** sharp gradient at edge or inside of plasma
**PF** poloidal magnetic field
**PF Coil** coil creating poloidal magnetic field
**PFM** plasma facing material
**PINI** positive ion neutral injector on JET
**plasma** collection of ions and electrons
**plug** electrostatic confinement region in mirror geometry
**ps** picosecond ($10^{-12}$ s)
**PWR** pressurized water reactor
**q** safety factor ratio toroidal to poloidal turns
**Q** ratio of fusion power to heating power
**q** electric charge
**QH** wide pedestal H-mode
**QHS** quasi-helical symmetric
**quasi-axisymmetry** symmetry when particles drift toroidally
**quasi-helical** where particle drift helically
**quasi-isodynamic** when particles drift poloidally
**quasi-symmetry** symmetric in absolute value of magnetic field

**R&D** research and development
**REBCO** rare earth barium copper oxide superconductor
**RF** radio frequency
**RMP** resonant magnetic perturbations
**RMP** resonant magnetic perturbation
**RMP ELM** RMP type ELM
**s** second
**sawtooth** oscillating signal with crash
**SCR** silicon controlled rectifier
**separatrix** magnetic surface intersecting x-point
**shear** radial gradient of rotational transform
**SMBI** supersonic molecular beam injection
**SPI** shattered pellet injection
**stellarator** toroidal device with helical windings
**symmetry** unchanged by a transformation of rotation or reflection
**system** describes connectivity between elements
**T** tesla (magnetic field)
**TAE** toroidal Alfven wave eigenmode
**tandem** magnetic mirror machine with end plugs
**TBM** tritium breeding module
**TBR** tritium breeding ratio
**Te** temperature of electrons
**TEM** trapped electron mode, also microwave transmission mode
**TEM** tearing electron mode
**TEM** trapped electron mode
**TF** toroidal magnetic field
**TF Coil** coil creating toroidal magnetic field
**thyristor** four-level high-power semiconductor
**Ti** temperature of ions
**tokamak** toroidal confinement machine with plasma current
**torsatron** see heliotron
**triple product** product of density, time, temperature
**Troyon limit** predicted maximum β formula
**VH-Mode** very high energy confinement mode
**VV** vacuum vessel
**WCCB** water-cooled ceramic tritium breeding blanket
**WCLL** water-cooled lithium lead tritium breeding blanket
**X-Point** point where poloidal field is zero
**Z** atomic number of element
**Zeff** effective atomic number
**κ** vertical elongation
ι rotational transform
**ι-bar** rotational transform divided by 2π
**β** ratio of plasma to magnetic field pressures

**βn** beta normalized to Troyon limit
**μs** microsecond
**ρ** rho minor radius
**τ** tau confinement time
**ω** omega circular frequency

# Index





F. B. Marcus, *Systems Approaches to Nuclear Fusion Reactors*, Springer Series in Plasma Science and Technology, https://doi.org/10.1007/978-3-031-17711-8

**F**

**J**



**K**



**L**

## S

The manufacturer's authorised representative in the EU is Springer Nature Customer Service Centre GmbH, Europaplatz 3, 69115 Heidelberg, Germany. If you have any concerns regarding our products, please contact ProductSafety@springernature.com

Printed and bound by CPI Group (UK) Ltd, Croydon, CR0 4YY
01/10/2026
02225872-0014